NUCLEAR RADIATION IN FOOD AND AGRICULTURE

THE GENEVA SERIES ON THE PEACEFUL USES OF ATOMIC ENERGY

Editor of the Series

JAMES G. BECKERLEY

Head, Engineering Physics Department,
Schlumberger Well Surveying Corporation;
Formerly Director of Classification
United States Atomic Energy Commission

NUCLEAR FUELS
by David H. Gurinsky and G. J. Dienes

EXPLORATION FOR NUCLEAR RAW MATERIALS
by Robert D. Nininger

NUCLEAR REACTORS FOR RESEARCH
by Clifford K. Beck

NUCLEAR POWER REACTORS
by James K. Pickard

SAFETY ASPECTS OF NUCLEAR REACTORS
by C. Rogers McCullough

NUCLEAR RADIATION IN FOOD AND AGRICULTURE
by W. Ralph Singleton

NUCLEAR RADIATION IN FOOD AND AGRICULTURE

Edited by

W. RALPH SINGLETON

Miller Professor of Biology
The University of Virginia
Director, The Blandy Experimental Farm

D. VAN NOSTRAND COMPANY, INC.
PRINCETON, NEW JERSEY
TORONTO LONDON
NEW YORK

D. VAN NOSTRAND COMPANY, INC.
120 Alexander St., Princeton, New Jersey (*Principal office*)
257 Fourth Avenue, New York 10, New York

D. VAN NOSTRAND COMPANY, LTD.
358, Kensington High Street, London, W.14, England

D. VAN NOSTRAND COMPANY (Canada), LTD.
25 Hollinger Road, Toronto 16, Canada

Published simultaneously in Canada by
D. VAN NOSTRAND COMPANY (Canada), LTD.

Library of Congress Catalogue Card No. 57-11590

PRINTED IN THE UNITED STATES OF AMERICA

". . . the United States pledges before you—and therefore before the world—its determination to help solve the fearful atomic dilemma—to devote its entire heart and mind to find the way by which the miraculous inventiveness of man shall not be dedicated to his death, but consecrated to his life."

President Dwight D. Eisenhower
address before the General
Assembly of the United Nations,
December 8, 1953.

FOREWORD

No matter how the Geneva Conference will be judged by historians of the future, there is one fact which will hardly be forgotten for some time to come: in ten days of August, 1955, an unprecedented volume of technical information on atomic energy was put into the public record. Not all the items of information were of equal value, to be sure, but relatively little was of a trivial nature.

Realizing that a verbatim record of this avalanche of data would be relatively indigestible to a large body of workers in the field, the Publishers requested me to undertake the task of organizing and editing a series of books which would present material of most urgent usefulness. Preprints of many of the conference papers were made available through the courtesy of the United Nations staff.

From a study of these papers it was concluded that six volumes of nominal length would be required. Each would cover a specific subject and include the key papers, suitably edited to eliminate duplication and secondary material, and arranged in a logical manner.

Because of the great interest in finding the raw materials for the growing nuclear appetite, Robert D. Nininger, Deputy Director for Exploration, Division of Raw Materials, United States Atomic Energy Commission, was asked to edit a volume on the geology of and the exploration for uranium and thorium.

All atomic energy projects to date have had research reactors at the base of their technical efforts. Dr. Clifford Beck, Professor of Physics, North Carolina State College, whose foresight and determination put the first research reactor on a university campus, consented to edit a volume on research reactors.

At the center of the stage, particularly for the energy-hungry nations of the world, is the power reactor, a device of many and varied forms. James K. Pickard, now a consultant engineer on atomic energy developments, for many years previously one of the stalwarts of the AEC Reactor Development Division, accepted the task of preparing a nuclear power reactors volume.

Because one of the most difficult and urgent problems in the power reactor program is assessing the hazard and providing safety in reactor operation, it appeared urgent to include a volume devoted exclusively to reactor safety. One of the pioneers in the field, Dr. C. R. McCullough of the Monsanto Chemical Company, has undertaken to edit a volume on the safety aspects of nuclear reactors.

The heart of the power reactor is the nuclear fuel, the energy source and the focal point for intense metallurgical and solid state research efforts. Dr. David H. Gurinsky and Dr. G. J. Dienes of Brookhaven National Laboratory, who have made substantial contributions to these efforts, have edited a volume on nuclear fuels.

In addition to the direct heat energy from fission, the ten per cent or so of reactor energy available as radiation is already exerting profound effect on civilization. One who has probed deeply into what might be termed the interaction of nuclear energy and agriculture, Dr. W. Ralph Singleton, Director of the Blandy Experimental Farm, University of Virginia, has edited a volume on nuclear radiations in food and agriculture.

These six volumes are not intended to cover all the technical areas of the Geneva Conference. Subjects such as radiobiology, nuclear physics, reactor physics, chemical processing—to name but a few—were not included (except where they touch on the subjects chosen) for a variety of reasons. A first consideration was that, if all subjects were properly represented, the series would presume an ungainly large number of volumes. Reducing the number of volumes could only be done by limiting coverage either to a few papers on each subject or to a superficial digest of a large number of papers. It seemed a better choice to cover fewer topics more completely. A secondary consideration was an assessment of the state of the literature—where a subject was rather well represented in the public literature, it could be argued that the need for a corresponding volume was less urgent. Certainly, for example, the literature of nuclear physics does not urgently require amplification. Moreover, although the nuclear physics papers published at Geneva were substantial, they did not disclose any striking new information. (In addition, many of the new nuclear data have been adequately disseminated by the United States Government handbooks.)

In any effort of this kind, arbitrary decisions are made in many places. Material is deleted, rearranged, abbreviated, and new material is added. It is hoped that these decisions have been wise and that the six volumes will be a fundamental reference work for the increasing number of scientists and engineers devoted to the peaceful uses of atomic energy.

J. G. Beckerley

PREFACE

The chapters of this book on Nuclear Radiation in Food and Agriculture are based on papers selected from those presented at the Conference on Peaceful Uses of Atomic Energy held in Geneva, Switzerland, in August 1955. Many of the papers presented here have also been published in Volume 12 of the records of the Geneva Conference; but papers on the genetic effects not covered in Volume 12 will be found in Part V of this volume, "Genetic and Biological Hazards of Nuclear Radiation."

Selection was exercised not only in the individual papers chosen but also in the topics. The volume will be of special use to plant workers. There were several excellent papers presented in the field of animal physiology at the Geneva Conference, but most of these were the reports of original research and are considered somewhat too specific for this volume.

Part I of the eight parts of this book concerns the general use of radioisotopes in agriculture. In addition there is a chapter on the biosynthesis of carbon-14-labeled plants.

Part II consists of two papers on photosynthesis, one by workers in this country and one from Russia.

Part III, on plant physiology, pathology, and cytology, contains one of the most controversial papers of the Conference. This describes stimulation of plant growth with ionizing radiation by the Russian investigator, A. M. Kuzin. The inclusion of the report does not necessarily mean that the editor endorses all its findings. However, it is well to bear in mind that the fact that we do not understand the reason for a possible stimulation by low doses of ionizing radiation is no justification for denying that such a stimulation exists. Stimulation from low doses has been found in several experiments by various competent investigators. Part III also contains interesting papers on root grafting, the foliar absorption of plants, the uptake of minerals, tracing fungicidal actions, the effects on plants grown under gamma radiation, and cytological and chemical effects of radiation.

Part IV consists of a single chapter on the use of radioisotopes in soil and fertilizer studies. It is a review paper with a rather extensive bibliography.

Part V, on genetic and biological hazards of nuclear radiation, will be of interest to all, since everyone is concerned with these effects. It is included because of the widespread interest in this subject.

Part VI consists of a single paper, on the eradication of the screw-worm fly in the island of Curaçao by the use of ionizing radiation. This is a unique method of controlling or eliminating an insect population. It was suggested, but not tried, a good many years ago by Dr. Richard Goldschmidt for controlling the gypsy moth infestation in the northeastern United States.

Part VII, on crop improvement, consists of one paper from Sweden, one from Norway, and two from the United States. The paper by the Swedish scientists is particularly pertinent, since it was in Sweden that the practical uses of radiation in plant breeding were first demonstrated by one of the co-authors, A. Gustafsson.

Part VIII on food sterilization represents the latest application of radiation to an old problem. Much more research in this field is necessary before radiation will take its place along with canning and freezing in food preservation, but rapid advances are being made in this as well as all fields of agricultural uses of atomic energy. Already atomic energy is having a profound effect on the lives of all of us, especially in the fields of food and agriculture.

W. Ralph Singleton

The Blandy Experimental Farm
University of Virginia

CONTENTS

PART V—GENETIC AND BIOLOGICAL HAZARDS OF NUCLEAR RADIATION

PART VI—GENETIC ERADICATION OF INSECT PESTS

PART VII—CROP IMPROVEMENT

PART VIII—FOOD STERILIZATION

Part I

THE USES OF RADIOISOTOPES IN AGRICULTURE

Chapter 1

RADIOISOTOPES IN AGRICULTURAL AND SILVICULTURAL RESEARCH *

It is now over thirty years since Hevesy introduced the application of radioactive isotopes as indicators in plant studies (Ref. 1). The discovery of artificial radioactivity by Curie and Joliot in 1934 greatly enlarged the scope of the method, but it is probably fair to say that it was not until the advent of the atomic pile that the use of radioisotopes in agricultural research became at all widespread. Today isotopes such as H^3, C^{14}, Na^{22}, P^{32} and Co^{60} are available in quantity and have been used to investigate a number of problems of interest to agriculture—problems of analysis, animal metabolism, the relation of the plant to its soil food, trace elements, fertilizers, photosynthesis and weed and pest control.

Fertilizer Utilization

General principles. Within the last few years radioisotopes have provided an invaluable tool for investigating the availability of plant nutrients under field conditions. While a number of different isotopes have been used, the economic importance of phosphate fertilizer and the relative ease of handling P^{32} have resulted in particular attention being paid to phosphorus. Qualitative soil and plant studies using P^{32} began in 1936 and were followed ten years later by quantitative field studies of phosphate fertilizer uptake by wheat plants (Refs. 2, 3, 4, 5). By combining the radioactive measurement of fertilizer uptake with the measurement by ordinary chemical methods of total phosphorus uptake, the uptake of soil phosphorus could be determined by difference. In the past, the recovery of phosphate fertilizer by a crop was determined by a comparison of the uptake of phosphorus by crops grown with and without fertilizer. The extra phosphorus in the fertilized crop was taken as the quantity coming from the fertilizer. This method assumed that fertilized and

* This chapter is taken from Geneva Conference Paper 10, "Studies of Special Problems in Agriculture and Silviculture by the Use of Radioisotopes" by J. W. T. Spinks of Canada. Numbered references are listed at the end of the chapter.

unfertilized crops take up the same amount of soil phosphorus, but tracer experiments indicate that this is often far from being the case.

Numerous experiments have been reported during the last decade covering such diverse topics as utilization of various phosphatic fertilizers by different types of crops at various stages of growth, grown on different types of soil, using different methods of placement and different rates of application. In the first field experiment, 1 millicurie P^{32} was used. Four years later, 30 curies was used in the United States alone; and now the labeled phosphate fertilizers are made by the ton.

Preparation and assay of labeled material. It is important that the P^{32} be in the same valence state and chemical form as the phosphorus in the phosphate fertilizer being studied. For example, radioactive ammonium phosphate ($NH_4H_2PO_4$) may be prepared by adding phosphoric acid of high specific activity to a solution of inactive ammonium phosphate and subsequently evaporating this solution to dryness with stirring (Ref. 6). Standard methods have been worked out for preparing various labeled phosphate fertilizers, and the procedures have in some cases been stepped up to pilot-plant scale (Ref. 7). It should be pointed out that the seemingly attractive neutron irradiation of phosphatic materials is likely to lead to difficulties since the P^{32} produced will almost certainly be present in several different chemical forms and give rise to ambiguity in interpretation of results (Refs. 8, 9, 10, 11).

In the early work, the phosphorus in the plant was converted to magnesium pyrophosphate and counted under an end-window counter (Ref. 4). Since then the solution counting technique has been used, utilizing either dip counters (Ref. 12) or immersion counters (Ref. 13). However, most workers now measure the activity of briquettes of dried plant material, using an end-window counter (Ref. 14), or the activity of hollow cylinders of dried plant material placed around a thin wall counter of the thyrode tube type (Ref. 15). By using as standard a cylinder made from an aliquot of the original labeled fertilizer mixed with inactive plant material, the uptake of the fertilizer is very simply and directly measured. The hollow cylinder method has increased both the sensitivity of the measurements and the ease of making them, so that it is now possible to work with much lower specific activities and still do many hundreds of analyses in a relatively short time.

Phosphorus is determined chemically by using the phosphomolybdate blue reaction with hydrazine as a reducing agent (Ref. 16).

Field experiments: typical results. In a typical field experiment each treatment plot consists of five rows, the two outside rows being guard rows, with no fertilizer. The three inner rows receive the designated fertilizer treatment, but only the center row is treated with radioactive fertilizer. The level of radioactivity in most field experiments is 100 microcuries P^{32} per gm of inactive P^{31}. The rows are 5 m long, the center 3.7 m being harvested. Six similar plots of land are used in a random arrangement, or a balanced lattice design. Except for placement experiments, fertilizer and seed are at the same level, the fertilizer and grain usually being sown together.

The following results were obtained in experiments with the soils of Saskatchewan, Canada, and may not apply in all details to other types of soil.

Stage of growth. There is little uptake of fertilizer or soil phosphorus in the earliest stages of growth of the wheat plant. The uptake of fertilizer phosphorus reaches a maximum at 2 to 6 weeks and begins to fall off at about 8 weeks as the plant reaches the heading stage. In the earlier stages of growth, the wheat plant obtains a larger portion of its phosphorus from the fertilizer than from the soil, but later on the situation is reversed (Refs. 17, 18).

Type and variety of crop. In a typical tracer experiment (Ref. 19) plots were set out at two locations, one on a less responsive brown heavy clay (Regina Heavy Clay) and the other on a highly responsive moderately heavy black soil (Melfort Silty Clay). Four varieties of wheat (Thatcher, Apex, Redman, and Rescue), three varieties of barley (Titan, Montcalm, and Vantage) and three varieties of oats (Ajax, Exeter, and Fortune) were included in the test. The tests were conducted on fallow land, and the fertilizer used was monoammonium phosphate at 24 lb P_2O_5 per acre. The order of yield increase in the two trials was wheat (least), barley, oats; and it is to be noted that barley utilized a greater portion of the fertilizer phosphorus and showed a greater uptake of total phosphorus on the fertilized plot. This was also the case for the uptake of phosphorus on the unfertilized plots. While, as might be expected, there were a number of significant differences in yield between varieties of the grain attributable to inherent varietal characteristics, it was only in barley that differential response to the fertilizer treatment was found. The varieties Montcalm and Vantage gave a greater response to the phosphate fertilizer than Titan at Birch Hills, but not at Rosetown on the less responsive soil. Titan is an earlier maturing variety, and this factor may be related to the result obtained. It is of interest that such differences betweeen varieties can exist, and conceivably such differences could be of considerable interest to the plant breeder (Refs. 20, 21, 22).

Type of soil. In a typical experiment, using wheat, barley, and oats with ammonium phosphate fertilizer, two types of soil were compared. One was a less responsive heavy clay (Rosetown, a chestnut soil on a black lake-bed base material) and the other a more responsive silty clay at Birch Hills (a thick black chernozem-like soil on lake-bed material). All three crops showed a greater utilization of fertilizer phosphorus on the more responsive soil at Birch Hills—33% as compared with 24% at Rosetown (Ref. 21).

Placement. In a typical placement experiment with Thatcher wheat and 11-48-0 fertilizer (monoammonium phosphate), seed was placed at depths of 7.5 and 11 cm, with fertilizer at depths of 5, 7.5, and 11 cm. The experiment was done at three locations, using fertilizer in the powdered and granular forms. The results indicated a maximum yield and maximum utilization of fertilizer with seed and fertilizer (granular form) both at depth of three inches (Refs. 21, 22).

Type of fertilizer. Numerous tracer experiments have been done with wheat, oats, and barley, and various phosphatic fertilizers such as ammonium phosphate, standard superphosphate, monocalcium phosphate, etc. The experiments all indicate that for the Saskatchewan prairie soils, ammonium phosphate provides the most readily available source of phosphate (Refs. 6, 22, 23). In other experiments, the effect of nitrogen addition on fertilizer phosphate availability has been studied in some detail (Ref. 24). Field tests were carried out in ten plots located in the Brown, Dark Brown, Black, and Gray soil zones of Saskatchewan. Fertilizers were monoammonium phosphate and monocalcium phosphate applied at 24 lb P_2O_5 per acre (lb/acre or kg/hectare), varying amounts of NH_4NO_3 being added to give ratios of N to P varying from 1:4 to 1:0.8. A marked increase in the uptake of fertilizer phosphate from both carriers occurred as the amount of nitrogen was increased. Stubble trash reduced fertilizer availability considerably. Later experiments in which the N:P_2O_5 ratio was varied from 1:4 to 4:1 have confirmed the above findings (Ref. 25).

Rate of application. The effect of rate of application has been intensively investigated. For example, in 1950 experiments were done at Melfort and Watson using Thatcher wheat and 11-48-0 applied at 6, 12, 24, 48 and 96 lb P_2O_5 per acre, or kg per hectare (Ref. 22). With increasing rate of application, the fertilizer uptake increases while the soil phosphorus uptake steadily decreases (Fig. 1.1). Before the advent of tracers, fertilizer uptake was esti-

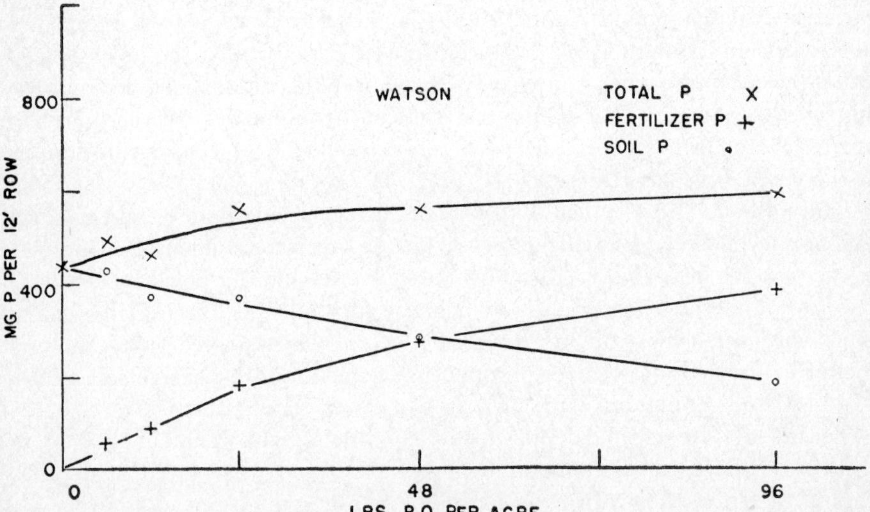

FIG. 1.1. Relationship between total phosphorus in plant and phosphorus uptake from soil and fertilizer, as influenced by rate of application at Watson, Saskatchewan, 1950.

mated by comparing the phosphorus uptake from a fertilized plot with that from a control plot, the assumption being made that an equal quantity of soil phosphorus is taken up from fertilizer and control plots. The values obtained in this way are shown by the broken line in Fig. 1.2. They can obviously be greatly in error.

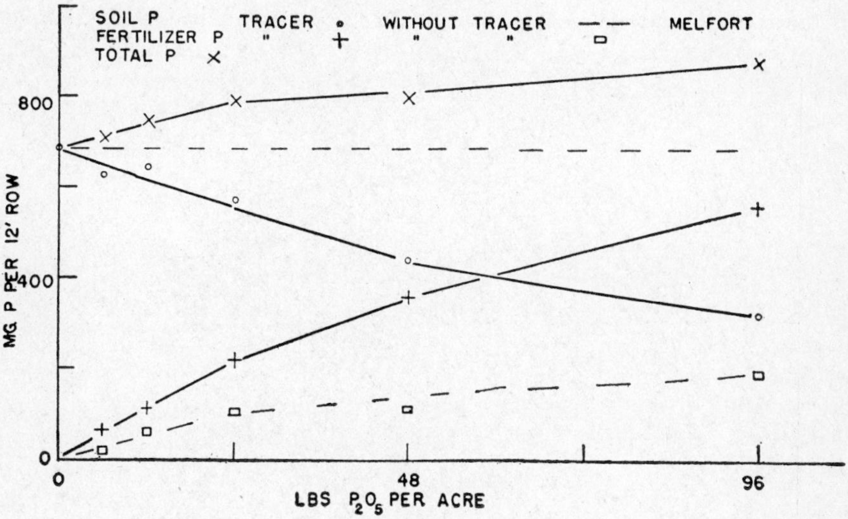

Fig. 1.2. Relationship of uptake of phosphorus from soil and fertilizer with increasing rate at Melfort, Saskatchewan, 1950. Broken lines indicate results as might be calculated without the information provided through the use of tracer phosphorus.

Available soil phosphorus: A value. In Fig. 1.3 the rate of application of fertilizer has been plotted against the ratio: (P from the fertilizer)/(P from the soil). It is of considerable interest that the points lie on a reasonably straight line, the slope of the line being different for different types of soil. This behavior has been found by a number of workers and appears to be quite general (Ref. 26). Having established that this behavior is quite general, the slope of the line, usually called A, can be determined by doing experiments for just one or two rates of application of fertilizer. In fact,

$$A = (\text{rate of application of fertilizer}) \times \frac{(\%\ \text{P from soil})}{(\%\ \text{P from fertilizer})}$$

From the figure for Melfort soil it appears that when 11-48-0 is applied at 60 lb P_2O_5 per acre, soil and fertilizer contribute equal amounts of phosphorus to the plant; in other words, the soil phosphorus made use of is equivalent to that supplied by 60 lb P_2O_5 as 11-48-0 per acre. The 60 lb per acre is just the slope of the line, and *thus A gives a measure of the available soil phosphorus* in terms of 11-48-0. The theoretical basis for this expression has been discussed by Fried and Dean (Ref. 26) and by Larsen (Ref. 27). Actually, this expression is just what we would expect if we were doing a quantitative analysis by means of the isotope dilution technique. In order to determine, for example, the amount of a given substance in a liquid sample by isotope dilution, a known amount of tracer of known specific activity and in the same valence form is added to the sample. After thorough mixing, a portion is pipetted out and the specific activity redetermined. In the soil-plant case, the sample is

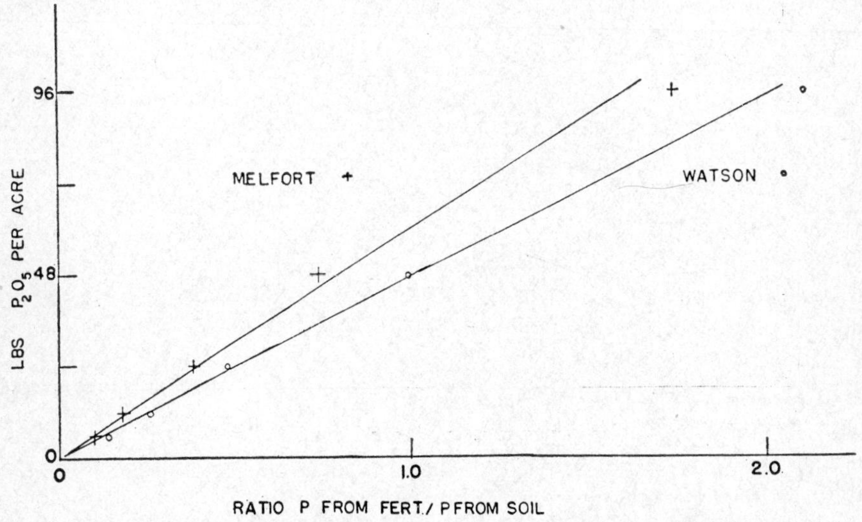

FIG. 1.3. Ratio of fertilizer phosphorus to soil phosphorus in the plant as related to rate of application of phosphate fertilizer. Melfort and Watson, 1950.

the phosphorus in the soil available to a given crop (plant or plants), the tracer is the tagged fertilizer, and the crop represents the pipette. It has sometimes been claimed that the occurrence of surface exchange between the phosphate fertilizer and the phosphate in the soil invalidates the method, but actually, as was explained in one of the earliest papers (Ref. 4), surface exchange will occur to the same extent for labeled and unlabeled fertilizer molecules and the mathematical result is that the conclusions drawn are not affected.

The available soil phosphorus, as indicated by the A value, has recently been used to help evaluate chemical methods used in determining soil phosphorus availability (Ref. 28).

Radiation damage. It is well known that the radiation from radioisotopes will inhibit the growth of plants. It is therefore important to use the tracer at a specific activity which is sufficiently low that radiation effects do not invalidate the method. Just what the safe level is has not yet been settled with certainty. There is an added difficulty that the safe level will depend on the particular design of the experiment, and it is unfortunate that different workers have used different techniques in trying to establish the critical level. Russell and Martin claim that in nutrient solution, 10 microcuries P^{32} per liter may interfere with the growth of young barley plants (Ref. 29). The same authors have claimed that P^{32} produces variations in the ratio of soil phosphorus to fertilizer phosphorus, although the variations seemed quite irregular (Ref. 30). Bould *et al.* (Ref. 31) reported effects on plant weight with as little as 5 microcuries P^{32} per pot, but failed to find any effect on the ratio of soil phosphorus to fertilizer phosphorus with 300 microcuries P^{32} per pot! Blume (Ref. 32) and Bould *et al.* both concluded that in experiments in which activity levels were much greater than those normally used, the smallness of the radiation effects was the most startling feature of the experiment. Numerous field experiments have failed to show any radiation effect on yields of fertilized crops grown with and without P^{32} up to quite high levels of P^{32} activity (Refs. 33, 34, 35).

In an elaborate series of greenhouse experiments conducted by Penner (Ref. 36), no deviations were found when radioactive $NH_4H_2PO_4$ was applied in granular form with specific activities from 0 to 3600 microcuries P^{32} per gm P^{31}. In particular, no alterations in the ratio of soil P to fertilizer P occurred in any of the experiments even when the fertilizer was applied in solution. Small radiation effects on plant weight were thought to occur when the P^{32} was applied in solution at 12 microcuries per gm P^{31}, but an examination of Table 1 of Ref. 36 makes it questionable whether this effect was really due to radiation, since 120 and 1200 microcuries P^{32} per gm P^{31} showed no effect, and even the 12 microcuries showed no effect in the ratio of soil P to fertilizer P, per cent fertilizer absorbed, total fertilizer absorbed, etc. Field experiments using 0 to 3600 microcuries P^{32} per gm P^{31} produced erratic results: at one location no effects were found for three harvest dates, whereas at another location (where there was, incidentally, a considerably smaller uptake of fertilizer) radiation effects were claimed at the second harvest, and even here the criteria showing an effect varied from time to time (Ref. 37)! A further field experiment at two locations and two harvest dates showed no effect in three of the trials, and the effect in the fourth was such as to be quite suspect—a slight decrease in fertilizer P uptake at 120 microcuries as compared with 12, 60, 240, and 720 microcuries per gm P^{31} (Table 7 in Ref. 37). While a good deal more work on radiation damage is desirable, it seems that activity levels in the usual field experiments, with less than 200 microcuries P^{32} per gm P^{31}, are low enough to avoid appreciable radiation damage (Ref. 38).

Diffusion of ions in the soil. Another factor which has to be considered in carrying out field tests is the amount of phosphorus which might be obtained by the plant from adjacent rows. In our experiments the row spacing is 15 cm, three rows being fertilized, only the fertilizer in the center row being tagged. For ammonium phosphate, the maximum "piracy" occurring between rows was less than 1% of the applied phosphorus, indicating that this effect can be neglected in our experiments.

The actual movement of ions in soil is, of course, of considerable interest and can be studied with radioactive tracers (see for example Ref. 39).

Comments. Considering that the whole development in this field has taken place in the last decade, it is perhaps not surprising that there is still some discussion, at times acrimonious, over the exact interpretation of some of the results (Refs. 40, 41). However, it cannot be denied that tagging provides the *only* means of determining the path of the fertilizer phosphorus as distinct from that of the soil phosphorus. It is also quite clear that the active research of the last nine or ten years has led to a very considerable increase in understanding of the complex relationship existing between plant and soil.

Neutron Measurement of Soil Moisture

The amount of moisture in the soil is of obvious importance to plant growth, as is also the problem of moisture movement in soils. Measuring changes in soil moisture in a given sample of soil involves using a non-destructive method of sampling, and heretofore no completely satisfactory method has been available. The neutron method seems to offer certain advantages over those previously suggested in that it is non-destructive, is adaptable to all types of soils, is independent of the state of the moisture (whether vapor, liquid, or solid), is able to indicate rapid changes in moisture content, and is relatively independent of salt content of the soil. In the neutron method, fast neutrons emitted by a neutron source are scattered by the atoms in the soil, thereby losing their energy and becoming slow neutrons (apart from loss by capture). The slow neutrons are then measured by means of a slow neutron detector such as an indium or rhodium foil or a BF_3 chamber. The possible fractional loss in energy when a neutron collides with an atom is at a maximum for hydrogen (Ref. 42), and thus soils with a high moisture content are particularly effective in slowing down neutrons. The activity induced in a slow neutron detector is correspondingly high and is in fact proportional to the moisture content of the soil when standard conditions of neutron irradiation are adopted (Refs. 43, 44, 45, 46). It can be shown theoretically that the distribution of slow neutrons about a point source of fast neutrons is a function of the ratio of the slowing-down length to the diffusion length (Ref. 47). The slowing-down length is defined as the average distance a neutron must travel in a medium before it is slowed down to thermal velocity. It is made relatively small by hydrogen in comparison with other common elements in soil. The diffusion length is one-sixth the average distance from point of origin (of the thermal neutron) to point of capture of the neutron. It does not vary with changing moisture content as strongly as does the slowing-down length. The net effect is that if a slow neutron detector is placed near a source of fast neutrons in soil, the activity will be largely a function of the hydrogen (or moisture) content of the soil.

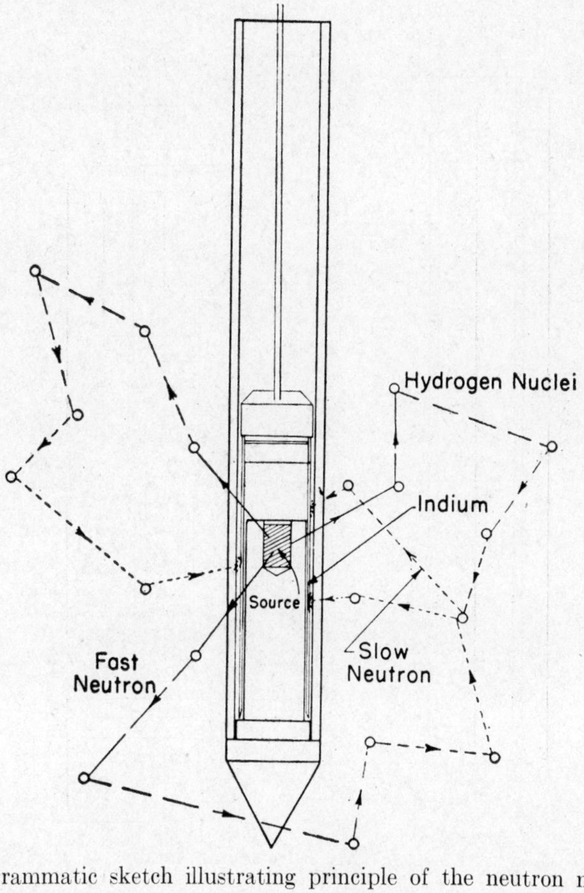

FIG. 1.4. Diagrammatic sketch illustrating principle of the neutron moisture meter.

The underlying principle and general experimental arrangement of the moisture meter is shown in Fig. 1.4, while mechanical details are given in Fig. 1.5. The apparatus consists of a probe head, D (see Fig. 1.5), which is lowered into a 5-cm-diameter vertical hole in the soil by means of an aluminum probe cylinder *A*. A 50-millicurie Ra-Be neutron source is then lowered into its seat, using an electromagnet. A foil holder, *C*, containing indium foil is lowered into its seat by means of a cord attached to the foil holder cap, *B*, thus placing the indium foil around the neutron source. After exposure of the foil for a definite time (measured with a stopwatch) the foil in its holder is withdrawn and placed around the Geiger tube of a portable beta-gamma rate meter, which is read at a definite time after removing the foil from the neigh-

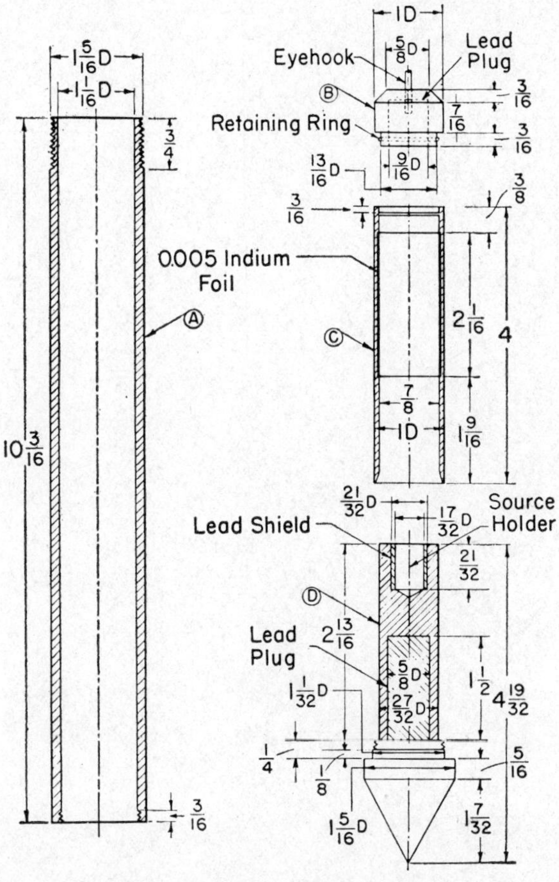

FIG. 1.5. Moisture meter details.

borhood of the source. After some experimentation, a 10-min exposure followed by a 1-min delay before measuring the activity was chosen as standard procedure.

Field experiments with a neutron moisture meter. In a typical field experiment, a series of holes 6 ft deep and 2 in. in diameter were drilled at various levels on the banks of the South Saskatchewan River to give a variety of soil types, densities, and moisture contents. The soil types encountered in these holes ranged from sandy silts to medium plastic alluvial clays. The holes were drilled with a 5-cm auger and lined with 5-cm-diameter aluminum pipe (0.125 cm thick). The depth of the source could be varied by adding lengths of aluminum tubing to the source holder and lowering it down to the desired depth in the hole. Samples of the soil were taken every 15 cm during excavation of the holes, for determination of moisture content by the standard oven-drying method. At each depth, three readings for moisture content,

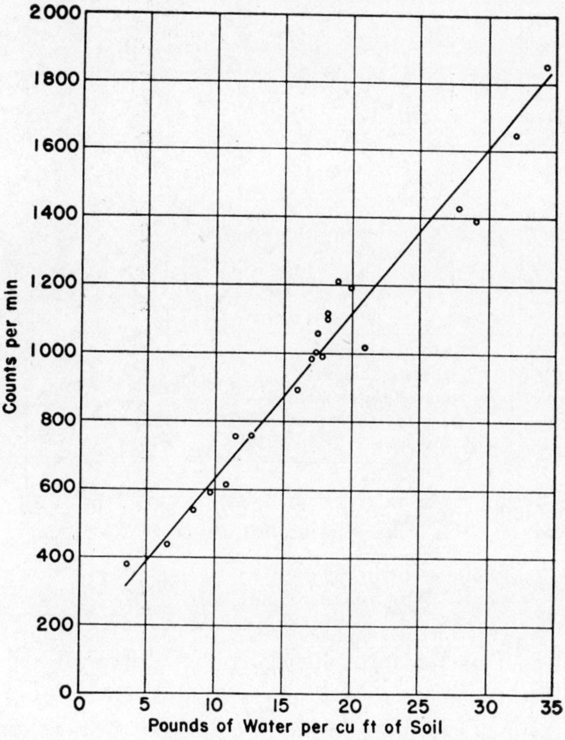

FIG. 1.6. Calibration curve for neutron moisture meter using a 50 millicurie radium beryllium source.

using the rate meter, were averaged to give a mean value. This was plotted in Fig. 1.6 against the average measured moisture content in the 15 cm above and below the source, since the zone of influence of the source is primarily a sphere of about 15 cm radius. Fig. 1.6 shows a reasonably linear relation between count rate and mass of water per unit volume. This curve was used as a standard calibration curve for subsequent tests. Fig. 1.7 shows a comparison of moisture content determinations in a vertical hole on the campus of the University of Saskatchewan for the neutron meter and the usual oven-drying method. All determinations are within 3% moisture, and most are within 2% moisture or less.

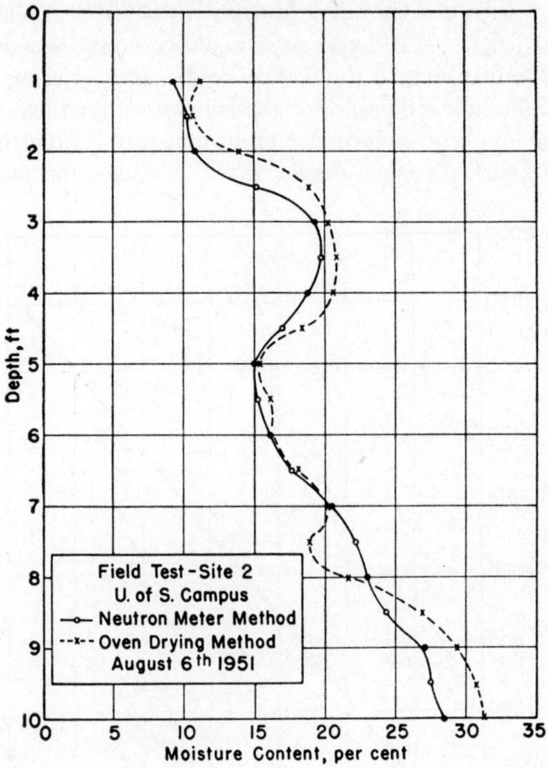

FIG. 1.7. Moisture profile comparison between standard oven-drying method and neutron meter.

An improved neutron moisture meter. It has already been noted that the zone of influence of the neutrons is a sphere of about 15 cm radius, and that as a consequence the neutron moisture meter readings represent an average value for a sphere of about 15 cm radius, the average being weighted somewhat in favor of points close to the neutron source. For some agricultural purposes it is desirable to have a somewhat more detailed knowledge of the variation of moisture content with distance. After varying a number of the parameters involved, the arrangement illustrated in Fig. 1.8 was adopted.

The neutron detector now consists of a thin strip of rhodium foil 2 mm wide and 66 mm long, bent into a ring and mounted in a lucite holder. During irradiation the lucite holder is slipped over the neutron source so that the center line of the rhodium foil is at the same elevation as the center line of the neutron source. After 4 min of irradiation time, followed by a lapse of 30 sec, the activity of the rhodium strip is measured by slipping it over a

"Thyrode" thin-walled Geiger counter tube (connected to a scaler) and counting it for 4 min. The neutron beam is collimated to some extent by using two shields, each consisting of a hollow Cd cylinder 5 cm long and 4.75 cm in diameter, with a central hole 2.5 cm in diameter, the hollow cylinders being filled with paraffin. The Cd shields fit snugly onto the probe cylinder. It has been found that optimum results are obtained with a shield spacing of about

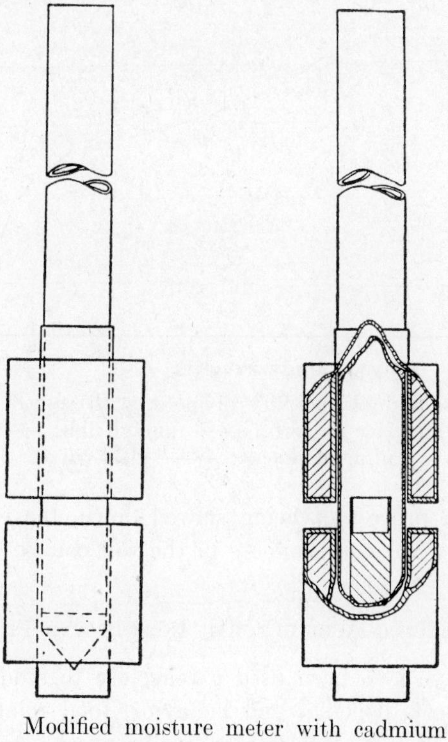

FIG. 1.8. Modified moisture meter with cadmium shields.

1 cm. The sensitivity of the device was tested using a simulated soil profile consisting of a 23-cm layer of dry sand topped by a 12.5-cm layer of paraffin wax, the whole being contained in a box of rectangular cross section (25 cm square). A 5-cm-diameter vertical aluminum tube down the axis of the box made it possible to take meter readings at various positions with respect to the sand-wax interface. The results are recorded graphically in Fig. 1.9. They indicate that the modification does effect an improvement, but that the device could probably be improved still further, possibly by using a cadmium shield of somewhat larger diameter. It has been suggested by Gueron (Ref. 48) that a further improvement can be obtained by measuring the epithermal neutrons, using indium foil sandwiched between two pieces of cadmium foil. By measur-

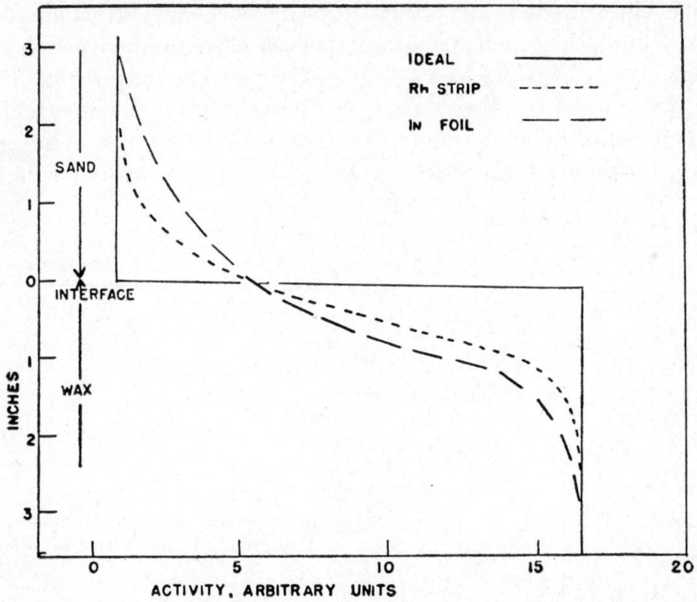

FIG. 1.9. Variation of induced activity in meter with respect to distance in inches relative to wax-sand interface as zero. —— indium foil; ---- rhodium strip and cadmium shields; ——— ideal curve.

ing the neutrons in process of being slowed down, the effects of capture of thermal neutrons by various elements in the soil can be avoided.

TRACING AGRICULTURAL INSECTS AND PESTS

Radioisotopes have also been used extensively to study agricultural pests with respect to insect dispersal and behavior, food relations, disease transmission, mode of action of insecticides, including systemics, dispersal of insecticides, and deterioration of insecticides in the soil. (For a general reference, see Ref. 49.)

Desirable characteristics of tags. In tracer studies of insect dispersal and behavior, insects or groups of insects are tagged with a suitable radioactive tracer and then followed after release using a radiation detector. Among the desirable characteristics of the tracer are ease of application, minimal effect on the insect, ease of recognition of the tag, persistence, suitable half-life, and availability. These points will be illustrated by reference to tagging experiments with mosquitoes, blackflies, grasshoppers, wireworms, and cutworms.

Dispersal studies of insects. *Mosquitoes.* During the last few years, experiments have been reported in which highly radioactive mosquitoes were reared from larvae left in a solution of radioactive phosphorus (Refs. 50, 51).

Subsequent field experiments in Saskatchewan have shown that large numbers of adult mosquitoes can be economically tagged by keeping larvae at the fourth instar stage of development in a 0.1 microcurie per ml solution of P^{32} (in the form of $PO_4^{\equiv}$) for 24 hr at a density of 1 larva per ml, and then allowing the tagged larvae to complete development to the adult stage in their original or normal habitat. In a typical experiment, about 45,000 larvae were placed in 45 l of pond water in a tub, together with 4.5 millicuries P^{32} as H_3PO_4. After 24 hr the larvae were placed in the pond and a further batch of larvae added to the tub together with a sufficient quantity of P^{32} to maintain the level of P^{32} activity. Approximately 450,000 larvae were tagged in this way. In the following month about 500,000 mosquitoes were caught in the neighborhood of the pond. Radioactive mosquitoes were found as far as 7 miles from the release point, but the majority were within 1/8 mile of the point of release. A total of 84 radioactive mosquitoes, 19 flies, 3 parasites, and 1 leaf hopper were found at various points. It was found that larvae lost 50% of the absorbed activity within 2 days after removal from the tagging solution, but after that the loss was relatively slow. The average retention of activity through to the adult stage was about 15%.

Control experiments indicated that larval mortality was no greater with 0.1 microcurie P^{32} per ml than with inactive media. For the first 60 hr the uptake of radioactivity is proportional to the time of exposure to the P^{32} solution. There is also a very strong dependence of absorption on temperature (Refs. 52, 53).

Blackflies. The method successfully used to tag mosquitoes had to be modified somewhat for application to the blackfly. In laboratory experiments, blackfly larvae were reared in 5-l jars containing water circulated and aerated by compressed air. As before, P^{32} was in the form of H_3PO_4. It was found that from 2 to 4 larvae per ml could be kept in a solution at 0.2 microcurie P^{32} per ml without excessive mortality and that an adequate amount of activity was absorbed in 24 hr. After transferring the larvae to an inactive solution, there was a relatively large loss of radioactivity in the first two days, but the loss thereafter was relatively small (Ref. 54). Field tests were made in the Torch and Battle Rivers in Saskatchewan in 1950 and 1951, the larvae being tagged in a tube aerated with a stream-driven paddle wheel. As before, the same solution was used to tag successive batches of larvae, the level of activity being brought up to 0.2 microcurie per ml by addition of P^{32} as necessary.

On being released into the stream, the tagged larvae attached themselves to rocks and leaves in the stream. They were detected as far as 520 yards down stream from the point of release. The recovery of tagged blackflies was extremely poor, mainly, it is thought, owing to inadequacies in collecting methods. In one experiment 300,000 larvae were tagged. Subsequently, 16,000 adults were caught in the neighborhood, only 1 being tagged. In another similar experiment, 19 tagged predators were recovered. It is believed, however, that the method for mass tagging is basically sound and that it could be used, with minor modifications, in similar studies of stream-inhabiting insects.

Grasshoppers. In a grasshopper dispersal study, 20,000 grasshoppers (nymphs and adults) were allowed to feed on wheat seedlings 6 in. high, in a cage 4 sq ft in area of cross section for several hours. The wheat seedlings had been sprayed with 0.5 millicurie P^{32} in 50 ml solution. About 14% of the activity was taken up and retained by the grasshoppers. The loss in activity by the grasshoppers was high for the first few days, but was very small thereafter. Loss of activity through molting was negligible, and survival was normal (Ref. 55). The method was used to measure the dispersal of nymphs from the second instar and adults of *Camnula pellucida* (Scudd) and *Melanoplus mexicanus mexicanus* (Sauss). When released on bare cultivated fields, they showed no ability to orient themselves and move toward a food supply. The average rate of movement at 70° F was about 7 yards per hr. In these particular experiments, the position after 7 days corresponded to random movement plus a response to wind direction (Ref. 56).

Behavior of wireworms and cutworms. In making behavior studies, it is necessary to tag individual specimens. Where the species is of the soil-burrowing type, the tag must be a gamma emitter in order for it to be detected through several inches of soil. The feeding of activity to the animal has its limitations, since any material excreted will contaminate the soil and make it difficult to follow. In studies of wireworm movement, this difficulty was avoided by sticking a minute piece of radioactive cobalt wire to the wireworm with vinylite plastic. Once tagged, the larva may be followed by using a Geiger probe connected to a rate meter. By moving the Geiger probe over the soil surface and determining the position of maximum counting rate, the wireworm is localized in the horizontal plane. By previously calibrating the instrument for varying soil depths, the position in a vertical plane is also determined. Twenty microcuries Co^{60} per tag is a suitable amount of activity. No harmful effects are observed over some months, and the long half-life of Co^{60} (5.3 yr) minimizes corrections for radioactive decay. The external tag is of course lost at each molt. This difficulty is avoided by tagging the wireworm internally. Successful insertions of cobalt wire have been made in both wireworms and cutworms. No apparent abnormality or loss of movement was evident up to 3 months after treatment. When using highly active cobalt wire, it was found that the body fluids of the wireworm slowly attacked the wire so that activity was excreted by the insect. This difficulty could probably be avoided by using gold-plated Co wire.

Tagged insects have been used to measure the response of the prairie grain wireworm to moisture, food, and temperature, and information on the rate of movement has been obtained. Radioactive larvae of the red-backed cutworm have been used to determine the fate and underground activity of larvae placed in open cages utilized for chemical control studies and observation on habits of the larvae (Ref. 57). *Ctenicera destructor* larvae moved very quickly at temperatures above 90° F, a little slower between 80 and 90° F, and were found to prefer temperatures in the range 72° to 80° F. *C. destructor* larvae were observed to avoid dry soil when offered a choice of moist or dry. However, larvae sometimes entered the dry soil region when food was placed there. For the types of food tried, the movement of larvae to food was found to be a random process. A great deal of information on the rate and extent of movement was obtained. The larvae were capable of moving several yards in the space of a few days. A maximum rate of 12.5 cm in 3 min was observed, although the usual rate was much slower.

Other observations have been made on cannibalism in wireworms and the eating of eggs of other insect pests (e.g., grasshopper eggs).

Automatic plotting of the position of a tagged insect. It has been suggested from time to time that an automatic device for continuous following and recording the position of a suitable tagged insect would be of some value (Ref. 58). Tracking or following continuously requires, first of all, a sensitive head and recorder which will follow the insect's movements. The sensitive head must be such that when it is not directly over the insect, a definite type of error signal is produced. The error signal is amplified, and the amplified signal applied to a servo mechanism which causes the head to move in such a way that the error signal is minimized by restoring the head to a position directly over the insect. The movable parts must be carried on a suitable movable framework. If the surface over which the follower operates is reasonably smooth and solid, the movable framework can consist of a cart or trolley mounted on wheels. Here again considerable mechanical variation is possible. One of the simplest schemes is to have two main coaxial traction wheels capable of independent rotation in both forward and reverse directions, the third point of support for the framework being merely a caster. The path taken by the insect could be marked directly on the surface traversed or recorded remotely, using a suitable remote recording device.

In the present apparatus, the sensing element is a "Thyrode" tube (Victoreen Co. 1B85) which is rotated about a vertical axis at the end of a fixed arm. The axis of the tube lies in a plane parallel to the operating surface (soil). The axis is tangential to the circumference of a circle lying in this plane and about whose center the arm supporting the tube rotates (Fig. 1.10). If a radioactive source lies on the axis of revolution, the distance between the source and the Geiger tube will not vary during a revolution, and the counting rate of the Geiger tube per unit of revolution will be "constant" except for variations arising from the usual statistical fluctuations associated with radioactive disintegration and background.

If, however, the source does not lie on the axis of revolution, the distance between the tube and source will vary during a revolution. The count rate

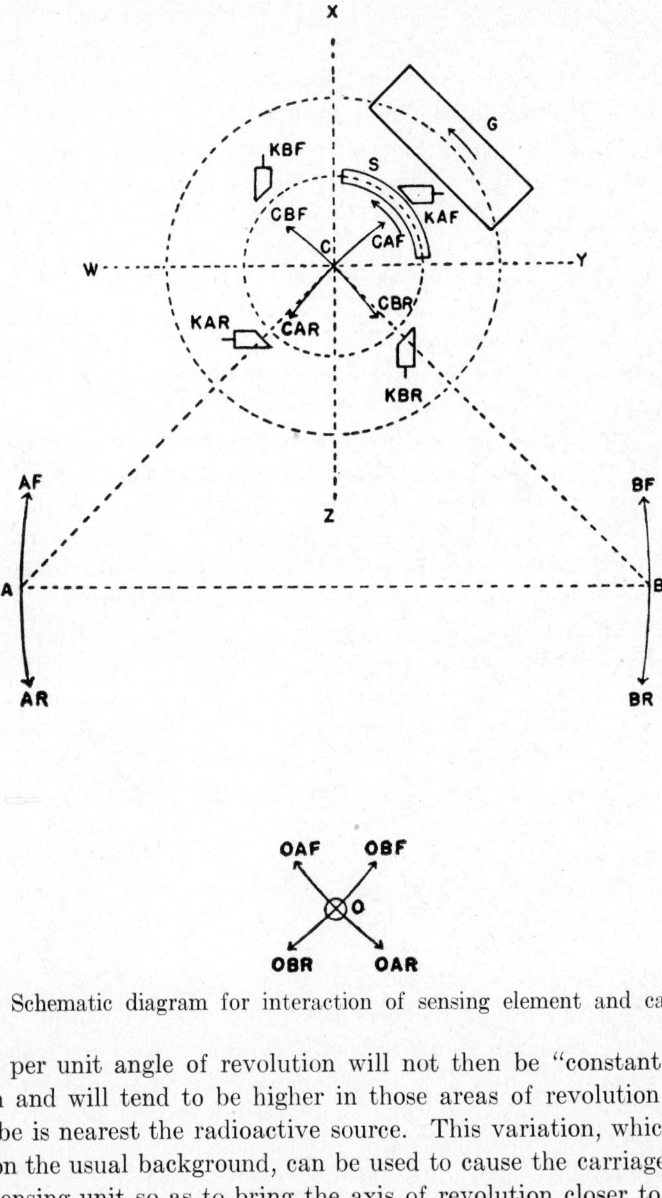

FIG. 1.10. Schematic diagram for interaction of sensing element and carriage unit.

generated per unit angle of revolution will not then be "constant" during a revolution and will tend to be higher in those areas of revolution where the Geiger tube is nearest the radioactive source. This variation, which is superimposed on the usual background, can be used to cause the carriage to reposition the sensing unit so as to bring the axis of revolution closer to the radioactive source. Shielding may be used to intensify these variations.

In the present model of the following unit, the Geiger pulses for each quadrant of revolution are separated by mechanical commutation. These pulses, in effect, are *stored* temporarily in order to develop voltages which are a function of the *average* count rate for each quadrant. The voltages resulting from collections from opposite quadrants are balanced against each other. These two different voltages are used to control the carriage movements (Fig. 1.11). The carriage chosen for the present tests involves a "free" unit which is capable of a crab-like motion. This machine is free to move over a horizontal area limited only by the boundaries of a suitable operating surface and the length of the attached electrical cable.

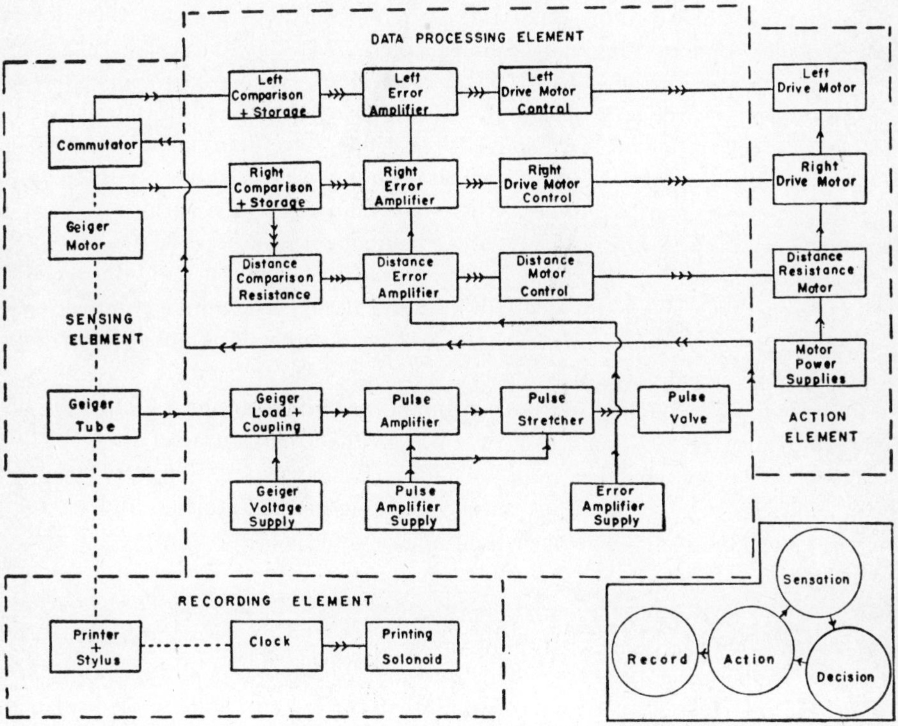

FIG. 1.11. Block diagram for automatic following device.

There appear to be at least two possible crab carriage mechanisms which can utilize the error information provided by the sensing element for explicit position correction. In the one chosen, the axis of revolution of the Geiger tube holder passes through the vertex of a triangle which is isosceles and approximately right-angled, the axis of revolution being at right angles to the plane of the triangle. This triangle lies in a plane approximately parallel to the plane of revolution of the Geiger and in or near the operating surface. Tractive effort produced by reversible electric motors is applied at each end of the base of this triangle. The direction of application of this effort is approximately at right angles to the base of this triangle. Most of the weight of the unit is distributed equally on the traction members. The remaining weight is carried by a small polished surface (coaster) which can glide freely to any position. Small movements by either tractor result in the machine pivoting about the opposite stationary traction point. The vertex of the triangle describes a small arc of a circle for movement of either traction unit. The arcs meet approximately at right angles. The quadrants of collection for pulses from the moving Geiger tube are approximately bisected by these arcs (Fig. 1.10).

The motors which produce the tractive effort may operate in either direction and may operate singly or simultaneously. If the source position does not lie on or close to the axis of Geiger revolution, error signals develop and the appropriate corrective action by one or both motors is initiated. This method of steering has proved "dodge" free. There are many general methods of converting the measured error signals into corrective tractive efforts, and for each of these general methods there are many possible variations in the arrangement of electronic and electrical components. The electrical circuits were selected mainly on the basis of availability of components. The model in its present form is not the ultimate in performance or simplicity, but it does demonstrate that a simple, economical, and practical unit is possible, and its present performance is equal to expectations.

Control in the model is of the "on-off" type, and traction is supplied by relay-controlled series-wound electric motors. The relays are driven from the error signal voltages by d-c amplifiers using hard vacuum tubes. The error signal voltages are obtained by amplifying, lengthening, storing, and balancing the Geiger pulses. Control of hunting was obtained by limiting the speed of correction. One method used was to limit position correction to small "parcels." These "parcels" were approximately half the size of the error which could be detected. The application of these "parcels" was followed by a forced inoperative interval. This interval allowed the storage and balancing circuits to readjust and accumulate reliable error data as a basis for each succeeding parcel of correction.

All translatory movements of the axis of Geiger revolution are recorded on the operating surface by a stylus which is almost coincident with the axis. A record of the time at which the machine and active source were in a given position is obtained from a clock-driven printing wheel carrying on its periphery suitable marking symbols representing the time. At chosen intervals this wheel is inked and momentarily pressed against the operating surface. The printer is at present activated by a solenoid controlled by the clock through a switch, a sensitive relay, and a slowly charged condenser.

A record of the distance of the radioactive source from the operating surface is also available. This is obtained by measuring the mean of one pair of the voltages collected for position error determination. The magnitude of these mean voltages depends on the distance of the radioactive source from the operating surface. It is measured with a self-balancing bridge. The collected voltage is compared with the voltage across the slider and ground end of a potentiometer by a third d-c amplifier. If these voltages are not in close agreement, a relay-controlled meter repositions the potentiometer slider so as to obtain agreement. Rotation of this potentiometer shaft moves a crank which changes the distance between the time printing wheel and a similar and angularly synchronized distance printing wheel. The printing operation, therefore, produces two identical characters whose character indicates time and whose separation is a measure of the distance of the radioactive source from the operating surface.

The apparatus described has been built and is able to home onto a wireworm tagged with 20 microcuries of Co^{60} from a distance of about 40 cm. It is able to follow such a tagged wireworm through 10 cm of soil, with an average position error of about ½ cm. It is capable of continuous operation at the usual speed for wireworms, a few cm per hr. It can, if necessary, travel much faster than this.

The principle of a rotating Geiger would seem to be capable of numerous other applications.

Use of Isotopes in Canadian Silvicultural Research

Plant nutrients. Although the translocation and ultimate disposition of nutrients absorbed by plant roots have long been of interest to plant physiologists, there are still differences of opinion as to the mechanism of transfer, the path of movement, and the functions of the different nutrients in the metabolic activities of the plant. Radioactive tracers have been used in investigating a variety of such problems related to silviculture. One of the earliest of these studies concerned the movement of phosphate in trees (Ref. 59).

A recent Canadian study describes the location of radiophosphorus in various parts of red pine seedlings after being absorbed from the surrounding soil (Ref. 60) by a part of the root system. The most noticeable feature observed was the uneven distribution of the radioactivity in the active plant. Often the activity would be concentrated in one of the branches, suggesting that certain roots provide nutrients to certain branches. Phosphorus was accumulated in the base of the leaf where the leaf meristem is located. Phosphorus is generally thought to concentrate in meristematic regions.

In other studies, radioisotopes such as Ca^{45} and Rb^{86} were injected into the trunks of yellow birch and white pine (Ref. 61). The movement of the radioisotope was followed using a newly developed portable scintillation counter. The maximum rate of upward movement of the Rb^{86} in the xylem of yellow birch approximates 30 cm per min along a narrow channel spiraling upwards (usually dextrally) from the point where the isotope was first introduced.

Movement in decadent yellow birch was very slow, with an apparent increase of permeability of the bark tissue as indicated by lateral diffusion of the isotope. In October no upward movement was discerned in healthy trees, but rather an active downward translocation in the phloem.

Radioarsenic, As^{76}, has been used in debarking studies of the Douglas fir (Ref. 62). The initial rate of rise of arsenic in the tree was 120 cm per hr. The arsenic is absorbed immediately by the sapwood, rises through the outer layers of the wood and apparently diffuses into the bark. Other tracer studies of interest to silviculture are the use of C^{14} and P^{32} in uptake studies of stomatal cells and parasitized leaves (Refs. 63, 64).

Plant pests. Radioisotopes can obviously be used to study forest insect pests. In one such Canadian study, the white-pine weevil was tagged with Co^{60} as nitrate dissolved in cellulose acetate and acetone to form an adhesive (Ref. 65). The Co^{60} was applied to the elytra and the tagged weevils were then released in a white pine plantation. Sixty-four were tagged with 0.2 to 0.5 millicurie each, and after 2 months, 21 were still alive and tagged. This is about the usual percentage of survival for untagged specimens under similar conditions; but it was thought that a smaller amount of radioactivity might have been better, since in the following spring only 19% emerged as compared to 56% in a control group. The tagged insects could be detected from a distance of about 2.75 m.

Other tracer studies are in progress in Canada on the use of radioactive carbon in metabolic studies of forest insect viruses (Ref. 66). Radioactive carbon has also been used in studies of the action of DDT on houseflies (Refs. 67, 68, 69), while radioarsenic has been used to tag mealworm larvae and tomato hornworms (Ref. 70).

Conclusion

While this chapter has been restricted to a consideration of applications of radioisotopes to studies of plant nutrients and pests in agriculture and silviculture, numerous other biological applications have been made in Canada. These cover such diverse topics as radiation-induced mutations in *Drosophila* and barley, cobalt deficiency in sheep, phosphorus and calcium metabolism in hens, and the mode of action of the anti-blood-clotting agent, dicumarol. All in all, we can say that Canada is pursuing these studies actively, and that her atomic scientists are trying to live up to the ideals expressed in Swift's Brobdingnag: "He gave it for his opinion that whoever could make two ears of corn, or two blades of grass to grow upon a spot of land where only one grew before would deserve better of mankind and do more essential service to his country, than the whole race of politicians put together."

REFERENCES FOR CHAPTER 1

1. G. Hevesy, *Biochem. J.* 17:439 (1923).
2. J. W. T. Spinks and S. A. Barber, *J. Am. Chem. Soc.* 68:2748 (1946).
3. S. A. Barber, J. Mitchell, and J. W. T. Spinks, *Can. Chem. & Proc. Ind.*, 1947, p. 757.
4. J. W. T. Spinks and S. A. Barber, *Sci. Agr.* 27:145 (1947).
5. J. W. T. Spinks and S. A. Barber, *Sci. Agr.* 28:79 (1948).
6. H. G. Dion, J. E. Dehm, and J. W. T. Spinks, *Sci. Agr.* 29:512 (1949).
7. W. L. Hill, E. J. Fox, and J. F. Mullins, *Ind. & Eng. Chem.* 41:1328 (1949).
8. J. G. A. Fiskell, *Science* 113:244 (1951).
9. J. G. A. Fiskell, W. A. DeLong, and W. F. Oliver, *Can. J. Chem.* 30:9 (1952); and *Can. J. Chem.* 30:185 (1952).
10. J. G. A. Fiskell, W. A. DeLong, and W. F. Oliver, *Sci. Agr.* 32:480 (1952).
11. W. D. E. Thomas and D. J. D. Nicholas, *Nature* 163:719 (1949).
12. C. D. McAuliffe, *Analyt. Chem.* 21:1059 (1949).
13. N. Veall, *Brit. J. Rad.* 21:347 (1948).
14. A. J. MacKenzie and L. A. Dean, *Analyt. Chem.* 22:489 (1950).
15. A. M. Kristjanson, H. G. Dion, and J. W. T. Spinks, *Can. J. Tech.* 29:496 (1951).
16. W. R. Shelton and H. J. Harper, *Iowa State Coll. J. Sci.* 15:403 (1941).
17. J. W. T. Spinks and H. G. Dion, *J. Chem. Soc.* S410 (1949).
18. H. G. Dion, J. E. Dehm, and J. W. T. Spinks, *Can. Chem. & Proc. Ind.* 34(11):905 (1950).
19. Descriptions of soils may be found in *Saskatchewan Soil Survey Report No. 12*, University of Saskatchewan, Saskatoon.
20. J. Mitchell, "Tracer studies of the utilization of phosphates by cereals in Saskatchewan, 1953" (to be published).
21. J. Mitchell, H. G. Dion, A. M. Kristjanson, and J. W. T. Spinks, *Agr. J.* 45:6 (1953).
22. J. Mitchell, A. M. Kristjanson, H. G. Dion, and J. W. T. Spinks, *Sci. Agr.* 32:511 (1952).
23. H. G. Dion, J. W. T. Spinks, and J. Mitchell, *Sci. Agr.* 29:167 (1949).
24. D. A. Rennie and J. Mitchell, *Can. J. Agr. Sci.* 34:353 (1954).
25. D. A. Rennie and J. R. Soper, *Tracer Fertilizer Research Report*, Soil Science Department, University of Saskatchewan, 1954.
26. M. Fried and L. A. Dean, *Research Report 197*, U.S. Department of Agriculture, 1950.
27. S. Larsen, *Den kgl. vet og Land*, Copenhagen, 1950.
28. R. S. Olsen, C. V. Cole, F. S. Watanabe, and L. A. Dean, *Circular 939*, U.S. Department of Agriculture, 1954.
29. R. S. Russell and R. P. Martin, *Nature* 163:71 (1949).
30. R. S. Russell, S. N. Adams, and R. P. Martin, *Nature* 164:993 (1949).
31. C. Bould, J. D. Nicholas, and E. D. Thomas, *Nature* 167:141 (1951).
32. J. M. Blume, *Soil Sci.* 73:299 (1952).
33. H. G. Dion, C. F. Bedford, R. St. Arnaud, and J. W. T. Spinks, *Nature* 163:906 (1949).
34. S. B. Hendricks and L. A. Dean, *Proc. Soil Sci. Am.* 12:98 (1948).
35. J. W. T. Spinks, M. A. Reade, J. E. Dehm, and H. G. Dion, *Sci. Agr.* 28:309 (1948).
36. E. Penner, *Can. J. Agr. Sci.* 34:41 (1954).
37. E. Penner, *Can. J. Agr. Sci.* 34:214 (1954).

38. M. Fried, *Report TID 5115,* p. 452. U.S. Atomic Energy Commission, 1953.
39. J. Govaerts, A. Lecrenier, C. Corin, E. Derinne, J. Trejinsky, and O. Liard, *Radioisotope Techniques,* H.M. Stationery Office, London, 1953, Vol. 1, p. 395.
40. R. S. Russell, J. B. Rickson, and S. N. Adams, *J. Soil Sci.* 5:85 (1954).
41. R. S. Russell, *Radioisotope Techniques* 1:402 (1953).
42. S. Glasstone, *Source Book on Atomic Energy,* D. Van Nostrand Company, Inc., Princeton, N. J., 1950, p. 295.
43. J. W. T. Spinks, D. A. Lane, and B. B. Torchinsky, *Can. J. Tech.* 29:371 (1951).
44. D. A. Lane, B. B. Torchinsky, and J. W. T. Spinks, *ASTM Special Technical Publication No. 134* (1952), p. 23.
45. D. J. Belcher, R. C. Herner, T. R. Cuykendall, and H. S. Sach, *ASTM Special Technical Publication No. 134* (1952), p. 10.
46. W. Gardner and D. Kirkham, *Soil Sci.* 73:391 (1952).
47. P. R. Wallace, *Nucleonics* 4:30 (1949).
48. J. Gueron, *J. de Physique et de Radium* 15, Supp. 5:5 (1954).
49. D. W. Jenkins and C. L. Hassett, *Nucleonics* 6(3):5 (1950).
50. J. C. Bugher and M. Taylor, *Science* 110:146 (1949).
51. D. W. J. Jenkins, *Econ. Ent.* 42:98 (1949).
52. J. A. Schemanchuk, J. W. T. Spinks, and F. J. H. Fredeen, *Can. Ent.* 85:269 (1953).
53. A. M. Kristjanson, *Report,* University of Saskatchewan, 1953.
54. F. J. H. Fredeen, J. W. T. Spinks, J. R. Anderson, A. P. Arnason, and J. G. Rempel, *Can. J. Zool.* 31:1 (1953).
55. R. A. Fuller, P. W. Riegert, and J. W. T. Spinks, *Can. Ent.* 86:201 (1954).
56. P. W. Riegert, R. A. Fuller, and L. G. Putnam, *Can. Ent.* 86:223 (1954).
57. R. A. Fuller, J. W. T. Spinks, A. P. Arnason, and H. MacDonald, *Eighty-first Annual Report of the Entomological Society of Ontario,* 1950, p. 7.
58. B. C. Green and J. W. T. Spinks, *Can. J. Tech.* (to be published).
59. P. R. Stout and D. R. Hoagland, *Am. J. Bot.* 26:320 (1939).
60. J. L. Farrar, *Silvicultural Leaflet No. 78,* Forestry Branch, Dept. of Northern Affairs and Natural Resources, Ottawa, 1953.
61. D. A. Fraser and C. A. Mawson, *Can. J. Bot.* 31:324 (1953).
62. D. G. Wort, *Can. Pulp & Paper Ind.* 7(7):8 (1954).
63. M. Shaw and G. A. MacLachlan, *Nature* 173:29 (1954).
64. M. Shaw, S. A. Brown, and D. R. Jones, *Nature* 173:768 (1954).
65. C. R. Sullivan, *Can. Ent.* 85:273 (1953).
66. G. Bergold, Forest Insect Laboratory, Sault Ste. Marie, Ont.
67. E. J. LeRoux and F. O. Morrison, *J. Econ. Ent.* 46:1109–1110 (1953).
68. E. J. LeRoux and F. O. Morrison, *J. Econ. Ent.* 47:1058–1066 (1954).
69. F. O. Morrison and E. J. LeRoux, *J. Agr. Sci.* 34:316–318 (1954).
70. F. O. Morrison and W. F. Oliver, *Can. J. Res.* D,7:265–269 (1949).

Chapter 2

THE ATOM AND THE WORLD FOOD PROBLEM *

The present world population of about 2.5 billion is growing at the rate of nearly 1½% each year. Every day nearly 100,000 additional hungry mouths appear at the breakfast table, and twenty-five years hence it is expected that we shall be 3.5 to 4 billion in number. Evidently one of the many major problems facing us, probably basically the most important for the future welfare of mankind, is how we shall be able to provide food, clothing and shelter for the world's peoples at the ever higher standards that they so rightly expect.

The Neo-Malthusians believe that we are engaged in a losing battle and that it will not be possible for world resources to meet requirements. The Food and Agriculture Organization of the United Nations believes that technically it is possible to achieve the necessary increases in production of foodstuffs and basic raw materials to meet world needs for the foreseeable future, and has recommended to governments that they aim at increases in production 1% to 2% greater than their anticipated growth in population. For many countries this will mean production increases of 3% to 4% a year. This admittedly will do little more than maintain the *status quo;* but even so, the social, political, and administrative problems involved will be of immense magnitude. Their solution alone, quite apart from the further urgent need to make a real improvement in the inadequate subsistence levels which at present characterize so many parts of the world, will require determination and fortitude on the part of governments. Although freedom from want must depend essentially upon the efforts made by each country on its own behalf, success in achieving this is of such general concern to all peoples that these national efforts should be supplemented by pooling the knowledge and experience of all countries on an international basis for the common good. In doing so, it is imperative that all possible advantage be taken of the contributions to increased production

* This chapter is taken from Geneva Conference Paper 780, "The Uses of Atomic Energy in Food and Agriculture" by R. A. Silow of The Food and Agriculture Organization of the United Nations, and M. E. Jefferson of the United States Department of Agriculture.

that might come from scientific and technical advances in other fields. Among those advances the development of atomic energy is of great significance for agriculture and the related industries of forestry and fisheries.

From the technical standpoint the solution of the problems of providing more of the primary necessities of life can come in three main ways. The easiest method of increasing supplies is by taking adequate measures to reduce the heavy losses which at present occur in all segments and stages of production, storage, and distribution. Secondly, the productivity of land now under cultivation could be much increased and the current utilization of our fisheries and forestry resources intensified through development of improved technical methods. Finally, supplies could be increased by developing new areas and resources; but this is the most difficult course since the areas and resources most easily developed are already being utilized.

The ways in which atomic energy can help in feeding, clothing, and housing the world's growing population will be considered in relation to these three main ways of improving production and making better use of our natural resources.

The Significance of Atomic Energy for Food and Agriculture

When relatively inexpensive power from nuclear reactors becomes generally available it will undoubtedly make a substantial impact on agriculture, if only by cheapening costs of production and distribution, improving conditions of work, and making modern conveniences and comforts more widely available in rural communities. This is particularly true in areas where other sources of power are deficient. These and other at present more speculative developments must, however, await the widespread application of nuclear power in industry, and its possible contributions will be discussed a little more fully later. At present and for the immediate future the radioactive isotopes and radiation that have become available as by-products of nuclear reaction are of greater importance to agriculture.

It was at one time thought that radiation and radioactive isotopes might prove of direct value in the stimulation of plant growth. To date the only stimulatory effects in the irradiated generation which have been observed in extensive investigations conducted in the United States of America, the United Kingdom, and Canada, have been those associated with damage to the plant, much as growth is stimulated, in a sense, by the use of the pruning knife, and there has been complete failure to substantiate the earlier hope for useful stimulation of growth.

Neither radiation nor radioisotopes therefore can make any direct contribution to increased production in the sense that an application of fertilizer leads to increased crop yields. Their contribution to food and agriculture is indirect, but nevertheless of immense potential. The value of radiations lies partly in their ability to induce inherited changes in the germ plasm and partly in their sterilizing effects on biological tissues. The ability to induce mutations is being used in plant breeding programs and in other ways, while the sterilizing effects have promising applications in food preservation and pest control. The value of radioisotopes in agriculture arises from the ease and accuracy with which they can be identified and measured in extremely minute amounts, which makes it possible to use them as highly refined research tools in so-called tracer studies in a wide range of nutritional, metabolic, developmental, and pathological investigations in plants, animals, and man. In this way they are giving information which at the present time could be obtained in no other way or only at much greater expense in terms of time and money. Thus, by giving a clearer insight into basic biological processes that have hitherto been obscure, the use of radioisotopes in tracer studies is already leading to greater efficiency and economy in the production and utilization of agricultural products. Bearing in mind that it was only unfettered scientific investigation of the nature of the atom—the pursuit of knowledge for its own sake—that made nuclear energy available to mankind, the potential value of the contributions which radioactive tracer studies can make to food and agriculture may similarly be almost unlimited.

Applications of Radioisotopes in Agriculture and Forestry

Food preservation and storage. Potential food supplies and other agricultural products are subject to heavy losses in all phases of production, distribution, and storage through fungal and bacterial infection and the ravages of insect pests, and it is by combatting these that the most immediate and spectacular improvements in supplies can be achieved. No valid estimate of total world losses is possible, but they are undoubtedly of immense magnitude. A very conservative estimate that has been brought forward for losses in stored grain, largely from the depredations of weevils and other insect pests, is 10%, but the losses are undoubtedly much greater in the hot and humid areas of the world, where the figure of 25% to 50% loss of harvested cereals and pulses which has been estimated for Central America is probably generally applicable in most of the less advanced countries. In addition, perishable foodstuffs such as fruits, vegetables, meat, and fish are particularly subject to spoilage in distribution and storage. Similarly the deterioration resulting from fungal infection and the attacks of various insect pests is of major concern in timber utilization.

Evidently there is great scope for the adoption of control measures, and these can be applied relatively easily at a cost which is usually but a small fraction of the value of the returns. Already radiations and radioactive isotopes show promise of making important contributions to the development of improved control measures which would markedly improve the supply situation. Thus the destructive power of radiation has been used to eliminate insect infestations in grain and cereal products at costs which compare favorably with those of the more conventional procedures such as fumigation. The sprouting of potatoes has been successfully inhibited as a result of the effects of radiation on the enzyme system, thus permitting transportation under less stringent conditions than are usually required and extending storage life by many months. Much attention is also being given to the possibility of food preservation through cold sterilization by irradiation at normal temperatures. While full success has not yet been attained, significant improvements in the keeping quality of meat and meat products have been achieved without off flavors or color changes arising from detrimental side reactions. In some countries trichinosis is an important health problem and it has been shown that the irradiation of pork can kill or sterilize the trichinae, thus rendering infected meat safe for human consumption.

Evidently radiation may have a very useful field of application in food processing and preservation, although much more exploratory work must be done to evaluate its full potentialities. One of the most attractive features of this application is the wide scope it offers for the useful employment of the radioactive residues arising as a by-product of the operation of nuclear reactors.

Reducing losses in growing crops. Every year a large part of the potential harvest is destroyed by diseases and pests which attack the growing crop. It has been said that even in such relatively developed countries as the United Kingdom the work of over 51,000 skilled farm workers is lost each year, and that in the United States of America losses caused by insects, weeds and plant diseases account for losses in farm output equivalent to $13 billion per year. Most of these losses could be avoided by the timely application of control measures, and radioisotopes are proving to be exceptional tools in studies which are leading to the development of improved materials and methods for safeguarding our harvests. Adequate control of a destructive insect, for instance, generally requires a thorough knowledge of its life cycle and habits, and tagging insects with radioisotopes provides a much more efficient means of determining their flight range, migrating routes, and overwintering habits than methods such as painting which were previously used. In Canada cobalt^{60} has been used to label wireworms, making it possible to follow their underground meanderings, while in both the United States of America and Canada the flight and overwintering habits of a number of forest insects are under investigation, using similar techniques. These studies will undoubtedly be important factors in better control.

Extremely effective insecticides have become available in recent years but one of the problems associated with their continued use is that insects frequently build up resistance to these poisons. Studies with insecticides labeled with a radioisotope permit comparison of uptake and metabolism by normal insects and by those that have developed resistance. This work may be a step in determining the nature of resistance to the poisons, concerning which so far little is known except that resistant insects take up the poisons just as susceptible ones do, but do not react to it in the same way.

Similarly radioactive tags are playing an important part in the development of fungicides, insecticides, and weed killers, for use either as direct applications or as systemic poisons. These latter are substances which are applied, usually through the soil, and taken up by the plant without harm to it but are toxic to pests feeding on the plant. It is important to know that such substances are not altered in use to products harmful to man or animals. For instance the weed killer 2-4-D when used to kill weeds in the bean field is readily absorbed by the bean plant also and has been found by the use of labeled material to be distributed throughout the plant together with at least two additional products derived from 2-4-D. It is obviously important in the case of edible plants to be able to trace such compounds and their metabolic products because of their possible effects on the consumer, animal or man, and the use of radioactive labels is playing an important part in the development of safe materials and methods.

Breeding improved crop varieties. Although control of crop pests and diseases by chemicals can be extremely effective, their use involves additional expenditure and usually requires considerable care. Undoubtedly the most satisfactory safeguard, and one that is particularly applicable in the underdeveloped areas, is the growing of crop varieties resistant to the prevalent pests and diseases. The plant breeder has long been engaged in developing such varieties, by well-known conventional methods involving selection and hybridization, and has achieved marked success; but this is a never-ending task, for existing varieties do not satisfy all needs and new pests and diseases or new forms of old ones are constantly appearing. In his task of developing disease- and pest-resistant varieties as well as other improved types with better agronomic characteristics and higher productivity, the plant breeder is already being greatly assisted by the new potent sources of radiation which have recently become available, and their use in this connection may well ultimately prove to be one of the most significant contributions of atomic energy to agriculture. It has long been known that radiations cause heritable mutations in plants and animals. With the advent of atomic energy further kinds of radiations and more potent sources have become freely available for experimental use, and in many countries extensive programs have recently been initiated for crop improvement through irradiation, with a view to accelerating the normal rate of mutation and so increasing the variability available to the plant breeder for selection.

Radiations are applied either in the early flowering stages in order to affect the developing gametes or to seeds in which somatic mutations are first induced, and such affected cells may subsequently give rise to germinal tissue, in which case the mutations are transmitted to later generations. Although as in the case of spontaneously appearing mutations the vast majority are deleterious, desirable types occur in low proportion and have been selected in a wide variety of crops. Improved types that have been obtained include higher-yielding or disease-resistant strains of cereals and other crops such as peanuts, stiff-strawed types of cereals resistant to lodging, types better adapted to mechanical harvesting on account of particular size or shape characteristics, types with extended or reduced maturity period, and types with changed ecological requirements, for instance in relation to higher or lower rainfall or soil fertility. An example of a particularly significant outcome of work of this type may be cited from Canada, where among some twenty barley mutants produced by irradiation and now under field trial are some maturing sufficiently early to extend the area in which barley might be grown in that country.

Plants such as some orchard crops which are normally propagated vegetatively are also receiving radiation treatment for the induction of bud mutations. Mutants have similarly been induced in algae, and in this way types have been obtained which are adapted to high temperatures and thus more suitable for growth in mass culture. Such types may one day become an important source of food and industrial raw materials. In forestry increasing emphasis is being given to the breeding of fast-growing types and varieties resistant to pests and diseases, and radiation may play an important role in the production of such improved strains of trees.

In the case of disease control the ability of radiations to induce mutations in the disease-producing organism itself may also prove to be a tool of exceptional value. Such pathogens as the rusts that affect wheat and oats or the smut attacking maize undergo spontaneous mutations in nature at a rate rapid enough to cause constant trouble for the crop breeder. If he develops, for example, a wheat resistant to black stem rust, he may find that within a comparatively short time a new mutated form of the rust organism appears to which his variety is no longer resistant as it was to the original form. Current work has shown that radiation will produce new races of such disease organisms with increased virulence. By developing these new races artificially and under controlled conditions, the breeder may be able to anticipate the resistance requirements of his crop and to breed adequate resistance prior to the appearance of new strains of the pathogens in the field.

Increasing productivity through improved practices. In addition to the adoption of adequate measures for the control of plant and animal diseases and pests and the use of higher-yielding crop varieties developed by the plant breeder, the productivity of land now under cultivation can be increased by the adoption of a wide range of improved methods of crop and animal husbandry. Here again through tracer studies using radioisotopes, atomic energy is making notable contributions to the development of improved techniques through the advancement of fundamental knowledge in animal and plant nutrition and physiology and a better understanding of the complex relationships between animal, plant, soil, water and sunlight on which man is dependent for his existence. The study of such a dynamic biological system is normally particularly difficult and the tracer technique has immeasurably facilitated such investigations. In many cases it provides the only practical approach to the solution of a problem and without it further progress in certain directions would at this time have been impossible.

Soil fertility. The major limiting factors in the productivity of land now in cultivation are the inherent yielding potential of the crops and livestock on them and the amount of nutrients available to enable those plants and animals to produce to their capacity. Contrary to general belief, some of the most productive soils in the world were originally of rather poor fertility and have been brought to their present capabilities by man's careful tending of the soil and the development of various principles of good husbandry, which through the ages have built up the fertility of the soil. The potentialities for increasing production through adoption of those principles on a wider scale are very great. As just one example, the yield of rice in Japan is about four times that of India's one ton per hectare. Much, though admittedly not all, of this increased productivity is due to the much greater use of inorganic fertilizers in Japan, and evidence from experiments indicates that through the use of only a moderate dressing of 30 kg of nitrogen per hectare, with other plant nutrients where necessary, India's annual rice production of about 35 million tons could be raised by more than 10 million tons—which would mean much for the welfare of the population and for the economy of the country. However, the present limited use of fertilizers in such countries is almost entirely a question of economics, so that it is essential that the best use be made of the natural fertility of the soil, and that added fertilizers be employed in such a way as to ensure the fullest possible return from them. Here tracer studies with isotopes in many countries are providing a wealth of fundamental information and practical hints. Radioisotopes of most of the important plant nutrients are now available and they have opened up new avenues of attack on important problems of soil fertility.

The fact that phosphorus is to so large a degree retained by soils in forms that are not readily available for use by plants is one of the major concerns of soil science. Advances in the study of this problem are being made through such work as that in the United States of America with radioactive phosphorus and calcium on the factors influencing the fixation of phosphorus in calcareous soils. Canadian workers, who were among the first to point out the advantages of the tracer technique in soil fertility and plant nutrition problems, have shown that the usual procedure for evaluating the contribution of a phosphorus fertilizer by comparing the total phosphorus absorbed by the plant from fertilized and unfertilized plots could be misleading. Before the use of tagged material it was thought that the increased uptake from the fertilized plot all came from the fertilizer. Through the use of radioactive phosphorus it has been found that when the fertilizer is applied, the plant takes up additional phosphorus from the soil itself as well as from the added nutrient.

Rice, the basic food of nearly one-half of the world's population, is typically grown under irrigation, in submerged soil. In Japan the radioisotopes of phosphorus, sulfur and iron are being employed in preliminary investigations of the chemistry of such submerged soils, which is quite different from that of ordinary soils, and valuable information is to be expected on the factors affecting the uptake and translocation of these nutrients by rice. Similar work on the determination of the phosphorus fertility status of tropical and submerged soils is under way in India. Other investigations are giving information on the most economic source of specific nutrients and the best placement and time of application of added fertilizers, particularly in relation to the growth period when the plant can best use them and their placing in relation to the main feeding zones of the roots. In this connection a good deal of attention is being given to investigation of the characteristic rooting habits of different plants through use of labeled nutrients, and this ability to define the zones from which particular crops obtain the major portion of their nutrients and moisture requirements should make a valuable contribution to better agricultural practices.

Soil moisture, drainage, and irrigation problems. Many other soil problems are amenable to attack by this new tool. It makes possible, for instance, rapid determination of the moisture content and the density of soils, a matter of interest alike to the agronomist, the soil conservationist, and the engineer. Moisture content of the soil may be estimated by a method dependent on the degree of neutron scattering by hydrogen atoms contained in the soil water. A similar method involving gamma rays instead of neutrons is used in measuring the density or degree of compaction of the soil. One application of these methods in the United States of America is the study of the effects of tillage and harvesting machinery in compacting the soil.

A different application to soil problems is the addition of phosphorus-32 or rubidium-86 to surface waters such as streams or ponds to determine the rate and direction of drainage into the soil, and to irrigation waters to find out whether they reach the farthest points of the field as well as the nearest points —that is, to evaluate the efficiency of distribution of the water in the soil. Radioisotopes are also being used, for instance in Japan, to detect leakage in irrigation dams and to survey supplies of underground water, and tritium may have a particularly valuable application in large-scale research in hydrology, for example over entire watersheds.

Plant nutrition and metabolism. In plant nutrition the basic mechanisms involved are the uptake of nutrient elements, their transport through cell membranes, and their subsequent translocation or movement throughout the plant. Such mechanisms are of course basic also in the case of animals.

Investigation of these fundamental mechanisms through the use of a considerable variety of radioactive elements is under way in many countries. Fuller knowledge of the phenomena involved is essential to the understanding of plant nutrition and growth, and discoveries that were hitherto impossible in this field are now being made. Amongst other practical aspects the mechanism of transport plays an important part in the utilization of growth-affecting or growth-regulating factors—the so-called plant hormones which are used to regulate development in crops, particularly in horticulture—and in the utilization of such compounds as the previously mentioned 2-4-D which are used as weedkillers, and the systemic insecticides which are administered to the pest by way of the host plant. An interesting example of the place of such translocation studies in relation to practical problems is the work on mistletoe in Australia, where this parasite is a serious menace of eucalyptus. In work directed towards its control, the isotopes of cobalt, iron and zinc are being used to obtain information on the efficiency of movement of toxic compounds from the host tree to the parasite.

Many of the plant hormones and weedkillers are used as leaf applications, and essential nutrients may also be applied to aerial parts of the plant as well as by way of the soil. Thus under certain conditions, especially with trees, temporary deficiencies of such vital elements as iron and zinc may occur, and these can often most readily be corrected by spray applications either in the dormant stage or when in leaf. Radioisotopes have been particularly helpful in demonstrating that some plants can absorb such nutrients efficiently through the foliage and that nutrients so absorbed are rapidly translocated throughout the plant. This principle is already being applied rather widely in practice, urea for instance being used fairly extensively as a leaf spray by fruit and vegetable growers in the United States of America. Additional information on the absorption and translocation of such substances is needed and through the use of urea labeled with carbon-14 it has already been found that crops differ markedly in their ability to utilize this compound as a source of nitrogen, cucumbers using it more than four times as fast, for instance, as do cherries and potatoes. In the case of strawberries, calcium is readily absorbed, but it has been found that it is not translocated into daughter plants and hence foliar feeding as a major source of calcium is not adequate in this plant.

In Puerto Rico studies of the formation of *Hevea* rubber are being made in tissue cultures, using possible chemical precursors of rubber tagged with carbon-14. Such investigations should lead to increased knowledge of the basic reactions and mechanisms involved in the secretion of gums and resins and will contribute in the long run to more efficient production of these economically important materials.

Radioisotopes offer a means of determining the rate of movement of water in plants and rather direct evidence of the structural pathways involved. Canadian forests are suffering from a die-back in birch, and radioactive phosphorus and rubidium have been used in the current investigation of the disease. These isotopes have shown that the normal narrow-band upward-spiral path of solution movement is broken in the affected tree and replaced near the dying region by a confused irregular pattern. Although no answer to the die-back problem is as yet forthcoming, it may be anticipated that the new techniques will play their part in the ultimate solution.

Well in the forefront amongst the contributions of radioisotopes to the advancement of fundamental knowledge of plant nutrition and metabolism has been the remarkable progress made through their use in the elucidation of the highly complex mechanism of photosynthesis, the process by which green plants use the sun's energy in the formation, from air and water, of compounds essential to life. Using radioactive carbon, the early principal pathways of this element have been established, and subsidiary pathways and the specific biochemical reactions involved in the synthesis of carbohydrates are being investigated.

The efficiency of energy conversion by photosynthesis is low, probably not more than 1% of the total available energy in the sunlight falling on a green leaf is effectively used. This low value arises at least in part from the fact that the process is limited somehow by the plant itself. Investigations are now under way with the aim of identifying the compound or compounds that limit the photochemical reaction. Inefficient though this process is, so far in the earth's history it has been the primary source of all food used by men and animals to sustain life, and the source of all our fuels including wood and the fossil fuels, coal and oil—those to which we now refer as conventional fuels in contradistinction to the nuclear fuels of the atomic age.

The remarkable advances that are being made in the understanding of photosynthesis may well lead to important methods of increasing the efficiency of this conversion of energy from the sun into chemical compounds. How much this might contribute to the enlarging of the world's food supply is at this stage a matter of speculation, but the work so far seems to hold much promise.

Another field of investigation of major importance concerns the part played by enzymes in life processes. More detailed knowledge about them is necessary if we are to have a true understanding of the metabolism and the synthesis of organic products. In this field earlier British work on cell metabolism is being paralleled by Canadian studies on the role of enzymes in the synthesis of sugars and amino acids in the living organism, while in the United States of America fundamental investigations are under way in the basic mechanisms and the factors influencing reaction rates and the dynamic equilibria characteristic of living cells.

Animal husbandry. Tracer studies are also giving much impetus to investigations on basic problems involved in animal production. As in the case of plants, understanding of enzyme action and other metabolic processes is of prime importance in animal nutrition. Some of the more fundamental investigations concern the amino acids, which are combined in complex ways to make the proteins of the body. Among the amino acids essential in the animal diet are two that contain sulfur—methionine and cystine. It is known that cystine can partly replace methionine in animal metabolism. Cows, sheep, goats and other ruminants have the power of synthesizing these amino acids from inorganic sulfur through the action of microorganisms in the rumen, or first stomach. It had been thought that non-ruminants, which do not have multiple stomachs, could not carry out this process within their own bodies. But recent work with inorganic radioactive sulfate in the diet of poultry has shown, through the recovery of radioactive cystine, that non-ruminants can also synthesize at least part of the cystine they require, which in turn can partly replace methionine; and hence that inorganic sulfate should be considered an important mineral nutrient for poultry and for swine as well as for cattle and sheep.

Studies on the biochemistry of lactation provide another example of the effective use of tracer techniques in metabolic studies. In such work in the United Kingdom radioactive carbon and tritium, an isotope of hydrogen, have been used to label dietary components. Isolation of the milk constituents and determination of radioactivity indicates directly which milk components are derived from the labeled source and permits some well-founded speculation as to the metabolic route involved. In a typical experiment, sodium acetate labeled with radiocarbon was administered intravenously to a lactating goat. Analysis of the fatty acids in the milk clearly indicated that the synthesis of acids from the acetate took place in the udder itself, not earlier. Such studies, evaluating dietary components and elucidating the steps by which they pass into the sugars, fats (lipids), proteins, and other milk constituents, should lead to more efficient feeding and management for production.

Radioisotopes are especially effective in studying the efficiency of food utilization because it is possible with this technique to arrive at more accurate values than are obtainable by the usual methods employed in food balance studies, which permit only the determination of the total intake and loss of nutrients from the body. In materials found in the waste it has not hitherto been possible to differentiate those that passed through the intestinal tract without being absorbed from those absorbed and subsequently excreted. The use of tagged nutrients permits ready identification of the source of the excreted materials and in such investigations in animals it has been possible, for instance, to distinguish endogenous (body) and exogenous (feed) calcium and phosphorus.

As an example, the value of alfalfa as a source of phosphorus for lambs had long been considered, on the basis of the usual evaluation procedures, to be of the order of 20%; that is, only a fifth of the phosphorus in the alfalfa was utilized by the body. The tracer technique has shown that the phosphorus from alfalfa is actually far more effectively absorbed, about 90% being utilized.

The low value previously estimated was due to the fact that phosphorus was being rapidly involved in the body chemistry, and the large amount excreted was not phosphorus from the feed but phosphorus that had been absorbed earlier, used, and returned to the intestinal tract. Studies of this nature on food utilization have been extended to the egg, the fetus, and the milk, permitting evaluation of the relative contribution of the diet to each.

All of this newer knowledge contributed by radioisotopes work on the metabolism and nutrition of animals is of major concern for the world food problem, since with rising standards of living there is a demand for a higher proportion of animal proteins in the diet.

Just as in the case of crop production, advances in the control of animal pests and diseases and the development of improved breeds also constitute important methods of increasing animal production. In the case of animals, however, experimentation is usually more costly and time-consuming than with plants, with the result that (with some important exceptions) the utilization of radioisotopes and radiation has not progressed so rapidly. Thus while irradiation for inducing mutations in plants is being quite widely used for crop improvement, this technique has not as yet been employed to the same extent with livestock nor has it yielded comparable results. Irradiation has also been shown to induce mutations in fungi causing crop diseases, and these mutants are being used in experiments to develop greater resistance to those diseases. It may be expected that this kind of work with animal disease-causing organisms will have comparable important practical applications. Thus variants of reduced virulence developed by radiation may prove to be of value for use as vaccines in the production of immunity to commonly occurring more virulent forms of the pathogen, and irradiation may also be of help in developing polyvalent vaccines. These are possibilities which, though so far untested, may have great significance, since vaccines constitute the most effective way of combatting the many widespread animal diseases which cause losses of millions of animals throughout the world.

A somewhat different use of radiation has been made within a limited area in the case of at least one insect pest of livestock, the screwworm fly (*Callitroga americana* C. and P.), which is responsible for damage to the cattle industry in the United States of America to the extent of some twenty million dollars annually in killed or crippled animals and damaged hides. An ingenious method of control has been successfully demonstrated by the United States Department of Agriculture in cooperation with the Dutch authorities on the island of Curaçao in the West Indies. Large numbers of male flies were bred in the laboratory, sterilized by radiation from radioactive cobalt, and released in numbers far in excess of the normal population of males in the area so that females were much more likely to encounter a sterilized male than a fertile one. The female fly mates only once in her lifetime and hence there was no possibility that she would subsequently produce fertile eggs. This unique undertaking was carried out after a close study of the life cycle of the insect, the determination of the susceptible developmental stage and dosage requirements, and careful estimation of the screwworm population of the area. The possibility of eradicating other insects by radiation undoubtedly merits further attention following such fruitful results with the screwworm fly.

Applications of Radioisotopes in Fisheries

The world's fisheries at present contribute over 26 million tons of high-quality food. Although this represents but a small proportion of the animal protein consumed by man, fish is an important element of the diet of many peoples, especially in certain countries in the tropics where proteins of animal origin may otherwise be almost absent from the diet. Up to the present, however, the potentialities of the ocean and inland waters as sources of food have been utilized to only a relatively limited extent; but with the application of modern scientific and technological advances revolutionary developments in man's exploitation of the world's aquatic resources may be anticipated.

As in agriculture and forestry, radioactive isotopes have many uses in fisheries research and their employment will accelerate the conclusion of certain investigations which are essential for the evaluation of fishery resources and the understanding of those resources that will permit efficient exploitation. Probably the most important application has been the use of carbon-14 in estimating the productivity of ocean waters. Carbon-labeled sodium acetate has been added to water samples taken at various depths and the samples incubated under controlled conditions. The growth of the phytoplankton or plant life under these measured conditions gives an estimate of the basic productivity of the water from which the samples were taken. The use of this method is rapidly being extended and will contribute significantly to the achievement of a realistic estimate of the basic productivity throughout the seas. The importance of this contribution will be clear since the production from the plant life in the oceans determines the quantity of organic material in subsequent links of the food chains and more particularly in those links that furnish material of economic value to man.

Applications similar to this have been made in fresh-water studies. Radioactive phosphorus has been used in several investigations in fresh-water lakes to study the efficacy of mixing, or the distribution of the phosphorus throughout the water and to the flora and fauna. Recent experiments in Canada have indicated that the turnover of phosphorus in lake waters is much more rapid than had previously been estimated, and further that the rapid turnover under natural conditions appears to be caused primarily by bacteria. This clearly indicates the need for further work, since it had not hitherto been thought that planktonic bacteria play an active part in the phosphorus cycle. On the contrary, the primary organisms involved were considered to be algae, which are highly important in the food cycle of fishes. What contribution, if any, the bacteria make to productivity in these waters is therefore unknown, and it is possible that their high consumption of phosphorus may adversely affect the growth of algae. The importance of radioisotopes in this connection is that they will accelerate the analysis of the ecological systems in inland waters and add precision to the available methods. It may be emphasized that this is especially important in inland waters, where the opportunities for human intervention to modify ecological systems are much greater than in marine systems.

Radioactive tags have also been used in fisheries work for tracing and measuring more massive movements of material. For example, they have been employed in studies of the movement and population density of fish, where they have some advantages over the usual marking methods. There is also the possibility that radioactive tags might be used for following the movements of oceanic currents and for measuring water transport in marine systems. Evidently radioisotopes can play at least as great a part in increasing the productivity of our fisheries resources as they can in agriculture and forestry.

Applications of Radioisotopes in Human Nutrition

Reference has already been made to the particular value of radioisotopes in analyzing the movements of specific nutrients in a dynamic biological system and the important part that such tracer studies can play in elucidating problems of animal nutrition and metabolism. Much of the knowledge concerning human nutrition and metabolism is gained from animal studies and therefore much of what has been discussed earlier with reference to animal nutrition applies equally in the field of human nutrition. Thus work in progress on the metabolism of amino acids and proteins in mammals is especially significant in connection with certain deficiency diseases such as kwashiorkor in man. Current studies also include work on the absorption and excretion of fat and cholesterol, which are considered important factors in atherosclerosis, a type of hardening of the arteries.

Use of the tracer technique with human beings is rather limited because of understandable reluctance to submit them to chronic internal exposure to radiation; but such studies are being made in the United Kingdom, the United States of America, and France on fat and cholesterol metabolism under normal and abnormal conditions. A limited amount of work has also been done with human beings in the field of amino acid and protein metabolism, using carbon-14 and radioisotopes of sulfur and iodine. In connection with mineral nutrition, work has been carried out in the United States of America on the metabolism of calcium in young male children.

An excellent example of international cooperation in the study of an important health problem was one involving the use of the tracer technique in an investigation of iodine metabolism in areas of endemic goiter in Argentina, where with the help of radioactive iodine Argentinian and American doctors made a complete study of uptake and metabolism under conditions of minimum iodine supply. Although the work did not result in any startling new discoveries, it was possible to observe the patterns of iodine metabolism, the changes imposed by deficiency, and the adaptation of the body to deficiency. In addition this investigation contributed new knowledge of the dynamics of iodine transfer within the body.

It is evident that tracer studies can make valuable contributions to a better understanding of the physiology of human nutrition. This in turn will affect the pattern of utilization of available food supplies and help in the development of adequate balanced diets which are so important for the maintenance of health and efficiency.

Future Outlook

Research. It is obvious from the foregoing discussion of past and present work that radiations and radioisotopes will play an increasingly important part in research and development in agriculture and related fields. In fact, the vast field of potentialities of the peaceful uses of atomic energy has hardly been touched, and it may be said that the applications of radioisotopes and radiation to problems of concern to agriculture are limited only by the imagination and ingenuity of the investigators.

The adaptability of radioisotopes to such a wide range of research arises from the fact that they may be quantitatively determined in such minute amounts. As we have noted, this makes it possible to introduce an identifiable component into a complex system such as the soil, a plant, or an animal, and follow its path to determine its fate in the dynamic system. The techniques involved, however, require carefully trained personnel and a considerable investment in laboratory facilities and instrumentation. Because of these limitations it would be well, for the immediate future at least, to confine the use of this tool in general to those problems in which it is the only applicable method, or to cases in which the desired information may be obtained by the use of isotope techniques with a marked saving in time and effort.

The supply of the more important radioisotopes is now generally adequate for present needs and the rate of progress in research in the immediate future will be determined largely by the number of investigators with the necessary training and experimental facilities. Hence, it will be necessary to make arrangements as rapidly as possible for the training of investigators in the handling of radioactive materials and the research techniques involved. The governments of those countries with experience in atomic energy are already generously making their training facilities available to the nationals of other countries; and the Food and Agriculture Organization of the United Nations, like other international agencies, is prepared to assist in facilitating arrangements on request and hopes from time to time to be able to award fellowships for training and research in the applications of radioisotopes to problems in agriculture and related fields. To make the best use of the available knowledge, skill, and facilities, it may be desirable to develop cooperative programs of research on the more important problems of general interest. Many of the fundamental investigations using radioactive isotopes can be pursued at any centers with adequate facilities irrespective of location, but studies of the applicability of the results in specific regions must often be undertaken locally, and here again it is likely that progress will be accelerated through cooperation between neighboring countries.

The encouragement of promising long-term investigations might well be a suitable function of the projected atomic energy agency in consultation with the appropriate Specialized Agencies of the United Nations, which have much experience in stimulating international cooperative investigations where such procedures can lead to more effective use of relatively restricted resources in trained manpower and facilities.

Power aspects. Little reference has so far been made to the potential benefits to agriculture and related industries that may be expected from the more abundant supplies of cheaper electric power that will eventually become available from atomic energy developments, other than to indicate in a broad way the general lowering of costs of production and distribution, and the improvement in the conditions of work and in domestic amenities for rural populations that would result. With the reservation that such benefits must undoubtedly await the widespread application of nuclear power in industry, mention may appropriately be made of some of the further though admittedly still speculative developments that might be possible in agriculture, forestry, and fisheries. Although at this stage any discussion of these possibilities may seem to be purely visionary, some reasonable speculation appears to be fully justifiable in view of the spectacular technological advances in the past few decades that have revolutionized the human way of life in less than the span of a generation. Already an atomic-powered submarine is in existence and in some quarters atomic-powered locomotives and aircraft are said to be only a few years away. Much, therefore, of what appears to be fantastic today may be a commonplace of tomorrow.

It has already been pointed out that while the reduction of losses due to diseases, pests, and spoilage, and the adoption of various improved techniques which lead to more efficient exploitation of areas already cultivated are the most immediately promising ways of increasing production, attention should also be given to the benefits that might arise through bringing new areas into cultivation, although this is a more difficult and expensive undertaking. Probably the greatest contribution in this direction would be through the provision of irrigation facilities, particularly in the desert and semi-arid areas of the world. Many areas suitable for such development exist in North and South

America, Africa, Central Asia, and North China, to name but a few. There are also areas where drainage and reclamation of marshlands and deltas, especially in the tropics, could render land fit for cultivation, though there are fewer opportunities for this type of action than in the case of irrigation. The availability of more abundant and cheaper power from atomic reactors might well make possible the development of such irrigation and drainage projects. In addition work is now under way on the partial desalting of water, and abundant and cheaper power would make it possible to do this on a scale adequate for irrigation. Such applications would be of particular significance to the underdeveloped countries, especially where conventional power sources are deficient.

As in the case of agriculture, in forestry more abundant and cheaper power would have far-reaching implications. It would affect not only the primary, often small, rural forest industries as well as the larger ones such as the manufacture of pulp and paper, but also forest policy and management in general. For instance at the present time about one half of the world's harvest of wood is used for fuel, which forms a particularly important source of energy in rural areas. Past experience has shown that where a more convenient, and sometimes more efficient, source of power is available, the use of wood as fuel drops very rapidly. This may be of advantage or disadvantage to good forest management, according to the circumstances. In areas where wood is scarce and the forests badly abused by destructive harvesting to meet the needs of the population, as in some Mediterranean and Asian countries, the replacement of wood by an alternative source of power would greatly facilitate the much-needed reconstruction of an adequate forest cover. On the other hand in some regions the use of wood as fuel is often an important factor in good silviculture, offering the sole or one of the few markets for small and rough assortments that would otherwise be unmarketable. This shows the complexity of the problems involved and, particularly in view of the length of time needed to plan and execute changes in forest management practices and policies, explains the unusually keen interest of forestry in the long-time changes in power sources and costs that may be affected by atomic energy developments.

The availability of cheaper power would also make possible the more efficient exploitation of large forest areas and would have a particularly marked effect on the practicability of establishing economically sound and integrated wood industries in the remote and less developed parts of the world. Looking even further afield, it might play a most important part in opening up some of the world's last untapped forestry resources in the tropical areas, particularly in South America and Africa.

Similarly atomic power could be of considerable significance to the fishing industry, where the development of nuclear power units of a size suitable for installation in ships would be of particular interest for possible use in mother and factory vessels of fishing and whaling fleets operating over long periods at great distances from their bases, especially in the antarctic, since they might in various ways lead to substantial reductions in operating costs.

This discussion of the possible implications of cheaper and more abundant power for agriculture, forestry and fisheries is by no means intended to be exhaustive but rather to give in broad outline an indication of what atomic energy developments could mean to these industries.

Agriculture is perhaps the most conservative of our arts; but when modern methods of farming and the multitude of highly technical services provided by an up-to-date department of agriculture are compared with primitive nomadic and pastoral systems of food production, it is obvious that the farmer is ever ready to adopt improved methods which increase the efficiency of his usage of land, water, capital and labor. It is therefore reassuring to know that this newest of our scientific advances, atomic energy, can contribute in so many ways to man's oldest industry and thus open up the way to improved methods of feeding, clothing, and housing the world's ever-growing population.

Chapter 3

RADIOISOTOPE USES IN U.S.S.R. BIOLOGY AND AGRICULTURE *

Introduction

The production of isotopes in the U.S.S.R. has provided new and broad possibilities for scientific research in biology and agriculture. Artificially produced radioactive isotopes and also stable isotopes have now become an available means of research extensively used by scientific workers in our country, who strive to perfect the methods used for nutrition of cultivated plants and to improve their economic qualities. At the present time biologists and agronomists are using radioactive isotopes, mainly as marked atoms. This method better than any other permits one to investigate the processes that take place in the soil, the ways in which plants consume nutritive elements, the movement of various compounds in the plant tissues, and finally, the most intimate metabolic reactions in plant cells.

There is no doubt but that the application of marked atoms (as well as any other method for that matter) calls for the careful appraising of the results obtained. For instance, the extreme sensitivity of the isotopic method sometimes allows the experimenter to forget about the absolute mass of the substance which he investigates and to consider secondary phenomena, constituting but a small part of the general process, to be the principal substance transformation. In other cases, biologists have to take into consideration the possibility of direct incorporation of radioactive atoms or whole groups of atoms in the substance, as a result of non-biological isotopic metabolism. These phenomena should not be confused with the true course of biological transformations.

But in spite of certain peculiarities of the tracer method it remains extremely important.

The possibility of applying the energy of radioactive disintegration to stimu-

* This chapter is taken from Geneva Conference Paper 618, "The Utilization of Radioactive Isotopes in Biology and Agriculture in the U.S.S.R." by A. L. Kursanov of the U.S.S.R.

late or in some cases to retard the development of plants is of great interest to agriculture. Serious attention is also given to the problem of applying the energy of radioactive disintegration to make more radical changes in the structure and inheritance properties of organisms.

In the Soviet Union, where agriculture is largely mechanized, there are extensive possibilities for a rapid practical application of new methods of introducing fertilizers and cultivation of plants, provided that these methods are substantiated by scientific research and checked by practice. At the present time, when extensive measures are being undertaken in our country for the further development of agriculture, there appear extremely wide possibilities for the practical application of new scientific achievements. They can be used on virgin lands and under conditions of irrigated agriculture as well as when cultivated plants are acclimatized in new districts, etc. This is the reason why the workers of agricultural experimental establishments are very interested in the possibility of applying radioactive and stable isotopes to solve the problems facing them.

It has not been long since radioactive elements were first applied in the practical work of biological and agricultural laboratories. Yet already during this short period of time, isotopic indicators have helped to discover many new aspects of plant life. Such knowledge makes it possible to improve nutrition and breeding of agricultural crops. Further study of plants by the application of radioactive elements will undoubtedly bring to light many new and unexpected aspects of plant life. The present paper, which describes some of the more important results obtained in the Soviet Union where plants were studied with the help of labeled atoms, is therefore by no means a summary of all that has been achieved. It merely gives a description of the scope and the general direction of these rapidly developing investigations.

This chapter describes only the investigations carried on in the Soviet Union, and will not mention investigations undertaken in other countries. The reason for this is our desire to give a fuller account of the application of isotopic tracers in the spheres of biology and agriculture that is undertaken in our country [Russia]. Giving this limited description we must not, however, forget that in other countries where the tracer method is applied many important results have already been achieved.

To continuously provide Soviet biologists with information about the successes achieved abroad in the application of isotopes, a special magazine, *The Action of Radiations and the Application of Isotopes in Biology,* is published in the U.S.S.R. It is a widely circulated magazine which is of great assistance in the coordination of our work. Comparison of numerous investigations leads one to the conclusion that scientists of various countries could successfully join their efforts to solve urgent problems of biology and agriculture by means of radioactive isotopes.

Instrumentation. To ensure the extensive development of work on the application of radioactive isotopes in the sphere of biology and agriculture it was necessary to construct new apparatus and apply new methods.

For example, instruments have been developed which, with the help of $C^{14}O_2$ automatically and accurately record the course of photosynthesis and respiratory metabolism in plants (V. Shirshov, V. Rachinsky and others); simple field methods for physiological experiments with radioactive carbonic acid have been worked out (V. Zholkevich and others); various dosimeters have been designed, which help to determine radioactivity on uninjured plants (S. Tselishchev) and thus make it possible to follow substances marked by the isotope in the plant organism.

The technique of radioautography and microradioautography has been thoroughly elaborated (Y. Mamul and others); this ensures fresh possibilities for visual investigation of the processes occurring in the plant as a whole and in its various tissues.

In their researches Soviet scientists are widely using the plants themselves to obtain preparations of sugar, amino acids, alkaloids, phosphoric ethers, and many other substances marked by the isotopes of carbon, sulfur, phosphorus, or nitrogen (V. Merenova, A. Kuzin, O. Pavlinova, and others). These preparations are successfully applied in physiological observations on the movement and transformation of substances in organisms.

Finally, in more refined investigations, whose aim is the detailed study of separate biochemical reactions, one may apply chemically synthesized compounds marked with carbon or other elements in a strictly defined position (M. Shemyakin, B. Savinov, N. Melnikov, K. Bokarev, V. Baskakov, and others).

Plant Nutrition Studies

Root nutrition is that aspect of the physiological development of plants which is most easily controlled. That is the reason why many scientists, while applying radioactive phosphorus, calcium, sulfur and other elements, concentrated their attention on the problems of distribution and transformation of nutritive substances in the soil and on their assimilation by plants. By the application of radioactive phosphorus it was possible to change the previously existing opinion that only 10–12% of phosphoric fertilizers were assimilated by plants. This opinion was formed as a result of the comparison of the general amount of phosphorus present in the yields of plants grown on fertilized and non-fertilized soils; however, no distinction was made between the phosphorus present in the soil itself and the phosphorus introduced with the fertilizer. But the experiments conducted by A. Sokolov demonstrated that when phosphate fertilizers marked by P^{32} of the double superphosphate are introduced into the soil, wheat and other plants first of all assimilate the phosphorus of fertilizers and that to the extent of 48–68%; at the same time the amount of phosphorus assimilated from the soil itself is reduced.

These results prove that plants assimilate far more phosphoric fertilizers than previously supposed and they compel us to look for ways of increasing the percentage of phosphates assimilated by plants directly from the soil.

Another problem of vital importance for agriculture that can now be easily solved is the rational distribution of fertilizers in the soil, ensuring a most rapid and complete assimilation of fertilizers by the roots of plants. This problem is particularly important for the granulated phosphoric fertilizers which are lately being widely applied in the U.S.S.R. If we place the granules of fertilizers marked with radioactive phosphorus into different sections of the soil, we can easily see that in 15 to 20 min after the contact of the rootlet and the source, P^{32} isotope appears in the leaves. This makes it possible, by observing the appearance of the first signs of radioactivity in the leaf lamina, to establish exactly the moment when the roots come in contact with the fertilizers. And watching a further increase in radioactivity, we can also determine the rate at which a given fertilizer is assimilated.

For instance, employing this method in his experiments with oats, E. Ratner showed that if phosphorus, marked with the isotope, is introduced into the soil and placed 3 to 4 cm below the seeds, "contact" between the roots and the fertilizer occurs as early as 2 or 3 days after the germination of the seeds; but in case the granules are displaced 5 to 6 cm sideways or placed much deeper than the seeds, this "contact" is postponed to 3 or 4 weeks. Consequently, the beginning of phosphorus assimilation, which is so vital for young plants, is also postponed.

Similarly, at the present time radioactive phosphorus is being used to test the efficiency of various methods of mechanized fertilization in planting cotton in the fields of the Tajik S.S.R. (I. Antipov-Karatayev and I. Lipkind), in the Uzbek S.S.R. (F. Reshetnikov) and in other districts of Central Asia. Undoubtedly, in the future, isotopes will be of still greater importance when the various methods of mechanized fertilization are appraised.

By extensively employing P^{32} in their field experiments V. Klechkovsky and N. Kashirkina were able to study in every detail how different plants assimilate phosphoric fertilizers in the presence of other elements, of soil humidity, liming, and other conditions. It is quite obvious that such information can successfully be used to improve the methods of nutrition of agricultural crops. Many experimental stations in the Soviet Union are already extensively using these possibilities. Suffice it to say that in 1954 dozens of experimental stations used superphosphate marked with P^{32} in their field experiments.

Root systems. The introduction into the soil of radioactive isotopes of several elements can also be utilized to observe the distribution of the root system, which was previously done by laborious and far from perfect methods of digging out roots and freeing them from earth. By introducing fertilizers, marked with radioactive isotopes, into various soil strata, the researcher can easily observe the distribution of roots in this or that strata without disturbing the plant structure but simply watching for the appearance of radioactivity in the plant's leaves. In this way it is possible to study the influence of soil cultivation, of methods and time of irrigation, of temperature and other factors on the development of root systems in plants, which is sometimes necessary for the proper organization of nutrition of plants in accordance with concrete conditions.

Application of radioactive phosphorus (P^{32}) to study the absorbing function of roots has permitted investigators to discover in this process, which seemed to be so well known, some new and interesting features which provide a theoretical basis for application of granulated fertilizers. By experimenting with spring wheat, E. Ratner and his assistants were able to show that when a granule of phosphorus comes in contact with a single rootlet constituting only 4–5% of the whole root system, the absorbing function of such a rootlet immediately becomes 20 or even 30 times stronger than ordinarily; as a result of this, one rootlet is able to satisfy to a considerable degree the whole plant's requirement in this element.

Thus we no longer consider the root system to be a uniformly acting mechanism which sends water and nutritive substances to the above-ground portions of plants. The root system is a highly labile organ whose activity in different sections rapidly changes under the influence of nutritive substances and the requirements of the plant. It is these peculiarities of the roots that enable them to utilize easily the local deposits of nutritive substances—in particular, granulated fertilizers.

Radioactive isotopes have made it possible to study more closely the role played by soil microorganisms in the nutrition of plants. For instance, employing vitamin B_1 marked with radioactive carbon, G. Shavlovsky showed that this vitamin, as well as certain other vitamins of the B group, which microbes form in the soil, is easily absorbed by the roots of buckwheat and enhances the general development of plants. At the same time, isotopic sulfur helped to discover that plant roots also utilize amino acids—methionine and cysteine—which microorganisms excrete into the soil.

Non-root nutrition. In certain periods of their development, for example, setting of fruits, the plants often become incapable of absorbing phosphorus and other nutritive elements through their roots. Yet the processes of movement and accumulation of plastic substances, which predominate at this given period, still require new stocks of nutritive salts and could be intensified if these salts easily penetrated into the plant.

Agricultural science has already solved this problem by applying the so-called non-root nutrition, which includes sprinkling, pollination, and sometimes even fumigation of the above-ground parts of the plants, with the required nutritive substances. For example, in many parts of the U.S.S.R., non-root feeding by phosphorus raised the yields of sugar beet (Yakushkin) and of cotton (Uchevatkin); non-root nutrition with ammonium salts also raised the yields of cabbage and other vegetables, particularly in the North where low temperatures of the soil prevent normal penetration of nitrogen through the roots (Dadykin).

The application of labeled atoms in this case too resulted in substantiating this useful undertaking and in bringing order to it. For instance, when superphosphate marked with P^{32} was utilized as a non-root nutrient, it became possible to observe how the fertilizer penetrated the leaves and into the plant and how it was distributed in its tissues. Radioautographs reproduce a clear picture of this distribution. On the basis of these documentary pictures, agricultural workers can easily check the efficiency of nutrition and determine more exactly when such feeding should be done and how much nutritive substances should be applied.

Carbon dioxide and plant nutrition. Utilizing radioactive carbon (C^{14}), A. Kursanov, A. Kuzin, N. Krukova, and co-workers discovered a new function of the root system, which consists in the following: The roots absorb carbon dioxide from the soil and bring it to the leaves and other green portions of the plant. It turned out that for synthesis of sugars and other products of assimilation, carbon dioxide of the soil can be utilized on a par with the carbon dioxide absorbed from the air, provided there is light. This fact has not only theoretical value but practical value as well. It proves the important role humus and the microbiological processes going on in the soil play in supplying carbonic acid to the plants. It also warns against the one-sided understanding of the tasks and possibilities of utilizing mineral fertilizers.

If we take plants marked in their organic part with the radioactive carbon (as siderites) and introduce them into the soil, we can observe the disintegration of organic residue by the rate at which $C^{14}O_2$ is liberated. Such observations are of great practical importance. They enable one to determine the intensity with which humus-forming processes take place in the soil under the influence of its cultivation, humidity, temperature and other factors. At the same time, when we grow plants in the soil containing radioactive organic residues, we are able to directly observe how they absorb and utilize various organic substances as well as the carbonic acid of decaying remains (V. Merenova).

Obviously, the relative role of carbonic acid, received by the plant through its roots, in the carbon nutrition of the plants as a whole may vary greatly; however, this quantity is usually not very large and constitutes from 2% to 5% of the amount of CO_2 absorbed by leaves (O. Zalensky and V. Vosnesensky). At present CO_2 fixation by roots is being investigated under field conditions.

The application of radioactive carbon (C^{14}) together with other up-to-date methods enabled us to study this new phenomenon in great detail and to determine its significance for plant life. It was shown that sugars, formed in the leaves when CO_2 is assimilated from the air, move downwards along the phloem and, on reaching the roots, penetrate in their thinnest and most active branchings. The speed of this descending movement, determined by means of radioactive carbon, is in the case of sugar beet about 70 cm per hr.

In the roots, sugars undergo glycolytic disintegration with the accompanying formation of pyruvic acid. It is this acid which, by means of a special enzyme, incorporates the carbonic acid of the soil. The latter, in the form of carboxyl, combines with the pyruvic acid and transforms it into oxaloacetic acid, which is easily reduced and transformed into malic acid—the first comparatively stable compound containing carbonic acid of the soil. Later on, as a result of mutual transformations of acids, malic acid is partly transformed into citric, ketoglutaric, and other acids.

It is necessary to say that when CO_2 is introduced into organic acids in the capacity of the carboxyl group, we do not observe any practical increase in the free energy of the substance, and consequently, we cannot consider beta-carboxylization, as such, to be equivalent to carbon dioxide nutrition of plants. But organic acids, formed in the roots and containing the carbonic acid of the soil, move upwards and penetrate into the green fruit, growing apices and leaf laminae. The speed of this ascending movement, even in the case of herbaceous plants, is about 3 cm per min or 2 m per hr. Therefore it takes very little time for the carbonic acid of the soil to reach the assimilating tissues where it can be reduced in the process of photosynthesis, forming carbohydrates, proteins, and other products rich in energy.

Certain portions of the sugars thus formed are, in their turn, translocated to the roots, where during the process of glycolitic disintegration they are transformed into pyruvic acid and accept from the soil fresh portions of CO_2 in order to transfer them to the leaves. Such is the principal cycle along which the process develops. But in addition to this, there are a number of subsidiary phenomena which condition the dependence of carbon dioxide nutrition in roots on other aspects of the physiological activity of plants.

By employing isotopes and chromatographic analysis, A. Kursanov, O. Tuyeva, A. Vereshchagin, I. Kolosov, and S. Ukhina have succeeded in discovering a new and important function of the roots, which consists in the initial synthesis of a number (about 14) of amino acids. These amino acids are utilized in the root itself as well as in other portions of the plant where they participate in metabolism and the synthesis of new proteins. So it turned out that besides its absorbing function, which has been drawing the attention of scientists and practical workers for a long time, the root system fulfills another very important role connected with the protein metabolism of the entire plant.

At first it was considered that this aspect in the activity of the root system is an independent function having no connection with the absorbing function of the roots, but later on it was discovered that this is exactly the mechanism by which roots assimilate ammonium nitrate from the soil.

In this process a very important role belongs to the carbonic acid of the soil. This acid brings about the carboxylization of pyruvic acid and some other products which are initial acceptors of the nitrate of ammonium fertilizers.

At the same time it has been proved that for the functioning of this system the roots should be well provided with phosphorus, since phosphoric acid participates in a number of stages of the glycolitic process as well as in the cycle of di- and tricarboxylic acids. Therefore, when there is a shortage of phosphorus the roots lose, as a rule, the ability to accumulate organic acids (A. Vereshchagin) and consequently, to fix soil carbonic acid (O. Kulayeva). All this inevitably results in difficulties of nutrition of plants.

Such is the picture of one of the most important aspects of plant nutrition which we have at the present time, one that it did not take long to discover by means of labeled atoms. In our opinion, this is convincing proof of the broad possibilities which the application of radioactive and non-radioactive isotopes opens before biologists.

Rates of Plant Processes

Our conceptions of the velocity of the processes taking place in plants have also been modified. The exact course of some of these processes has now been determined thanks to application of tracers; we have been compelled to reject the opinion that various processes occurring in the plant proceed very slowly. For example, F. Turchin, who used heavy nitrogen (N^{15}) which he introduced into the soil in the form of ammonium sulfate, showed that proteins of all the organs of the plant are constantly renewed and that this renewal proceeds at such a pace that a protein particle of vegetating rye sometimes lives no more than a few hours. Even the nitrogen of chlorophyll, forming the very center of this pigment molecule, is renewed so rapidly in actively vegetating plants that sometimes in the course of 2 to 3 days, more than half of it is replaced.

Labeled atoms enable one to approach experimentally the problem of biosynthesis of chlorophyll and carotenoids in plants. Thus in the Byelorussian Republic, T. Godnev and A. Shlik, utilizing in their experiments glycocol, labeled in carboxyl with C^{14}, and also radioactive glucose, obtained new evidence on the course of chlorophyll, phytol and carotene synthesis in plants.

Now with the use of radioactive isotopes it is very easy to study in detail the movement of plastic substances in the plant, and to determine not only the direction and speed of this movement but also the composition of the moving products. If we bear in mind that proper nutrition of various portions of the plant and the accumulation in them of valuable nutritive elements depend precisely on this process, then we shall understand the great importance for biology and agriculture of direct observations on the flow of plastic substances which have been made possible only thanks to radioactive elements.

Employing this method, M. Turkina, E. Vyskrebentseva, and N. Pristupa, working in our laboratory, demonstrated that substances move in plants more quickly than was thought heretofore. For instance, the products of photosynthesis travel from the leaves to the roots of sugar beet, pumpkin, and some other plants at the rate of 70 to 100 cm per hr and even more. Somewhat slower is the movement of assimilants towards fruit and growing shoots, but even in this direction the organic substances usually make not less than 40 to 60 cm per hr. Therefore, in the majority of agricultural crops, it takes no more than 20 to 40 min for the products of photosynthesis to reach the growing points and organs of accumulation.

Finally, according to A. Akhromeiko's experiments, water, especially in some of the trees, moves along the xylem at a rate of 6 to 8 or even more meters per hour.

Labeled atoms, therefore, showed that in plants, despite their outward immobility, physiological processes proceed at a very rapid rate. Consequently, agricultural workers should take this fact into consideration so as to coordinate their methods of tending plants and the "rates" of the latter's activity.

Identification of Plant Substances

Radioactive carbon (C^{14}) is now being successfully used to study the composition of substances that move in the plant. Thus, experiments conducted in our laboratory with sugar beet, pumpkin, cotton and other crops, demonstrated that along the conducting tissue usually moves a mixture consisting of saccharose, glucose, fructose, hexosephosphoric ethers, organic acids and amino acids, with the saccharose and hexosephosphoric ethers forming a greater part of this mixture. In addition, they move faster than the other components and this leads to the conclusion that they are the basic mobile substances in plants.

However, the composition of the nutritive mixture moving in the conducting tissue differs in various plants and may change depending on the stage of development or under the influence of light and mineral nutrition.

All these data show that using radioactive carbon as an observation method, we can approach the problem of arbitrarily changing the direction and composition of the plastic substances that move in the plant; this will provide agriculture in the immediate future with some new and still unused possibilities.

However, already V. Zholkevich, who utilized the tracer method on the irrigated lands of the Trans-Volga, has proved that assimilants formed in wheat leaves from labeled carbonic acid travel to the seeds far more rapidly in the case of irrigated plants than in the case of plants suffering from water shortage. With the aid of radioactive carbon dioxide, it was also shown that when the potato is cultivated under the conditions of the Far North, the products of CO_2 assimilation mainly travel from the leaves to the stems and growing apices, which results in abundant stems and leaves. The long day and low temperature of the soil caused a slow and retarded flow of nutritive substances to the tubers, and this was the reason for comparatively low content of starch in the local potatoes (Z. Zhurbitsky). At the present time agricultural workers of the North are searching for methods to combat this negative phenomenon, employing isotopic indicators to appraise the efficiency of the measures they undertake.

Radioactive isotopes enable one to observe not only the movement of organic substances but also the way nutritive salts move and are redistributed in the plant. In this respect interesting experiments are conducted in the Ukrainian S.S.R. with fruit trees. By means of P^{32} and Ca^{45} it was possible to study in detail the picture of the distribution and redistribution of these elements in the development of trees (N. Lubinsky and K. Garnaga). Similar experiments are conducted on other wood plants, such as scoompia (*Cotynus coggygria scop*), eucommia (*Eucommia ulmoides*) and others (S. Slukhai). These experiments also deserve special attention, particularly if we bear in mind that our information on the requirements of wood plants in mineral nutrients is far from complete.

Special chemical preparations, such as 2-4-dichlorophenoxyacetic acid, methyl ether of alpha-naphthylacetic acid, 4-iodophenoxyacetic acid, etc., have become widely applied in practical farming as a means of influencing growth and formation processes in plants. The biological aspect of this peculiar influence has been studied lately thanks to the utilization of labeled atoms.

Thus, for instance, if we introduce into a tomato plant a solution of 4-iodophenoxyacetic acid, marked with radioactive iodine (I^{131}), we can exactly determine its localization in plant tissues after some time has passed. The greatest part of the iodine preparation is concentrated in flowers and young fruit, thus creating a peculiar polarization: from leaves, roots, and other portions of the plant, nutritive substances travel to these organs, ensuring their intensive growth and accumulation of nutritive products (Y. Rakitin).

Plant Chemical Processes

The isotopic method not only enables one to observe the movement and distribution of substances in the plant; it also helps to observe the course of further transformations.

Thus, A. Krylov, Y. Rakitin, and N. Melnikov utilized as growth stimulants the methyl ether of alpha-naphthylacetic acid, marked in carboxyl or in other positions by the radioactive carbon. They discovered that while this substance intensifies the growth, the plant at the same time always strives to rid itself of this influence by splitting off carboxyl groups and by further rearrangement of the molecule of the physiologically active compound. In this way the radioactive isotopes enable us to approach the solution of the problem which is of great importance to biology and agriculture—the detoxication of alien substances in the body of plants.

In a similar manner we can observe the normal course of processes. For instance, A. Prokofiev, introducing the solution of saccharose marked with radioactive carbon (C^{14}) into the leaves of kok-saghyz, discovered that radioactive rubber very shortly appears in the latex of the plant. In this way it was proved once and for all that rubber originates from carbohydrates. This experiment is interesting from another point of view also, for it provides an explanation of the accumulation of rubber in kok-saghyz after harvesting. Practical workers have been interested in this problem for quite some time.

Developing these experiments on kok-saghyz, B. Savinov and his assistants utilized D-fructose, D-glucose, levulinic, pyruvic, and acetic acids marked with radioactive carbon in definite positions. In this way they were able to study more closely the processes of sugar transformation, leading to the rubber formation.

The isotopic method has also considerably modified our former ideas on the synthesis of saccharose in sugar beet. Utilizing C^{14}-labeled carbon dioxide, and glucose and fructose also labeled with radioactive carbon, A. Kursanov and O. Pavlinova demonstrated that the tissues of the root and shortened stem of sugar beet—the tissues which were formerly considered to be the organs capable of intensive synthesis of saccharose—actually cannot form this from glucose or fructose.

Further experiments proved that in sugar beet the synthesis of saccharose takes place mainly in the leaves, where saccharose is easily formed as the first free sugar of photosynthesis, or where it may be formed as a secondary product from glucose and fructose. From the leaves saccharose rapidly travels along the conducting tissues to reach the roots. This conclusion compels agronomists and selectionists striving to raise the sugar content in the industrial varieties of sugar beet, to pay more serious attention to the development of the leaf system, and to the processes occurring in it, while previous investigations underestimated the role played by the leaf system in the synthesis of saccharose.

Another example of successful application of the isotope method in plant physiology is furnished by interesting experiments conducted by V. Pontovich and A. Prokofiev. By means of C^{14}-labeled carbon dioxide they discovered that in the ripening fruits of poppy there occurs a cycle of substance transformation which is of great significance for the normal development of seeds.

This process begins in the following manner: as a result of intensive respiration of young seeds, large portions of carbon dioxide are evolved into the core of the fruit. However, this gas is not accumulated here but diffuses to the green surface of the fruit walls, where it is quickly reduced in the process of photosynthesis, forming sugar and other nutritive products which, through a placenta, are transmitted from the walls back to the growing seeds. Thus, in the green poppy fruit, this peculiar type of turnover ensures a quick removal of superfluous carbonic acid from its core, maintains a high concentration of oxygen necessary for the synthesis of fat, and results in the restoration of organic substances which the seeds use in respiration.

Photosynthesis Studies

The tracer method opens especially broad possibilities before scientists engaged in investigating photosynthesis.

Utilizing isotopes, Soviet scientists have achieved a number of successes in this sphere too. The discovery of photolysis of water, made almost simultaneously in the Soviet Union (Vinogradov and R. Teis) and in some other countries, is particularly important. It enables one to understand the mechanism of photoreduction of carbonic acid.

At the present time we are approaching the solution of the problem of primary organic substances formed in the process of photosynthesis. The utilization of labeled carbon dioxide in combination with paper chromatography ensures an extreme sensitivity of such experiments and enables one to detect labeled products of photosynthesis as early as 0.5 sec after irradiation. Thus N. Doman has recently discovered two new substances of low molecular weight whose nature is now under investigation. For the time being, we know only that the first of these substances does not contain phosphorus and consequently is not phosphoglyceric acid. These and many other investigations conducted by means of radioactive and non-radioactive elements are a preliminary to the decisive step in the direction of disclosing the mystery of photosynthesis and of mastering this process; and this, probably, will take place in the near future.

Since we cannot consider in detail a number of such investigations in this paper, we shall take up only the question concerning the direct products of photosynthesis.

A. Nichiporovich, N. Voskresenskaya, T. Andreyeva, and others, utilizing radioactive carbon (C^{14}) in the form of carbon dioxide and heavy nitrogen (N^{15}) in the form of ammonium salt, demonstrated that not only carbohydrates but proteins as well are the direct products of photosynthesis in the leaves of plants. The composition of the photosynthetic products may considerably vary under the influence of the species of the plant, its age, and conditions of its existence. This process is most strongly influenced by the spectral composition of light and its intensity. These factors, in combination with mineral nutrition, can considerably change the composition of the primary products formed in the leaves. Thus, carbohydrates are synthesized mainly in the red and yellow part of the spectrum, while proteins are formed under the influence of blue light. Besides its principal significance for biology, this fact also provides practical possibilities of influencing the development and properties of plants by changing the composition of the primary products of photosynthesis as they are being formed. There is every ground for the practical utilization of this discovery under hothouse conditions where it is possible to use sources of lighting with varying spectral composition and intensity enabling the agronomist to control not only the quantitative aspect of photosynthesis but the qualitative as well.

Conclusion

We have already said that the present paper does not describe all the new data in the sphere of metabolism and plant nutrition obtained in the U.S.S.R. by applying the method of labeled atoms. It is our opinion, however, that this paper proves that application of the achievements of nuclear physics for studying plant life opens great possibilities for the development of biology and agriculture.

But even at the present time, when we are only beginning to discover the practical possibilities of radioactive isotopes as a means of studying plant life, we can state with satisfaction that in the U.S.S.R. this means has already seriously influenced the development and improvement of practical farming methods. For instance, it was by means of radioactive isotopes that the problem of the effective application of granulated fertilizers was finally solved; this at once ensured wide application of this method of plant nutrition, and it is now successfully used on thousands of large farms in the Soviet Union.

Radioactive elements helped to bring about a similarly radical change in the attitude to non-root nutrition, whose expediency was for a long time disputed. At the present time non-root nutrition is successfully practiced in the U.S.S.R. on thousands of hectares. This method is applied to those crops and in those periods of vegetation when non-root feeding can give the greatest effect required.

Radioactive isotopes have also demonstrated the possibility of a more economical and at the same time more effective utilization of fertilizers for feeding plants. This resulted in a number of rational changes in soil cultivation, in reconstruction and improvement of sowing machines and in other measures that are being rapidly introduced in practical farming in our country.

But all this does not exhaust the great possibilities that are opened before biologists and agronomists utilizing atomic energy for the aims of peaceful construction.

Employing radioactive isotopes, we can now more profoundly study the laws of plant life, and it is our duty to show to the agricultural workers ways and means of obtaining abundant and stable yields, ways of securing a prosperous and tranquil existence for all.

Chapter 4

USES OF RADIOISOTOPES BY THE HAWAIIAN SUGAR PLANTATIONS *

Since 1946, radioisotopes have been in continuous use in Hawaii for several purposes. These may be classified as

(1) Essential elements as tracers of absorption, translocation, metabolism, and photosynthesis. (Examples: C^{14}, P^{32}, Rb^{86}, Ca^{45}, Mo^{99}.)
(2) Tracers of leaching and distribution of irrigation water. (Examples: P^{32} and Rb^{86}.)
(3) Sources of radiation for instruments weighing streams of bagasse and sugar, measuring light and temperature, measuring soil moisture, measuring crop growth. (Examples: Sr^{90}, Co^{60}, Cs^{137}.)
(4) Biological effects of radiation. (Co^{60}.)

The following examples have been selected because of their direct application to the growing and manufacture of sugar on the plantations.

Translocation of Photosynthate

Tagged elements are ideal tools for measuring translocation. A number of practical problems in sugar cane culture are associated more or less directly with the translocation and storage of soluble carbohydrates. Hence, factors affecting rates of movement as distinct from photosynthetic manufacture have been investigated in some detail.

For the more exact work, plants in culture solution have been used. A glass tube is slipped over a leaf and sealed with a split stopper and plastic, and $C^{14}O_2$ is introduced. After a short period of photosynthesis in sunlight, the tube is removed and the plant subjected to the measurements planned.

Experimental results. In the first experiment a large, single-stalk plant was fed through blade 5. Eighteen hours later the alcohol soluble fraction of

* This chapter is taken from Geneva Conference Paper 115, "Uses of Radioisotopes by the Hawaiian Sugar Plantations" by G. O. Burr, T. Tanimoto, C. E. Hartt, A. Forbes, G. Sadaoka, F. M. Ashton, J. H. Payne, J. A. Silva, and G. E. Sloane of the United States.

every part of the plant was strongly tagged. The fed leaf and the growing tip were particularly rich.

For the second experiment a large stool of cane with 16 stalks in all stages of development was fed $C^{14}O_2$ through blade 3 of stalk 9. The plant was harvested 44 hr after feeding. Fig. 4.1 shows the relative specific activity of

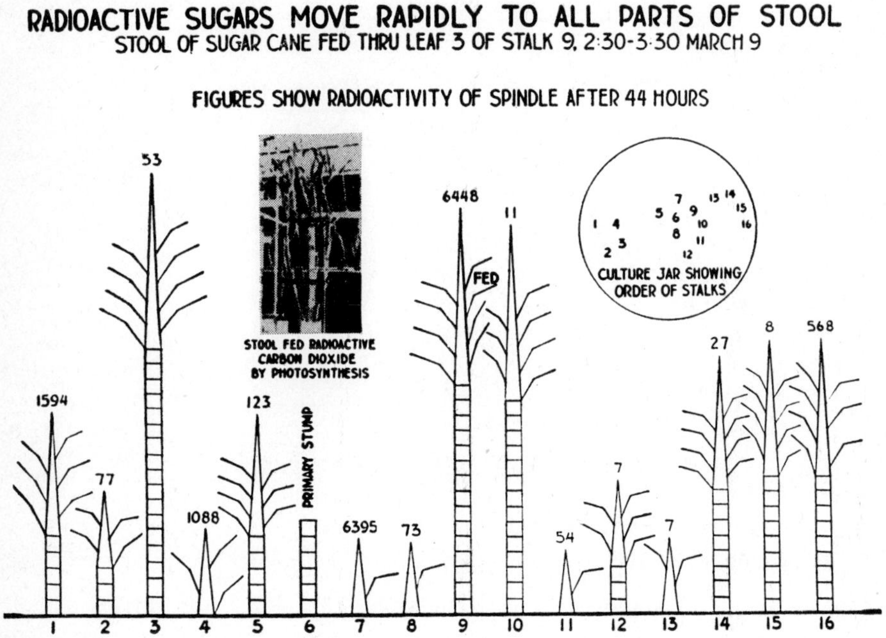

FIG. 4.1. Translocation of photosynthate in Exp. 2.

the spindle of each stalk. Table 4.1 gives the percentage distribution of the photosynthate among the parts of the stool.

TABLE 4.1. PERCENTAGE OF TOTAL COUNTS FOUND IN PARTS OF THE STOOL

Stalk	Percentage of Counts	Stalk	Percentage of Counts
1	7.2	10	0.05
2	0.2	11	0.006
3	1.0	12	0.08
4	0.7	13	0.003
5	1.4	14	0.3
6	0.1	15	0.1
7	0.5	16	1.2
8	0.003	Roots	17.2
9 (fed)	68.5	Stubble	1.3

In the third experiment the roots of a single tall stalk of sugar cane were enclosed so that the respired CO_2 could be absorbed from an air stream bubbling through the culture solution. The $BaCO_3$ was collected at short intervals and its relative specific activity determined. The first radioactive $BaCO_3$ appeared 6 hr after feeding blade 5 with $C^{14}O_2$ in sunlight. Total distance from fed blade to the top of the roots was 258 cm. Hence, the minimum rate of translocation from leaf to root is 43 cm per hr or 0.7 cm per min.

Maximum specific activity of CO_2 from roots was reached during the first night following the morning feeding (9:00 A.M.). Following this there is a decline until at 8 days a relatively constant level has been reached (Fig. 4.2).

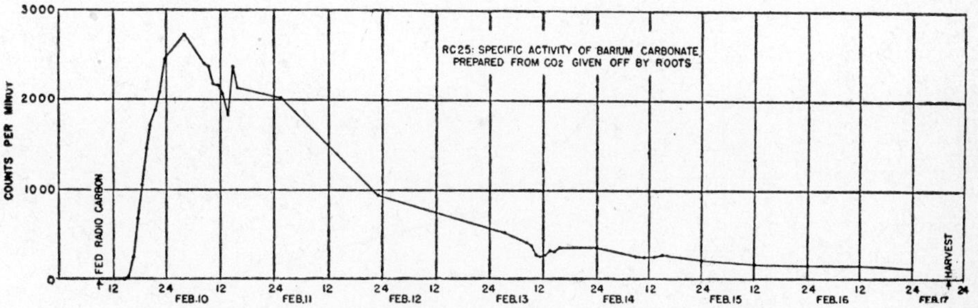

FIG. 4.2. Root respiration of photosynthate in Exp. 3.

Harvested after 8 days, the plant was found to have a general distribution of C^{14}. Even the oldest leaf, about ready to drop, received some C^{14} from leaf 5.

Discussion. The experiments described above demonstrate the high speed of movement of photosynthate in a general circulatory system. The wide range of specific activities shown by the same tissue of different stalks of a stool may be interpreted in terms of relative stalk vitality, i.e., the ability to take from a common pool more or less than its share. From a practical standpoint, this idea has been helpful in the study of the incidence of dead cane due to competition in a crowded, high-yielding field.

FEEDING SUGAR CANE THROUGH LEAVES

It has been known for some years that some minor (trace) elements can be successfully fed to plants by leaf spray. It is only in recent years that a serious effort has been made to supply some of the major elements in that way. Detailed studies have been made here of the uptake and distribution of urea tagged with N^{15}, P^{32}, Rb^{86}, radioactive sucrose, and other radioactive substances. All materials thus far tested are absorbed by the leaf and distributed more or less throughout the plant. In general the absorption curves are logarithmic, with half-times ranging from 1 day for sucrose to about 15 days for P^{32} and Rb^{86}. Absolute values depend upon the quantities applied and the environmental conditions.

Experimental results. Fig. 4.3 shows results of a single experiment. From a practical standpoint in the Hawaiian sugar industry, phosphate application via the leaf is of greatest interest. Many of the soils fix phosphate so quickly and completely that fertilizer applications lose their effectiveness. Hence a detailed study of utilization of leaf application was made. Fig. 4.4 shows the

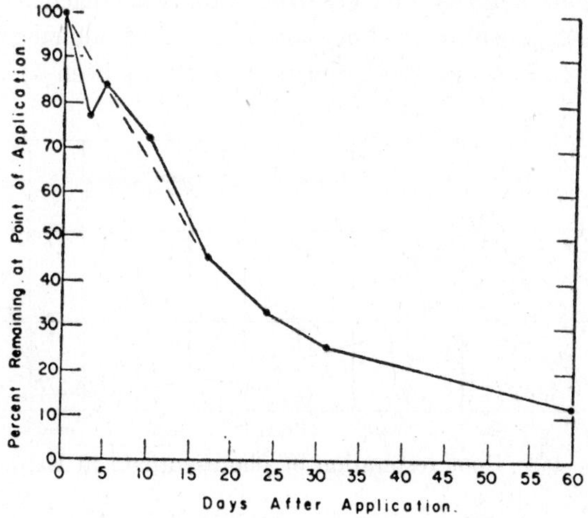

FIG. 4.3. Absorption of P^{32} applied in solution to the sugar cane leaf.

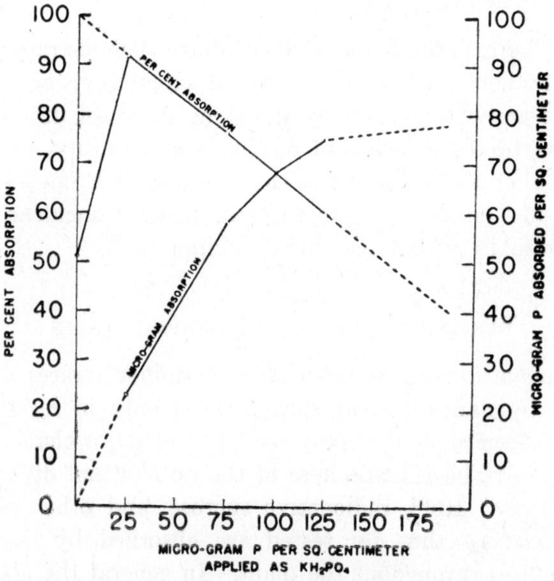

FIG. 4.4. Efficiency of absorption of P^{32} by the sugar cane leaf.

efficiency of absorption with different concentrations. From these data it can be readily calculated that the total required phosphate may be supplied in this manner.

The absorption of sugar, urea, and other organic compounds is also of great importance because a number of hormone-like substances are being used in control of weeds, control of sugar cane flowering, etc.

Fig. 4.5 illustrates the striking differences in uptake and translocation of radioactive 2-4-D in sugar cane and beans. The very low concentrations of

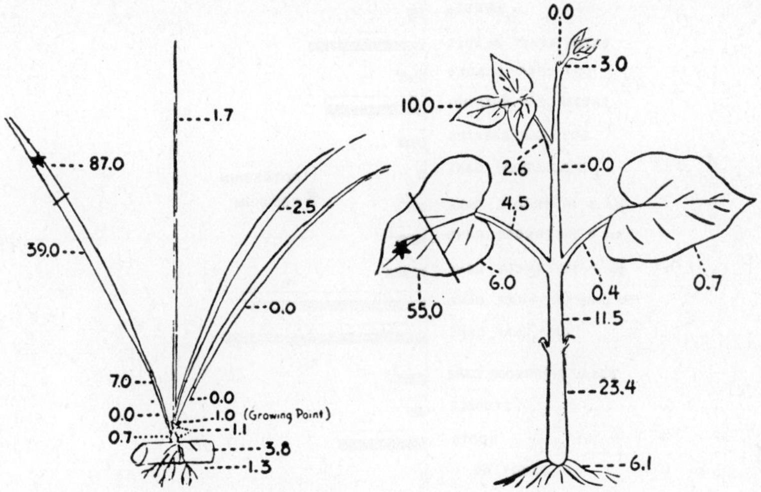

FIG. 4.5. 2,4-D is absorbed and translocated much more rapidly by the bean plant (right) than by sugar cane. Solution applied at star.

2-4-D reaching the vital growing parts of sugar cane may explain its high resistance to this chemical as contrasted with the bean plant. The point of application is shown by a star. A much larger proportion of the 2-4-D remained in the treated leaf of sugar cane.

RUBIDIUM-86 AS A SUBSTITUTE TRACER FOR POTASSIUM

The available radioactive isotope of potassium, K^{42}, has such a short half-life that it is not useful for much of the work with sugar cane. This is unfortunate since this element is a major fertilizer constituent and is used in large tonnages here. Therefore, a study of rubidium-86 was made to determine whether it would serve as a tracer for potassium. This element has a strong beta and gamma ray and convenient half-life of 19.5 days.

Healthy sugar cane plants 4½ months of age were fed a dose of Rb^{86} via the culture solution. After one week, when most of the rubidium had entered the plant, the plant was returned to the normal, rubidium-free culture solution. Two months later the plants were harvested, divided into many anatomical parts, and analyzed for potassium and rubidium. Potassium was determined with a flame photometer and Rb^{86} by Geiger counting.

Experimental results. Fig. 4.6 summarizes the results, expressed as percentage distribution of the two elements. Two conclusions are evident:

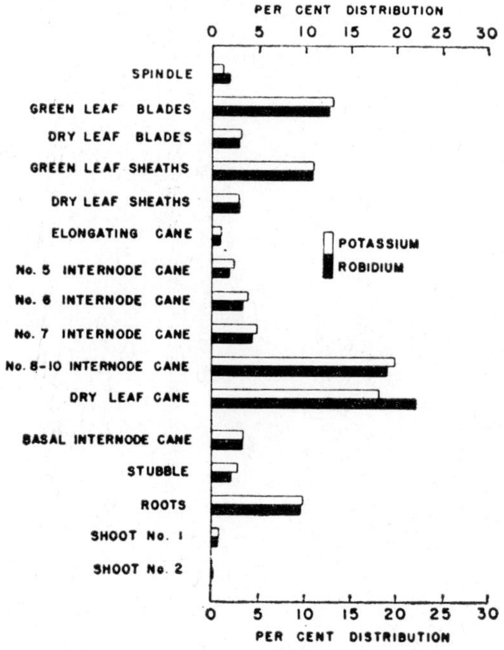

FIG. 4.6. Distribution of potassium and rubidium in different tissues of cane plant.

(1) Two months is sufficient time for rubidium to reach its final distribution in the growing sugar cane plant. (2) Rubidium is distributed within the plant in the same proportion as potassium.

Hence, Rb^{86} may be safely used as an index of potassium distribution. This relationship between the two elements is of great interest. For some time it has been suspected that rubidium may partially substitute for potassium as an essential element. Apparently the large difference in their atomic weights does not materially alter their "affinities" for different tissues. The forces which determine their relative concentration in different tissues apply alike to the two elements. This is all the more remarkable in this case, where the total potassium present was many times the total rubidium.

Measuring Sugar Cane Growth with Cobalt-60

The usual method of measuring growth of sugar cane is by tabulating elongation of the stalk above a fixed base mark. This has the disadvantages inherent in small selected samples as well as the obvious error of substituting length for mass. It is also difficult to reach stalks in the center of the plots.

The gamma-ray weighing method described here is very satisfactory for standard sugar cane plots and may well be useful for many other crops. Without disturbing the plants in any way, the entire vegetative mass is weighed, every stalk, large and small, being included.

Co^{60} slugs of 100 millicuries each are mounted in the pockets of two upright poles. A track and cart carrying a portable scaler or ion chamber are on the opposite side of the plot. The cart is rolled forward and backward at a uniform speed. This gives an integrated value which completely eliminates errors due to irregularities of stand in the lines. Sources and receiver are raised in steps of 1 foot. Each measurement with a scaler is run to a total of 10,000 counts. At a rate of 3000 counts per min the time at each level, including resetting the sources, is about 5 min. When the cane is 9 ft tall, total time per plot (6 lines 16 ft long) is 45 min.

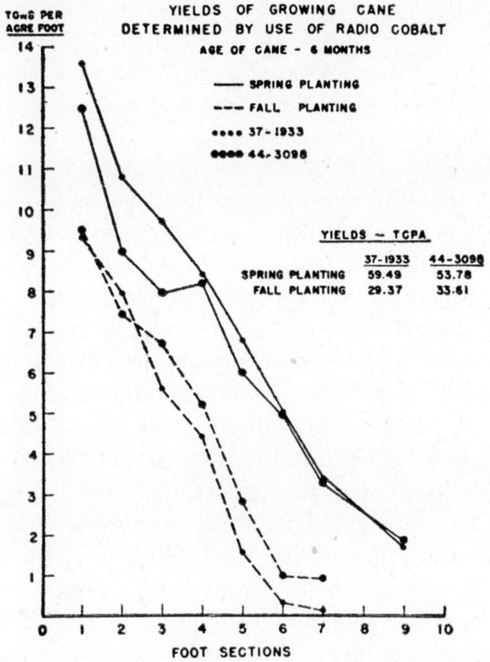

Fig. 4.7. Yields of 4 plots of sugar cane weighed standing in the field.

Experimental results. The results given in Fig. 4.7 are expressed as tons per acre-ft. The two varieties of sugar cane were planted in the spring and fall. Measurements at six months compare the winter and summer growth rates. Totals for the entire height of sugar cane show that summer growth rate is about twice that for winter.

The method has been applied to plots of sugar cane which were hand-harvested and weighed the following day. The two methods agree very well and the gamma-ray scale leaves the cane undisturbed for future measurements.

Tracing Irrigation Water Tagged with Radioisotopes

In Hawaii over 50% of the million-ton crop of sugar is grown on irrigated plantations. Uniform distribution of water in long irrigation lines has been of especial interest during times of water shortage. It has become of even greater importance in recent years as the practice of adding soluble fertilizer to the water has grown. In order to simplify the measurement of the amount of water percolating into the soil at points along the cane line, a method has been devised in which the water is tagged with a radioactive element which is later measured at chosen points. From the data thus obtained it is hoped that field layouts may be redesigned to give the best possible results on the various soil types.

The method. The method is based on the fact that some elements are so strongly adsorbed by most soils that they will be completely retained on the surface while the water percolates downward. The first element considered was radioactive phosphorus (P^{32}), which had the advantage of ready availability and cheapness, in addition to the fact that it has a convenient half-life and emits only a strong beta ray, making shielding easy. It works well on many soils. But the discovery that it does leach readily through some soils led to the search for an element of more universal applicability. Rubidium-86 seems to fill the bill. It also emits a strong beta ray and has a convenient half-life (19.5 days). However, it emits some gamma rays, which increases the cost of shipment and necessitates more care in handling in the laboratory. These problems are not too serious, and rubidium has been adopted for routine use in studying water distribution on Hawaiian sugar plantations.

The test lines are isolated for a hundred days until decay has reduced the radioactivity to a negligible quantity.

Approximately 1 millicurie of Rb^{86} per 100 ft of irrigation line is diluted to a convenient volume. A capillary tube is inserted in the outlet of an aspirator bottle which will give a constant rate of flow for a period in excess of the estimated irrigation time. The irrigation weir is adjusted to maintain a constant flow of water into the line throughout the entire period. The Rb^{86} and the irrigation water are fed into a mixing box which gives a thorough mix before the water starts down the line. Any desired number of sampling stations are set up where both water and soil samples are to be taken. The repeated water samples are required because Rb is so strongly adsorbed that it is picked up by the surface soil, thus reducing the activity of the water as it goes down the line. Each station is covered by four layers of cheesecloth about 15 in. wide extending across the line. This does not hinder percolation into the soil but does prevent surface pickup of Rb from passing water. Thus a pickup correction is eliminated, since all of the residual radioactivity in the soil must come from water which has gone down at that point.

Water flowing over each station is sampled continuously to give a true integrated mean activity. From the activity of the water at zero station, a few feet below the mixing box, the rate of water flow into the line may be calculated. For assay of Rb^{86} content, 50 ml of water in moisture cans 7 cm in diameter are counted at infinite thickness with a mica and window counter. The interval between time of arrival of irrigation water at each station and the time of stopping irrigation is recorded for each test. Record is also kept of the flow of water down the cane line for each 10-ft section.

When the last of the irrigation water has disappeared, the cheesecloth is removed and soil samples are taken with a cylindrical soil sampler 1 in. in diameter, which penetrates to a 1 in. depth. It has been found that sampling to this depth recovers all the Rb^{86}. Twelve samples are taken: two at the water lines which are marked during irrigation and the other ten equally spaced across the line. The total area sampled is 9.4 sq in. The soil sample is dried, weighed, and thoroughly mixed. Fifty grams of the sample is weighed into 7-cm moisture cans and leveled to an even layer for counting. From the activity of this aliquot, the total count in the whole sample is calculated.

A reference standard is prepared by adding to 100 gm of dry inactive soil the amount of Rb^{86} present in 9.4 cu in. of an approximately average water sample. This standard is dried and thoroughly mixed, and a 50-gm aliquot is counted. The calculated total activity is equivalent to 1 linear inch of water of known activity.

Thus from the total activity of the soil sample and the activity of the integrated water sample of a station in a cane line, the amount of water which has gone down at that station is

linear inches of water, $I =$

$$\frac{\text{total activity of soil sample}}{\text{total activity of standard soil}} \times \frac{\text{activity of standard water}}{\text{activity of water sample}}$$

Acre inches of water absorbed at any station are

$$\text{acre inches} = I \times \frac{\text{width of wetted line}}{\text{width of cane lines}}$$

Experimental results. Over a hundred tests, covering practically all the irrigated cane areas of the Hawaiian Islands, have been run to determine the efficiency of the various irrigation practices. The results of three of these tests are recorded in Table 4.2. These lines were 200 ft in length and five

TABLE 4.2. DISTRIBUTION OF IRRIGATION WATER TAGGED WITH RUBIDIUM-86

Test Number	Line Slope, %		Sampling Stations				
			A	B	C	D	E
0-16	0.75	Water Activity	16.3	15.8	15.4	14.7	14.3
		Soil Activity	5084	4782	4267	2745	2300
		Acre-Inch Irrigation	4.43	3.26	3.40	2.47	2.60
0-19	1.5	Water Activity	17.0	15.1	14.9	11.5	8.0
		Soil Activity	3726	4254	3546	4136	2698
		Acre-Inch Irrigation	2.58	3.04	2.56	5.11	6.61
0-13	2.7	Water Activity	21.3	17.8	14.3	9.7	6.3
		Soil Activity	3575	3984	5760	5412	2125
		Acre-Inch Irrigation	1.23	1.43	4.74	7.93	4.46

sampling stations were set up at 20, 60, 100, 150 and 190 ft (A, B, C, D, E) below the head of the line.

The very large differences in inches of water absorbed along the lines is due to line slope and rate of application of water. In order to get enough water into the soil at the points of lowest absorption, it is necessary to flood other sections of the line with three or four times the needed amount. This represents a great waste of water. The effects become even more disturbing when irrigation water is the medium for fertilizer application, a practice in wide use in Hawaii.

Measuring Soil Moisture with Gamma Rays

Some years ago it was found here that the gamma rays from Co^{60} could be used to measure soil moisture changes with sufficient accuracy for most needs. The method is now being used continuously.

In a field installation the Co^{60} source is lowered in one pipe and the ion chamber is lowered into another. Fig. 4.8 is a cross section of an 18-in. pot, showing plastic pipes, Cs^{137} source (A) and G-M tube receiver (B). With the source and tube each 4 in. long, the measurement is an integrated value for a rectangle of soil. Any depth can be chosen.

Fig. 4.9 shows the calibration curve for the pot. The width of the dots represents the error of about 1% moisture. The absolute moisture value is read in 5 min without disturbing the soil or roots.

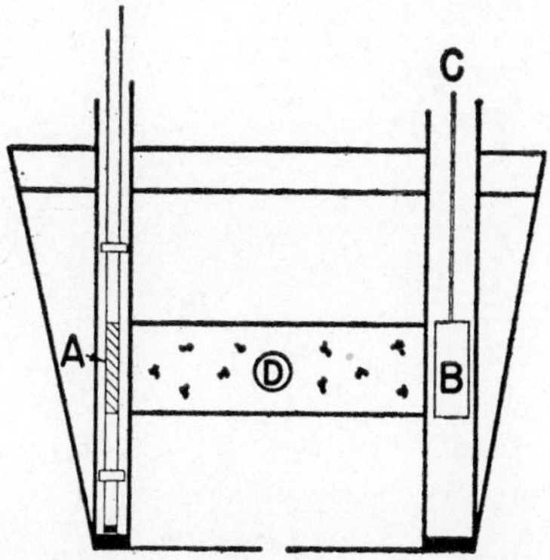

FIG. 4.8. Diagram of a pot of sugar cane with pipes for measuring soil moisture with G-M tube and Cs^{137}.

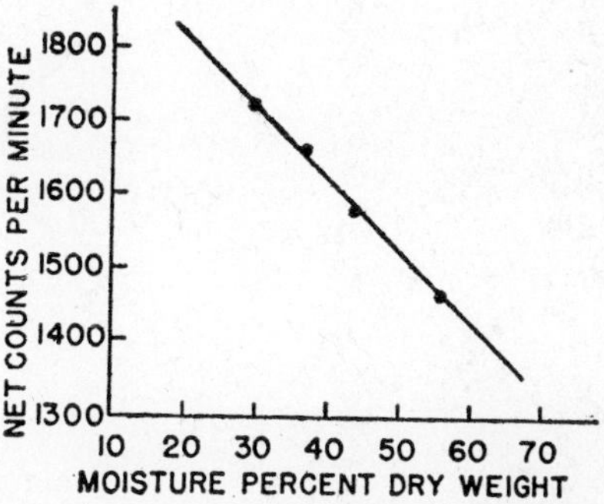

FIG. 4.9. Soil moisture calibration curve for pot shown in Fig. 4.8.

GAMMA-RAY SCALE FOR STREAMS OF BAGASSE AND SUGAR

Numerous uses may be made of radioisotopes within the sugar mill. The first continuous use to be made here is in the gamma-ray scale for streams of bagasse and sugar. This gives a continuous record of milling and production performance, as well as the integrated weight for any period of time.

The principle of the measurement is illustrated in Fig. 4.10. Although the system lacks the theoretical simplicity of a parallel beam of light passing through a homogeneous medium, in practice an entirely empirical calibration yields constants which are highly reproducible from day to day. In practice a sugar stream or a bagasse stream falls through the gamma-ray beam. The recorder is equipped with a precision integrator which totals the weight in each batch.

The source is a tube of cesium-137, which is ideal for this work, since it emits a soft γ ray and has a half-life of about 37 years. The receiver is a pair of ion chambers which are in balance when no sugar is flowing. A 10% absorption of gamma rays is spread across about two-thirds of the chart. This is the normal rate of flow and there is sufficient chart space to take care of the highest loads encountered.

Results and discussion. Calibrations against weighed loads of sugar indicate reproducibility under continuous operation. Reliability over very long periods is now being tested.

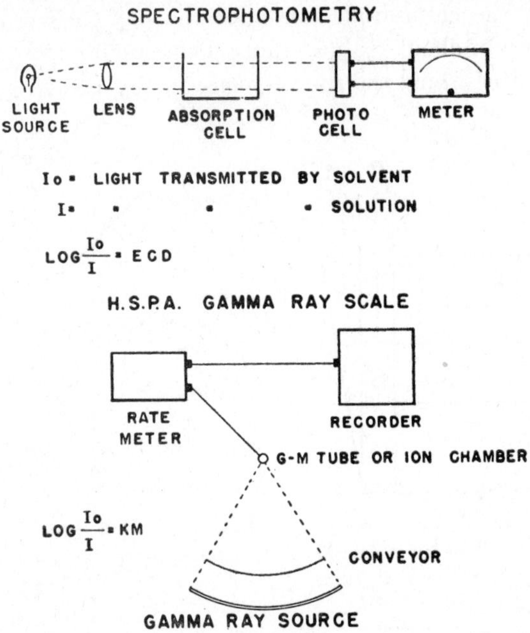

FIG. 4.10. Schematic drawing of gamma ray scale.

The records from these machines give very valuable pictures of mill operation. The uniformity of load at the crushers and the efficiency of output from the batch crystallizers are recorded. There are a number of operations within the mill where such records and integration may be well worth while. With minor modifications the same equipment is applicable to other operations.

Chapter 5

BIOSYNTHESIS IN CARBON-14-LABELED PLANTS *

The use of isotopes as tracer tools is unquestionably one of the most significant peacetime applications of the atom, one which will have particular and continuing utility in agricultural and biological research. Carbon-14, the long-lived isotope of carbon, has proved the most useful tracer isotope in biological research ever since it became readily available with its production in atomic reactors. The full tracer potential of radiocarbon (C^{14}) can be envisioned by simply noting that carbon accounts for 30–40% of the dry weight of living organisms.

The usefulness of any tracer isotope is limited by factors other than its availability in elemental form. The most limiting factor in applying C^{14} as a tracer in biological research is the requirement of incorporating it into the molecules of the particular native organic compound that is to be traced or studied. In many cases this incorporation can be most efficiently accomplished by means of chemical syntheses, in which case it is possible to place the isotope at specific carbon atom positions. In many more cases the mechanism of synthesis of naturally occurring organic compounds is not known and the preparation of a labeled molecule cannot be accomplished. It is only by use of biosynthesis that a radiocarbon-tagged form of all of the diverse organic molecules that occur in living organisms can be prepared.

In biosynthesis an actively metabolizing organism is supplied the carbon-containing substrates or "assimilates" normally utilized in its natural habitat. These substrates are supplied in C^{14}-enriched forms which essentially contain the specific activity required in the particular organic compound desired and which is known to occur in the organism. The initial labeled substrate supplied may be a carbohydrate, in the case of many bacteria, or, with chlorophylous higher plants, it can be carbon dioxide.

This chapter is predominantly concerned with describing in detail the re-

* This chapter is taken from Geneva Conference Paper 274, "Biosynthesis in C^{14} Labeled Plants: Their Use in Agricultural and Biological Research" by N. J. Scully, W. Chorney, G. Kostal, R. Watanabe, J. Skok, and J. W. Glattfeld of the United States.

quirements which must be met for efficient biosynthesis of C^{14}-labeled higher plants and their products. The various observations noted are the cumulative result of seven years of investigation of the ways and means of biosynthesizing radiocarbon-labeled plants as well as of their experimental tracer usefulness (Ref. 5).

Unless otherwise noted, the use of the term "labeled" refers to uniformly labeled plants or their specific products. In such instances a uniformly labeled plant is constituted of organic compounds of identical specific activity of radiocarbon as measured on a gram of carbon basis. Further, the uniformity of labeling is such that the percentage distribution of C^{14} is the same at the various carbon positions in a specific organic compound.

PHYSICAL FACILITIES: THE BIOSYNTHESIS CHAMBER

A number of diverse criteria were utilized in establishing the final design of the complete physical unit, termed a "biosynthesis chamber" (Fig. 5.1), required for culture of C^{14}-labeled higher plants. Included were the requirements associated with maintaining a hermetically sealed system and such factors as height and weight of plant species, length of culture period, control of

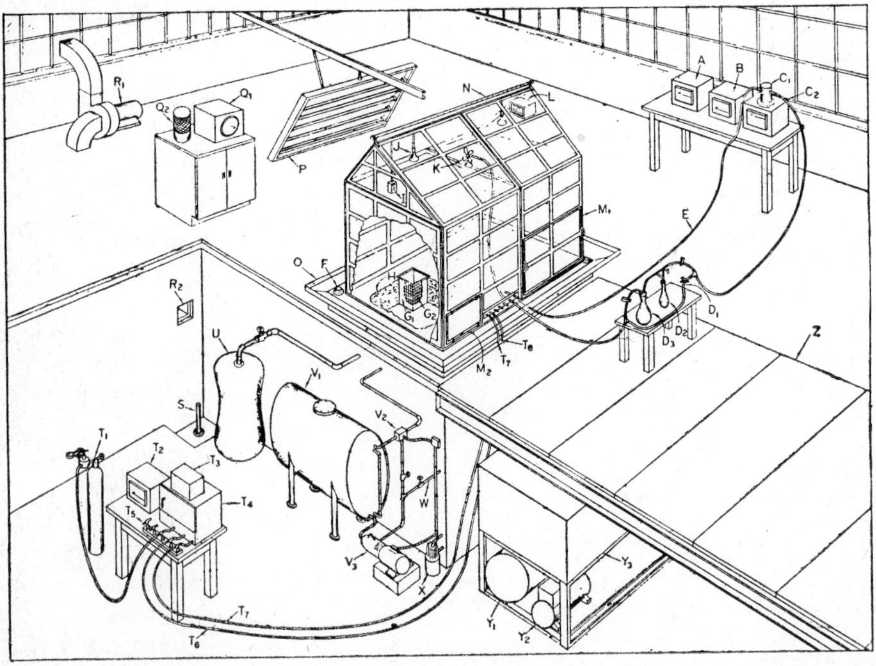

FIG. 5.1. Schematic drawing of a single, complete C^{14} biosynthesis chamber unit. The individual parts are indicated by the letters A–Z and are reviewed in detail in the text.

environmental variables such as temperature, humidity, and day length, mineral nutrition, and photosynthetic capacity, as measured by plant rate of assimilation of carbon dioxide. A two-year period was devoted to the construction and experimental testing of a pilot biosynthesis chamber (Ref. 6). The resultant experimental data were used to establish the final design incorporated in each of three independent replicate biosynthesis chamber units. The latter units have been in experimental use for five years.

Each of the various parts of a single, complete biosynthesis chamber unit are schematically designated by letter in Fig. 5.1 and identified in the text. The three units are located in a conventional greenhouse where plants can be grown with light intensities which substantially exceed those normally available from artificial light sources. Each unit has approximately a 10,000-liter internal free-air capacity when plants are being cultured experimentally under hermetically sealed conditions. The bulk of this volume is contained in the plant growth chamber proper as shown in Fig. 5.1.

The chamber is constructed of low carbon content steel, with standard commercial window sash continuously welded to the main supporting framework. The chamber dimensions in feet are 7⅔ long, 5 wide, and 5½ high to the eave of the roof. The chamber has two removable panels (Fig 5.1 M_1, M_2) for use as entrances at harvest times. The panes of standard, double-strength glass are glazed, using continuous glazing angles and a carbonate-, sulfide-, and sulfite-free mastic. Continuous with and subtending the base of the chamber is a 21-in. deep stainless-steel bed containing the inert substrate (H) used for nutriculture maintenance of plants.

Normally, eight stainless-steel wire-mesh baskets (Fig. 5.1 G_2) are equally spaced in the bed. The baskets set on the bed-floor, which has a gravel-free, recessed flooding-channel into which the nutrient solution is first pumped in the course of flooding the bed. Each of the baskets has a surface area 10×10 in. They are each contained within a solid, stainless-steel retainer (G_1) open at top and bottom, leaving a ½-in. free air space between the basket and the retainer wall. This arrangement facilitates recovery of roots by restricting their growth to the silica sand or quartz gravel substrate volume contained within the baskets.

All plants are supplied a complete nutrient solution which has been experimentally demonstrated to support good growth for the particular species to be cultured. The solution is contained in a nutrient reservoir tank (Fig. 5.1 V_1); by means of a centrifugal pump (V_3), activated by a time-clock, this solution is delivered by piping into the bed of the chamber. A solenoid valve (V_2) permits the solution to drain directly back into the tank without passing through the centrifugal pump. A vent connecting the air in the chamber with that above the solution in the tank avoids pressure change while the solution is pumped to or drained from the bed. If necessary, at the end of an experiment this solution can be pumped via a pipeline (W) to a hold-up tank. A plastic (Tygon) bag (U), having an air volume of 200 liters, is also connected by a vent pipe to the air in the growth chamber. This unit serves as an expansion bag, collapsing or expanding in response to change in pressure due either to temperature variation inside the plant biosynthesis unit or to change in outside barometric pressure.

The remaining parts of the hermetic system consist of ¼-in.-lumen plastic tubing (Fig. 5.1 D_1, E, T_6, T_7) connected to a 1-liter ionization chamber (C_1) of a vibrating diaphragm electrometer and to a 20-cm sample cell of an infrared, carbon dioxide gas analyzer. The concentration of C^{14} in the growth chamber atmosphere is continuously assayed by use of the electrometer and automatically indicated on a 30-day strip-chart recorder (C_2) in millivolts. The carbon dioxide detecting unit consists of an infrared gas analyzer (T_4) and a time-relay circuit (T_3) for activating a panel of 5 solenoid valves (T_5). An hourly assay record (T_2) of the carbon dioxide content is obtained in millivolts in each of two separate biosynthesis chambers. Every hour a 28-min record of the carbon dioxide in each of the two chambers is obtained. After each assay period the sample cell of the analyzer is flushed with nitrogen (T_1), thereby returning the sample to the chamber from which it was drawn. Air pumps located in the chamber continuously supply air samples at a rate of 1.5 liters per min to both the ionization chamber of the electrometer and the sample cell of the infrared gas analyzer.

Additions of labeled carbon dioxide to the chamber are accomplished by forcing a lactic acid solution, using pressure supplied by air pumps, from one flask (Fig. 5.1 D_2) into a flask (D_3) containing C^{14}-labeled carbonate. The generated CO_2 is expelled immediately in front of one of two air-circulating fans (K) in order to effect a quick and uniform distribution of the gas.

The air temperature within the plant growth chamber is regulated to $\pm 2°$ F over the range of 60–80° F by constantly flowing, temperature-regulated water over the four external chamber surfaces. A modulating thermostat (Fig. 5.1 I), centrally located inside the chamber, controls the temperature of water flowing out of numerous openings in pipes (N) located at the top of each external surface face of the chamber. The thermostat controls a three-way mixing valve connecting to water supplied from a chiller tank (Y_1) associated with a refrigeration compressor (Y_2) and to unchilled water supplied by a reservoir (Y_3) which receives all water that has passed over the chamber. Fractional quantities of the two types of water can be mixed, depending on the fractional heat-load recognized by the modulating thermostat. The water flowing off the chamber is caught in a splash pan (O) and returned to the unchilled water reservoir by means of a drain (F). A multiple-station, continuous-recording unit (B) measures the ambient temperature within the growth chambers as well as that of the greenhouse and outdoor air.

The relative humidity inside the chamber is regulated to $\pm 5\%$ over the range of 40–60%. This control is obtained by means of a dehumidifying unit

(Fig. 5.1 L) and a humidistat inside the chamber that activates a small compressor (X) supplying refrigerant gas to the dehumidifier. Condensed water is returned to the nutrient solution.

Two types of supplemental light are available, each independently operable by means of time-clock control. Three 75-watt incandescent, reflector-type lamps (Fig. 5.1 J) are used to extend the natural day length when required. Two movable, 18-tube fluorescent luminaires (P) provide supplemental, high-intensity light for growth of plants when natural light is deficient or not available. Light intensity is continuously recorded (A) in foot-candles as measured by a photocell.

The pit in which the nutrient reservoir tank and other equipment is located is normally completely covered with metal decking (Fig. 5.1 Z). An exhaust blower (R_1) and vent (R_2) provide continuous flush of the pit atmosphere as a safeguard against accumulation of C^{14}-dioxide in event of its leakage from the various parts of the biosynthesis unit located at this site. A drain, topped by a standpipe (S), provides a holdup reservoir for solution which may leak from the nutrient reservoir. This arrangement was particularly designed for biosynthesis of plants in which radioisotopes such as P^{32}, S^{35}, Zn^{65}, Fe^{59}, are also employed. With removal of the standpipe, the trapped solution is free to drain to holdup tanks for further processing. The normal greenhouse atmosphere is continuously monitored for $C^{14}O_2$ which may appear as a result of leakage from the growth chamber. A vibrating-diaphragm electrometer (Q_2) monitors a 5-liter air sample of the normal greenhouse and records (Q_1) with a full-scale sensitivity representing 50% of the $C^{14}O_2$ concentration considered safe for constant human exposure.

Performance of the facility. Certain of the steps as they occur in a typical C^{14} biosynthesis experiment are noted in a separate section. Here consideration is particularly given to various aspects of the performance of the plant biosynthesis facilities based on total experimental experiences. The facilities have proved to be hermetic, and no measurable loss of $C^{14}O_2$ has occurred during experimental usage. An internal pressure equivalent to that of a 10-in. column of light mineral oil could be maintained. A biosynthesis unit is not sealed until the ambient temperature in the growth chamber is at equilibrium, at which time the plastic breather-bag is partially inflated. When thus sealed, a maximum fluctuation of pressure equivalent to a ½-in. oil column has occurred as measured in an attached manometer. This is predominantly the result of barometric pressure changes. Pressure variations due to temperature variation are insignificant because of the precise control of this environmental factor.

A light intensity maximum of 8000 foot-candles has been observed inside the growth chambers when clear skies prevailed in summer, which is approximately a 25% maximum reduction of outdoor intensities.

In any green plant the rate of growth can be limited by the quantity of "photosynthates" available. In natural habitats and greenhouses the rate of photosynthesis may be limited by more environmental variables than in the plant biosynthesis chamber, where controlled environmental conditions exist. The variable that requires the closest attention in the biosynthesis units is the total quantity and concentration of CO_2. At normal concentrations (approximately 0.03%) of this gas, 3.0 liters are present in the system at 75° F. With 1.5 kg of fresh leaf tissue, the quantity accepted as maximal for the chamber under optimum light conditions, several species have assimilated this quantity of CO_2 in one hour. When the natural concentration is increased, still greater rates of fixation occur. Most plant species have not been able to effect a net assimilation of CO_2 from an atmosphere which has been lowered to 0.003%. On the other hand, it is detrimental to some plants if they are maintained for a few weeks with 0.20% CO_2 during winter light intensity and day-length conditions. In such instances alfalfa has been observed to exhibit severe epinastic responses and arrested growth rates.

Irrespective of seasonal period all species used in chamber biosynthesis experiments are cultured with a maximum CO_2 concentration of 0.10%. When assimilation has lowered the concentration to a third of the normal, or 0.010%, it is again raised. With the maximum quantity of leaf tissue noted earlier and with optimal photosynthetic conditions, several species have assimilated 24 liters of carbon dioxide during a single 12-hr photosynthesis period. In this instance three additions or generations of carbon dioxide had to be supplied to the growth chamber.

Usually a biosynthesis experiment which is designed to produce uniformly labeled plants is initiated with seedling stocks. These are then cultured to selected stages of maturity while exposed to atmospheres which are kept in the total carbon dioxide range noted earlier and always at the specific activity desired in the uniformly labeled plant or its particular products. It is maintained in the air phase by readjustment with additional quantities of carbon dioxide of the appropriate specific activity. Frequent readjustments are necessary early in the culture period, particularly if the experiment is started with plants well beyond seedling stage size. As much as a 50-fold increase in the initial seedling carbon pool has been obtained in a four-week culture of plants in the growth chamber. The air in the chamber is circulated at a 3-mph rate by the fans. When additional quantities of carbon dioxide are introduced the nutrient solution is pumped into the chamber bed to facilitate reaching the equilibrium concentration of the gas. Equilibrium is reached in a 15–20-min period.

The nutrient reservoir tank contains 750 liters of a complete mineral element solution freshly prepared at the start of an experiment. It is usually pumped into the bed twice a day, requiring 30 min per pumping period. A comparable drainage time is required. This large solution volume is necessary to flood completely the two tons of silica sand or quartz gravel substrate in the bed. This volume of nutrient has been found adequate to supply completely all the mineral elements necessary to the development of the plant tissue, irrespective of the length of culture period. No addition of any element has been necessary for any of 15 species of plants cultured in the chamber.

As a safeguard against the effect of overchilling by the temperature-controlling water passing over the chamber, two 1000-kw strip-heaters are controlled by a thermostat and serve as a heat source when the temperature drops several degrees Fahrenheit below the established mean ambient temperature.

With certain biosyntheses it is necessary to enter the chamber relatively frequently. At these times, as well as at final harvest of plants, the C^{14}-activity in the air phase is lowered to a safe health physics level by allowing the plants to reduce the total carbon dioxide concentration photosynthetically to levels (approximately 0.003%) at or near the compensation point, where respiratory loss is equal to the photosynthetic fixation rate of the gas. An activity of 0.001 microcurie of C^{14} per liter of air is accepted as safe in terms of continuous human exposure. When this concentration level is reached the large access panel is removed. The remaining C^{14} activity is removed from the chamber by a fan and the chamber is then entered. If a partial harvest is required a smaller panel can be removed. When the chamber is resealed after a harvest the activity of the air is then readjusted. If a harvest period is desired during a period of low natural light intensity it is necessary to use the fluorescent light source to expedite the lowering of the chamber atmospheric C^{14} to the accepted health-safety level.

A Typical Biosynthesis: C^{14} Tobacco

A biosynthesis was undertaken to produce uniformly tagged tobacco tissue from which tagged organic compounds could be isolated for use in tracer studies intended to clarify the role of the compounds in plant metabolism. The compounds selected for study were the flavonoid pigments and related polyhydroxy phenolic compounds. These compounds are apparently distributed throughout the plant kingdom, in concentrations varying from a few parts per million in some tissues up to as high as 20% of the dry weight of floral parts of some plants. Little is known of the exact role that they play in the over-all metabolism of plants. While these compounds are present in foods used by animals and man, our knowledge concerning their metabolic fate is sparse. It is likely that compounds exist in plants that have hitherto escaped isolation and identification because of their exceedingly low concentrations as well as a lack of suitable methods for detecting their presence by chemical, physical, or biological tests. The labeled tobacco tissue is being utilized in part to make such a natural product survey. The carbon-14 tag allows the detection of amounts about 10,000 times less than can be detected by the usual chemical methods. Since all the compounds of uniformly tagged material contain carbon-14, the isolated individual compounds can be detected with tracer sensitivity either by counting or by radiographic techniques.

Culture of C^{14} tobacco plants. Seed of *Nicotiana rustica* was planted in April in soil in the greenhouse and maintained on 18-hr day length. Six weeks later the soil was carefully washed from the seedling roots, which were then transplanted into fine quartz sand contained in two-quart glazed crocks having bottom drains. At this time the plants were supplied a complete nutrient solution (Table 5.1). Three weeks later eight of the more vigorous seedling

TABLE 5.1. COMPOSITION OF NUTRIENT SOLUTION

Macro Elements	Molarity
$Ca(NO_3)_2$	0.0045
$MgSO_4$	0.0023
KH_2PO_4	0.0023
NH_4SO_4	0.0007
Micro Elements	**Parts per Million**
B	0.50
Mn	0.25
Zn	0.05
Cu	0.02
Fe [a]	0.50
Mo	0.05

[a] Fe added as chelated iron salt.

plants, which had four small, partially expanded leaves, were carefully transplanted to quartz gravel in the C^{14} biosynthesis chamber.

The complete nutrient solution (Table 5.1) was supplied twice daily to plants. The temperature and humidity were maintained at 75° ± 2° F and 50–55% relative. Two days later the chamber was hermetically sealed and the atmosphere was charged with $C^{12}O_2$ to raise the concentration of CO_2 to 0.060%; at the same time enough $C^{14}O_2$ was released to adjust the specific activity of the air phase to 350 microcuries per gm C. After 4 hr of photosynthesis the plants had dropped the CO_2 concentration to 0.052% and the specific activity had dropped to 300 microcuries per gm C. The following morning the CO_2 concentration had risen to 0.062% while the specific activity of the carbon-14 in the air phase had dropped to 200 microcuries per gm C. These figures reflect the usual dilution effect which plant respiration has upon the specific activity of the atmosphere early in a biosynthesis period. As the biosynthesis period progressed, the plants became more uniformly labeled and the specific activity of the air phase was progressively less influenced by respiratory CO_2. Essentially the plants became uniformly labeled at the desired specific activity when it was observed that the activity of the air phase remained constant throughout a dark period.

At intervals, usually at the start of a photosynthesis period, a new charge was introduced into the chamber to raise the CO_2 concentration, and if necessary, to readjust the specific activity to the desired level. It was the intent of this experiment to tag the entire plant at a specific activity of 250 microcuries per gm C. Therefore in the early stages of the biosynthesis the specific activity of the air phase was adjusted to 350 microcuries per gm C, after two weeks to 300 microcuries per gm C, and in a few more days to 275 microcuries per gm C. Throughout the entire experimental period, the CO_2 level was maintained within the concentration range of 0.100% to 0.010%. When the plants lowered the atmospheric CO_2 concentration to 0.010%, a new charge was introduced to raise the CO_2 to the maximum level and adjust the specific activity.

Forty-seven millicuries were fixed by the plants and about 3.8 kg of fresh plant tissue were produced. Four of the eight plants were harvested on the twentieth culture day, two more on the thirty-first day and the final two on the forty-fourth or last day. The plants were divided into leaves, stems, roots, and flowers at harvest time and stored either in a deep freeze unit or in 85% isopropyl alcohol.

When a harvest was to be made, the plants were allowed to photosynthesize until they had reached a compensation point; at this point the rate of CO_2 fixation was equivalent to the rate of CO_2 respired. For tobacco at high light intensities (3000–4000 foot candles), the compensation point concentration was 0.003% CO_2. At this time the level of C^{14} activity had dropped below the maximum permissible health physics level for C^{14} in air. The large access panel was removed and the small amount of residual activity blown off before the chamber was entered.

Fixing and storage of plant tissue. The manner in which the plant tissue is fixed and stored after harvest is determined by its projected use. The most convenient method is drying in a forced air oven which is vented to the outside atmosphere. It is understood that a drying temperature is employed which does not destroy the desired compounds. Tissue can also be frozen and stored in a deep freeze or it can be fixed by lypholization (freeze drying). When one is interested in labile materials the tissues can be worked up fresh in suitable solvents like alcohol or boiling water. In the case of the tobacco tissue some was worked up fresh by fixing in boiling isopropyl alcohol and some was placed in plastic bags and stored in the frozen state.

Isolation and Identification of Flavonoids

The leaf tissue was transferred to boiling 85% isopropyl alcohol and boiled for 5 min to destroy enzymatic activity. The tissue was then disintegrated in a Waring blendor, transferred quantitatively to a Soxhlet extractor and extracted for 16 hr with 85% isopropyl alcohol. The flavonoids were in the alcohol extract. The alcohol was evaporated from this solution, care being exercised to prevent the solution from going to dryness. The fats, pigments, tannins and other water insolubles were removed by filtration. The flavonoids were in the water filtrate. They were further purified by extracting with isoamyl alcohol. The isoamyl alcohol was evaporated to dryness under reduced pressure in the presence of nitrogen.

The residue was taken up in a minimum of 80% ethyl alcohol for paper chromatographic analysis. A small aliquot of the 80% ethyl alcohol solution, containing 50–100 micrograms, was spotted on a large sheet of paper and a two-dimensional chromatogram run. The various components of the mixture separate on the paper because of their different rates of migration in the various solvent systems employed. Since most of the flavonoids and other polyhydroxy phenolic compounds exhibit fluorescences in ultraviolet light, these compounds can be visualized by examination under the ultraviolet lamp provided their concentration is sufficiently great. The minimum amount that can be detected is usually of the order of 5–10 micrograms, and partial identification can be established by the character of the movement of the compound on paper plus its fluorescent color. Colored derivates can also be made of the flavonoids by spraying the chromatographic paper with appropriate chromogenic sprays, such as aluminum chloride or lead acetate, and the colors observed in normal or ultraviolet light. For certain flavonoids as little as 0.5–1 microgram can thus be detected (Ref. 3). When the paper chromatogram obtained by the above procedures was examined, six spots or compounds were evident. Since all the compounds obtained from the randomly tagged tobacco plants contained C^{14}, a radioautograph was made by exposing the paper chromatogram to ordinary no-screen X-ray film. After such exposure and development of the film, additional compounds were detected and the number of compounds was then found to be twenty. Differences were noted in the detection sensitivity of the chemical as against the radioautographic technique.

Previous to this work three flavonoid compounds had been reported in tobacco leaves, namely, rutin, isoquercitrin, and quercetin. By use of the above technique at least four additional flavonoids previously unreported have been shown to occur naturally in tobacco leaves. In order that these compounds may be further identified and characterized by classical chemical procedures, much larger quantities than normally employed with paper chromatographic methods have to be processed. Such large-scale separation procedures can be worked out using small amounts of tagged compounds.

The biosynthesis of labeled forms of organic compounds, as well as their isolation and identification, completes the first major step in the application of radiocarbon as a tracer tool.

Discussion

Uniformly C^{14}-labeled intact plants have particular usefulness in the application of radiocarbon as a tracer tool in general agricultural and biological research. They can serve not only as the source of tracer forms of known natural organic constituents of importance in these and other fields, but can be uniquely used in studies designed to isolate and identify previously unknown natural products. It is likely that the latter use of radiocarbon plants can make the more significant contributions.

A great number of uniformly C^{14}-labeled organic constituents, particularly those classified as early products of photosynthesis, can be efficiently biosynthesized by use of simple bell-jar type facilities (Ref. 2). However, it is not possible to biosynthesize uniformly labeled, intact plants with such facilities. This is true mainly because it is not possible to provide adequately controlled conditions for normal plant growth in small-volume systems hermetically sealed for extended periods of time. A biosynthesis facility of the type reported here is needed to culture higher plants efficiently under these conditions. It should be noted that the usefulness of this type of facility extends to tracers other than radiocarbon, both stable and unstable types. In addition, the facility is adaptable to a variety of non-tracer, physiological studies concerned with the interaction effects of such environmental variables as carbon dioxide concentration, temperature, mineral nutrition and day length upon plant growth and development.

Plants of alfalfa, rye, artichoke, *Tradescantia,* onion, *Digitalis,* opium poppy, red kidney bean, *Hevea,* and several varieties of soybean, tobacco, buckwheat, and snapdragon have been cultured with $C^{14}O_2$ atmospheres in the plant biosynthesis chambers. The objective of the individual experiments, as well as the duration of the culture periods, varied from species to species. Many of these studies were conducted collaboratively with personnel of other research groups in the course of exploring and applying the usefulness of the biosynthesis facilities and the tracer products synthesized. Certain of these experimental studies are worthy of review for individuals interested in the scope of the research application which can be made with this approach.

The biosynthesis of uniformly labeled sucrose and dextran, a blood volume expander, has been reported (Ref. 7). In this study the biosynthesis of a total of 175 millicuries of uniformly C^{14}-labeled sucrose was accomplished with use of photosynthesis periods which were as short as 5 hr in duration. Of all the $C^{14}O_2$ supplied to excised *Canna* leaves an average of 99% was assimilated and an average of 57% was converted to sucrose. Microbiological degradation of the isolated sucrose indicated that the molecule was uniformly labeled (Ref. 7). Comparable efficiencies can be attained in the biosynthesis of a specific organic constituent when intact higher plants are cultured for periods as long as seven weeks. Uniformly C^{14}-labeled soybean oil, tagged at 70 microcuries per gm of carbon, was biosynthesized for use in tracer animal metabolism studies. Soybean plants were photoperiodically induced by short-day treatment to form microscopic terminal inflorescences at all meristems just previous to placing them in the biosynthesis chamber for a seven-week culture period. Such photoperiodic treatment prevents further meristematic differentiation of leaves in this variety of soybean. At harvest the tagged plants had mature pods which contained 56% of all the assimilated C^{14}. In the seed alone 39.6% of the assimilated C^{14} was present, while 12.6% was present in 12.2 gm of tagged oil isolated from the seed. This biosynthesis points out the efficiency which can be attained in converting C^{14} into either specific plant organs or their products by the physiological regulation of the character of growth of plants.

It is essential to emphasize the major limitation associated with the biosynthesis of C^{14} plants or their products. It is obvious that plants cannot be cultured with tissues which contain concentrations of C^{14} which will induce abnormal growth in response to the beta radiation emitted by this isotope. One of the continuing objectives of the C^{14} program has been to establish the concentrations of this isotope that induce abnormal plant behavior as judged by a number of response indices.

Roots from C^{14}-labeled onion plants cultured in the biosynthesis chamber for periods ranging from 7–14 weeks and tagged with 44–46 microcuries per gm of dry tissue (absolute activity) exhibited both chromosome and chromatid breaks and fragments in dividing cells, as well as bridging and micronuclei (Ref. 1). While no visible developmental anomalies occurred during the chamber culture period, the new growth of harvested dormant sets made in normal atmospheres was arrested, and mutant leaf areas (chlorophyll deficiency) occurred. At this C^{14} concentration, macroscopic anomalies in this species occurred only after long exposure to the beta radiation.

Several factors must be considered in evaluating the concentration of C^{14} occurring as constituent body carbon, which can make the biosynthesis of labeled plants or their products either impractical from an efficiency standpoint or an impossibility. The major biological factor to be evaluated is the relative radiation sensitivity of plant species, including not only the species sensitivity at various life-cycle stages, but also the relative sensitivity of the different tissue types that constitute the several organs.

The various organs of uniformly C^{14}-labeled plants are by no means exposed to identical beta-radiation dosages. In all cases a microcurie of C^{14} uniformly distributed in a gram of fresh tissue is calculated to result in approximately 3 rep (roentgen equivalent physical) per day (Ref. 4). This applies to a gram of tissue centrally located in an organ which is labeled at the same C^{14} concentration and whose average diameter exceeds the tissue range of the C^{14} beta particle. While a measurable variation occurs in the quantity of carbon per dry tissue weight, the most significant plant composition factor in effecting different radiation dosage is variation in moisture or dry weight. The freshly harvested seed of uniformly C^{14}-labeled mature red kidney bean and soybean plants were found to have radiation dose-rates that were 8 and 9 times greater, respectively, than that occurring in the fresh leaves. While no abnormal tissues appeared on the soybean plants while in the biosynthesis chamber, their seedlings exhibited arrested growth rates and abnormal tissues when grown immediately following harvest of the tagged seed.

As much as a tenfold difference in species sensitivity to radiocarbon has been observed in the course of culturing the various plants noted earlier. Red kidney bean has been predominantly used in evaluating the pattern of anomalies which occur in response to toxic concentrations of C^{14}. This species was cultured with atmospheres containing $C^{14}O_2$ with specific activities of 0, 50, 100, 1000 and 2000 microcuries per gm of carbon. In all cases plants were grown from seedling to adult stages. With the two highest specific activities, radiation-induced types of deformed leaves occurred (Fig. 5.2). In addition a chlorotic leaf symptom developed which was strikingly similar to that typically induced by nitrogen deficiency. With the 2000-microcurie

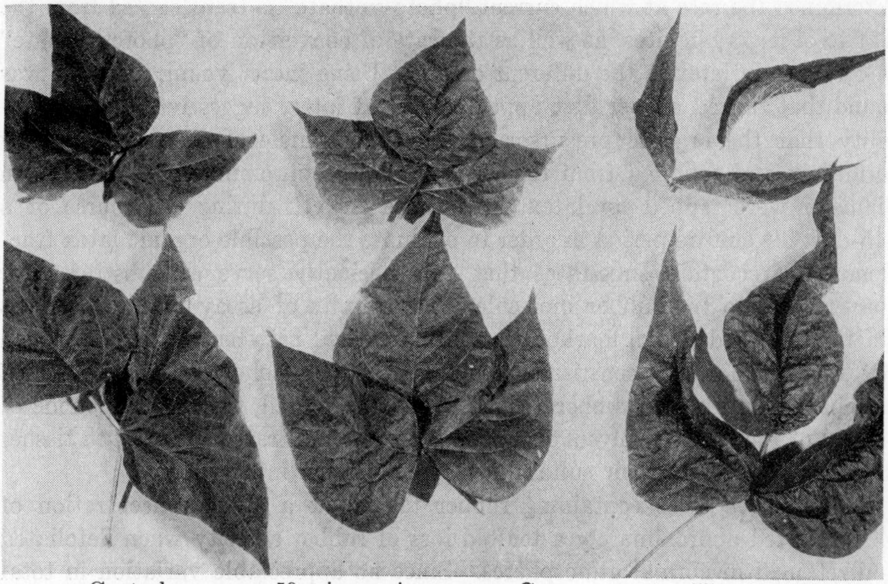

Control　　50 microcuries per gm C　　1 millicurie per gm C

FIG. 5.2. Leaves of red kidney bean plants grown in the biosynthesis chambers for 33 days in atmospheres containing indicated specific activities of $C^{14}O_2$. For each C^{14} concentration the upper young leaf was completely differentiated during the culture period while the bottom mature leaf had just emerged from the terminal bud at start of the experiment.

level the terminal, main axes meristems were aborted or partially necrotic when observed under the dissection microscope. Obviously in such instances the development of flowers or fruit cannot take place.

An effort was made to relate the "nitrogen deficiency" leaf response noted in the case of plants cultured with specific activities of 1000 and 2000 microcurie per gm of carbon to possible physiological or biochemical differences in the leaf tissues. Chemical assay for total nitrogen and ascorbic acid of both young and mature leaves of control and C^{14} plants was made. These assays indicated that a significantly decreased level of nitrogen and an increased level of ascorbic acid occurred beginning with plants cultured with the 1000-microcurie concentration of C^{14}. Although no other chemical constituents were assayed, it is likely that additional differences would have been found in the abnormal tissues of these plants and can be found in tissues of plants cultured at C^{14} concentrations which induce macroscopically visible growth anomalies.

The plant biosynthesis facility (Fig. 5.1) has been used in connection with a number of C^{14} tracer physiological studies. Relatively typical of the qualitative experimental data that can result is that recently obtained in the case of *Hevea braziliensis*. This tracer study was concerned with carbon assimilation and rubber formation and turnover in *Hevea*. These studies have determined the rate at which current "photosynthate" is translocated from the leaf to other organ sites, as well as the rate of conversion of "photosynthate" into rubber in latex in the different organs. Using intact young plants it was found that tagged rubber first appeared in leaf latex, suggestive of the possibility that the organic precursor of the rubber molecule is predominantly formed in and supplied from this organ. Trees approximately 3 to 4 in. in diameter were tapped for latex three times a week during the course of a 115-day C^{14} culture period in order to evaluate the possible organic latex fractions which contain substrates that can efficiently serve as substrates for conversion into the rubber molecule. The results of assay of latex solvent fractions showed that a marked correlation existed between the rate at which C^{14} appeared in latex constituents soluble in 80% alcohol and that at which it appeared in isolated rubber (Fig. 5.3). The isolated, tagged compounds of several of the organic solvent fractions are to be infiltrated into *Hevea* tissues in order to evaluate their suitability for conversion into rubber.

Other *Hevea* trees containing rubber tagged at a given concentration of C^{14} exhibited approximately a tenfold loss of rubber activity when defoliated and retained in normal atmospheres. Since no appreciable variation in total rubber concentration occurred in this experimental period the data are interpreted to be indicative of the fact that rubber, once formed, does turn over at a measurable rate, a point which has been in question for some time.

While the biosynthesis of C^{14} plants and their products is obviously restricted by the biological effects which beta radiation can exert at elevated dosages,

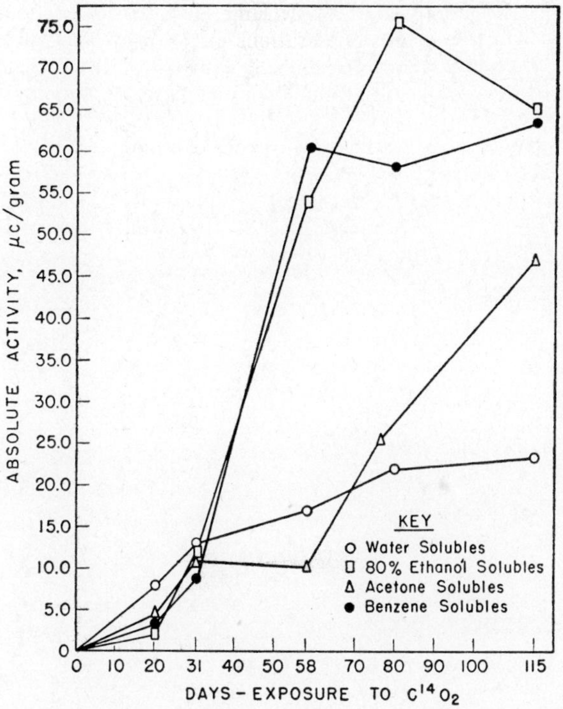

FIG. 5.3. Absolute activities of organic fractions of stock stem latex from a *Hevea* tree cultured in the biosynthesis chamber with $C^{14}O_2$ for 115 days. The solvent extraction sequence was water, alcohol, and acetone, followed by benzene. The benzene-soluble fraction represents rubber.

this method of tracer synthesis nevertheless offers relatively unlimited opportunity towards the profitable application of radiocarbon as a tracer tool where efforts are being made to obtain a better understanding of organic compounds, whether the concern be in the fields of agriculture, chemistry, biology, or medicine. It is hoped that the selected observations noted here will be useful to others as a basis for determining the value which the C^{14} plant biosynthesis facility and its products may be to those who foresee that radiocarbon can serve as a useful tool in their particular areas of research.

REFERENCES FOR CHAPTER 5

1. J. M. Beal and N. J. Scully, "Chromosomal aberrations in onion roots from plants grown in an atmosphere containing $C^{14}O_2$," *Botanical Gazette* 112:233–235 (1950).
2. M. Calvin, C. Heidelberger, J. C. Reid, B. M. Tolbert, and P. F. Yankwich, *Isotopic Carbon,* John Wiley & Sons, Inc., New York, 1949.
3. T. D. Gage, C. D. Douglass, and S. H. Wender, "Identification of flavenoid compounds by filter paper chromatography," *Anal. Chem.* 23:1582–1585 (1951).

4. L. D. Marinelli, E. Quimby, and G. J. Hine, "Dosage determination with radioactive isotopes: II. Practical considerations in therapy and protection," *Am. J. Roentgenol. & Radium Therapy* 59:260–280 (1948).
5. N. J. Scully, "Progress report of radiobiology experiment station," *ANL*-4205:164–165 (August 1948).
6. N. J. Scully and W. Chorney, "Progress report of radiobiology experiment station," *ANL*-4253:145–147 (February 1949).
7. N. J. Scully, H. E. Stavely, J. Skok, A. R. Stanley, J. K. Dale, J. T. Craig, E. B. Hodge, W. Chorney, R. Watanabe, and R. Baldwin, "Biosynthesis of the C^{14}-labeled form of dextran, 116:87–89 (1952).

Part II

STUDIES OF PHOTOSYNTHESIS

Chapter 6

THE PHOTOSYNTHETIC CYCLE *

Photosynthesis is usually defined as the biochemical reaction

$$CO_2 + H_2O \xrightarrow{nh\nu} (CH_2O)_x + O_2$$

This represents the conversion of carbon dioxide and water to carbohydrate and oxygen by green plants in the light. The reaction is separated both chronologically and chemically into two parts: the photolysis of water,

$$H_2O \xrightarrow{nh\nu} 2[H] + \tfrac{1}{2}O_2$$

and the reduction of carbon dioxide,

$$CO_2 + 4[H] \rightarrow CH_2O + H_2O$$

Each of these two reactions represents a complex series of reactions with many steps. The term [H] is used to denote reducing agents generated in the photochemical decomposition of water. These reducing agents probably undergo several transformations before they are used in the reduction of carbon dioxide.

The reactions involved in the reduction of carbon dioxide have been studied and the results of these studies have been reported in a series of papers on "The Path of Carbon in Photosynthesis" (Refs. 1, 2, 3).

The radioactive carbon isotope, C^{14}, was used throughout this investigation. To a lesser extent, radioactive phosphorus, P^{32}, was also employed. As a result of this work, it is now possible to write the complete path of carbon reduction in photosynthesis, with all intermediates and enzymatic reactions, from carbon dioxide to sucrose. The study of carbon reduction and its relation to respiratory transformations of carbon compounds has provided evidence regarding the nature of the reactions involved in the decomposition of water and the formation of the primary reducing agents and other energy-rich compounds required for carbon reduction.

* This chapter is taken from Geneva Conference Paper 259, "The Photosynthetic Cycle" by M. Calvin and J. A. Bassham of the United States.

First Products

The methods used in studying the path of carbon in photosynthesis are here described briefly. In nearly all cases the initial condition is an actively photosynthesizing plant in which photosynthesis has been maintained long enough to establish a "steady state." In this steady-state condition the concentrations of various intermediate compounds in the pathway from carbon dioxide to sucrose are constant. The plants commonly used in these experiments are the unicellular green algae *Chlorella* or *Scenedesmus,* but leaves of higher plants are sometimes used.

In the first type of experiment to be discussed, $C^{14}O_2$ is added to the unlabeled CO_2 that the plant has been using. After a measured short period of photosynthesis with $C^{14}O_2$, the plant is killed by sudden treatment with boiling ethanol. All enzymatic processes are thereby quickly halted. Extracts of the plant material are made, concentrated, and then analyzed by two-dimensional paper chromatography and radioautography. The techniques of two-dimensional chromatography and radioautography of plant extracts labeled with C^{14} have been described earlier (Ref. 1) as well as the identification of the numerous labeled compounds (Refs. 1, 4, 5, 6, 7). The radioautographs obtained from experiments of 10-sec and 60-sec photosynthesis with $C^{14}O_2$ are shown in Figs. 6.1 and 6.2. The 60-sec experiment illustrates the importance of various sugar phosphates and acid phosphates in carbon reduction. The 10-sec experiment shows the predominance of phosphoglyceric acid at short times. If the percentage of C^{14} in phosphoglyceric acid (PGA) of the total C^{14} incorporated during photosynthesis for various short periods of time

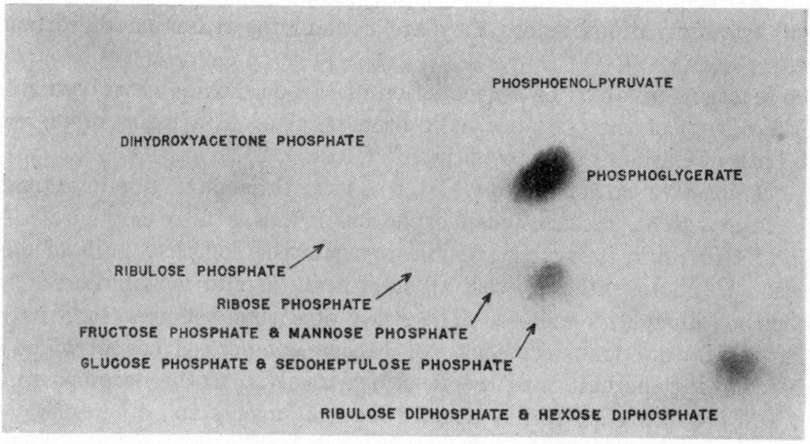

Fig. 6.1. Chromatogram of extract from algae, indicating uptake of radiocarbon during photosynthesis (10 seconds).

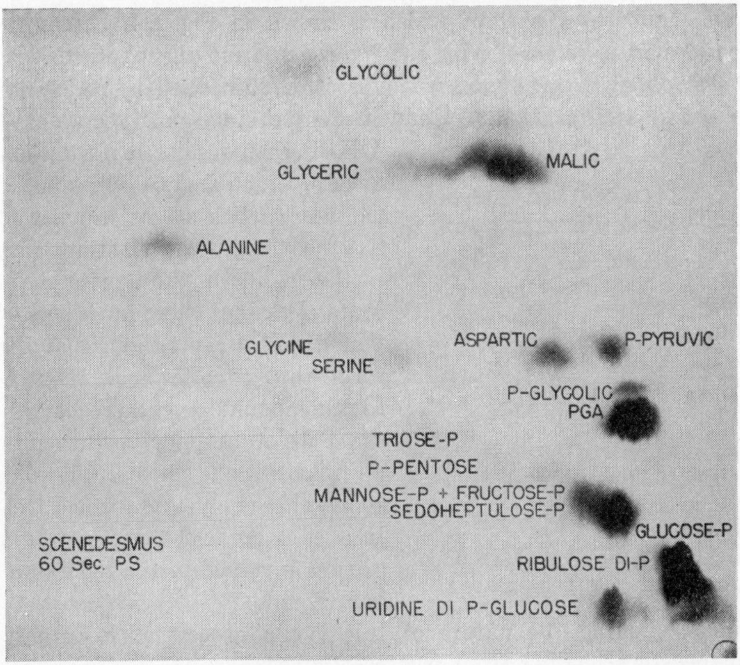

FIG. 6.2. Chromatogram of extract from algae indicating uptake of radiocarbon during photosynthesis (60 seconds).

is extrapolated to zero time, it is found that at zero time all the C^{14} should be in phosphoglyceric acid. This compound is therefore identified as the first compound into which carbon dioxide is incorporated in photosynthesis.

Fig. 6.3 shows the distribution of the labeled carbon in the three carbon atoms of the glyceric acid obtained from the phosphoglyceric acid in a 15-sec experiment. Half of the C^{14} is in the carboxyl group and the other half is divided equally between the other two carbon atoms. From the same experiment some hexose (fructose and glucose) was obtained and degraded. The distribution of carbon in the two three-carbon halves of the hexose was found to be very much the same as it is in the three carbons of glyceric acid. This result immediately suggests that the six-carbon piece is made from the two threes by joining the two carboxyl carbon atoms. This is simply a reversal of the well-known aldolase split of fructose diphosphate in the

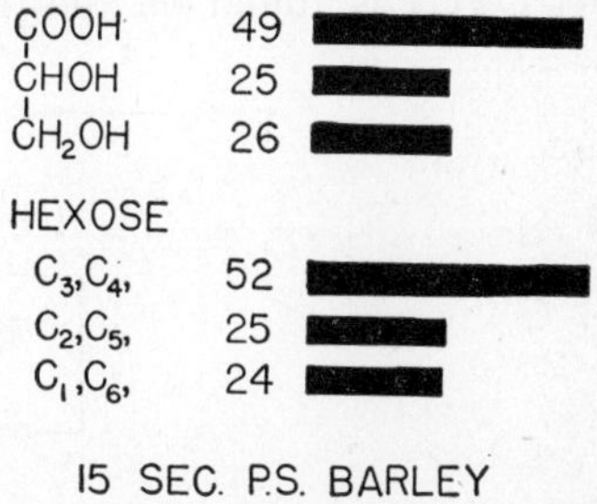

FIG. 6.3. Distribution of labeled carbon in photosynthesis experiments.

glycolytic sequence, a part of which is shown in Fig. 6.4. Here the phosphoglyceric acid is reduced with the hydrogen from the photochemical reaction to phosphoglyceraldehyde, which is then isomerized to form dihydroxyacetone phosphate (DHAP). Condensation of phosphoglyceraldehyde with DHAP then results in formation of the hexose, fructose–1,6-diphosphate. Thus, the two carbon atoms which were originally carboxyl-carbon atoms finally fall in the middle of the hexose chain. It is quite clear that there must be some compound that accepts the carbon dioxide to form the glyceric acid. Furthermore, that compound must be regenerated from the PGA (phosphoglyceric acid), triose phosphates, and hexose phosphates, or some other compound formed from them. It is thus evident that there is a cyclic process involved in the reduction of carbon dioxide.

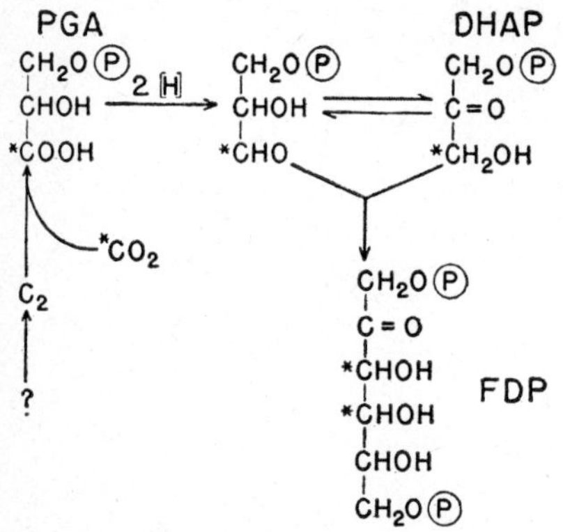

FIG. 6.4. Path of carbon from CO_2 to hexose during photosynthesis.

Before considering the nature of this cyclic process it is of interest to mention the steps leading from fructose diphosphate to the final product of photosynthesis, sucrose. These steps were identified after the intermediate compounds were isolated by paper chromatography and radioautography. Fig. 6.5 shows the relationship that was found. Here are shown the phosphoglyceric acid, fructose diphosphate, and the various transformations that lead ultimately to glucose–1-phosphate. This compound reacts with uridine triphosphate to make uridine diphosphoglucose. Uridine diphosphoglucose (UDPG) is found on the paper, with the glucose moiety labeled after very

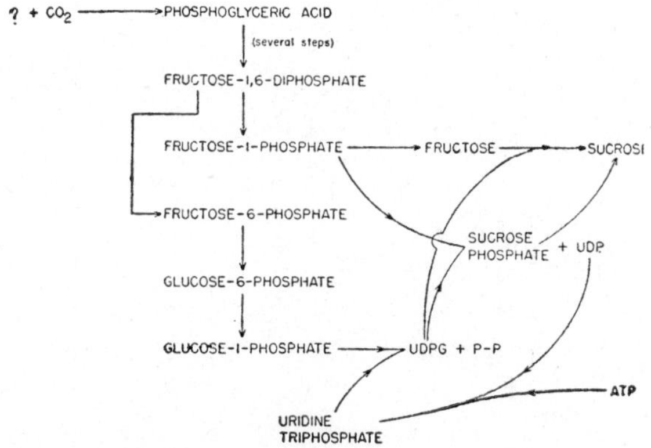

FIG. 6.5. Proposed mechanism for formation of sucrose with uridine diphosphoglucose.

short $C^{14}O_2$ exposures. UDPG can then react in one of two ways: either with fructose–1-phosphate to form sucrose phosphate which then is phosphatased to sucrose, or directly with free fructose to form sucrose in one operation. However, since one seldom finds any free labeled fructose, the first of these alternatives appears to be the major pathway for green leaves. An enzyme performing the reaction

$$\text{UDPG} + \text{fructose phosphate} \rightarrow \text{sucrose phosphate}$$

has recently been prepared in a partially purified state by Leloir in Argentina. Fig. 6.6 shows the structural formula for the UDPG and its reaction with

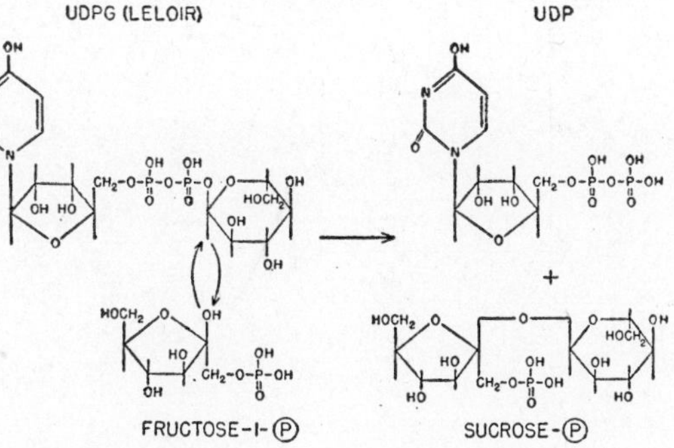

FIG. 6.6. Uridine diphosphoglucose reaction with fructose phosphate.

fructose–1-phosphate. This reaction gives uridine diphosphate and sucrose phosphate with the phosphate on carbon atom 1 of the fructose moiety. The phosphate is then removed to give sucrose. This appears to be the common route to sucrose and is therefore one of the major synthetic reactions in agriculture, since sucrose provides the substrate for a wide variety of other transformations.

C_5 and C_7 sugars. We now return to the problem of cyclic regeneration of the carbon dioxide acceptor. The roles of PGA, triose phosphates, hexose phosphates, UDPG, and sucrose have already been identified. Of the compounds labeled by short periods of photosynthesis, there were left only the seven- and five-carbon sugar phosphates. These were sedoheptulose–7-phosphate (SMP), ribose–4-phosphate (RMP), ribulose–5-phosphate (RuMP) and ribulose diphosphate (RuDP).

An attempt was made to determine the order of occurrence of these compounds by the same technique as was used to identify PGA as the first product of CO_2 fixation.

Since the reactions of carbon reduction are so rapid, a flow system was designed to obtain sufficiently short periods of exposure to $C^{14}O_2$ to permit observation of the relative rates of labeling of the various sugar phosphates

(Ref. 3). The system used is shown in Fig. 6.7. A suspension of algae was forced by means of a pump from a transparent tank through a length of transparent tubing into boiling methanol. An aqueous solution of $C^{14}O_2$ was injected at a constant rate into the tubing. The time of exposure of the algae to $C^{14}O_2$ was determined by the rate of flow of algae through the tubing and the

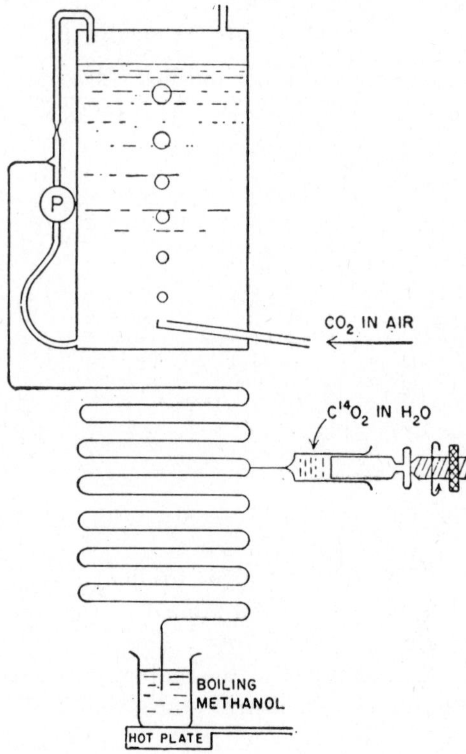

FIG. 6.7. Schematic diagram of flow system for short exposures of algae to $C^{14}O_2$.

length of tubing between the point of injection and the killing with methanol. In this way exposure times ranging from 1 to 20 sec were obtained. When the radioactivity found in each of the sugar phosphates was extrapolated to zero time of exposure, however, no choice could be made between the pentose, hexose, and heptose phosphates. It appeared that all were formed at the same time. It was necessary, therefore, to turn to degradation studies of these various sugar phosphates labeled in the very short exposures.

A detailed analysis of the distribution of radioactivity among the carbons of these sugars is shown in Fig. 6.8. Here, besides PGA, are the five-carbon sugar, ribulose diphosphate (RuDP); the seven-carbon sugar, sedoheptulose phosphate (SMP); and the skeleton of a six-carbon sugar, corresponding either to glucose or fructose (these are the major six-carbon sugars that we find). The stars give some indication of the order of appearance of radioactive carbon in these compounds, and it was from an analysis of these data that it became possible to deduce relationships between the various compounds.

In much the same way as we deduced the relation between the three-carbon PGA and the six-carbon sugars we were able to deduce the relationships be-

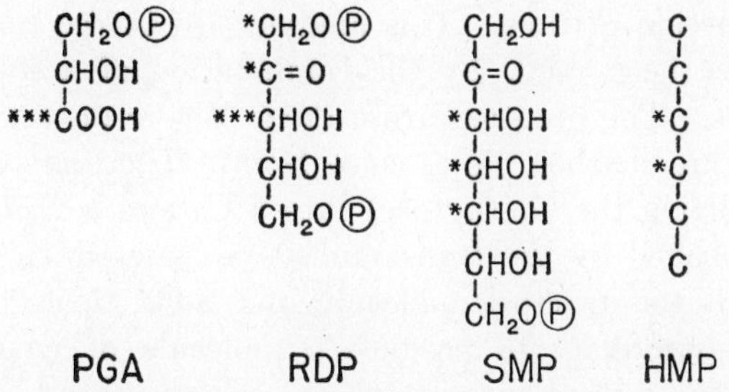

FIG. 6.8. Distribution of radioactive carbon in certain sugars.

tween the five-, seven,- six- and three-carbon compounds that are shown here. It is quite clear at a glance that there is no simple structural relationship between the five- and the seven-carbon compounds and the other sugars. At least, there is nothing as simple as the relationship between the three-carbon PGA and the six-carbon hexose. There is no sequence of carbon atoms in the C_5 or C_7 sugars that could be considered as simply the intact C_3 or the intact

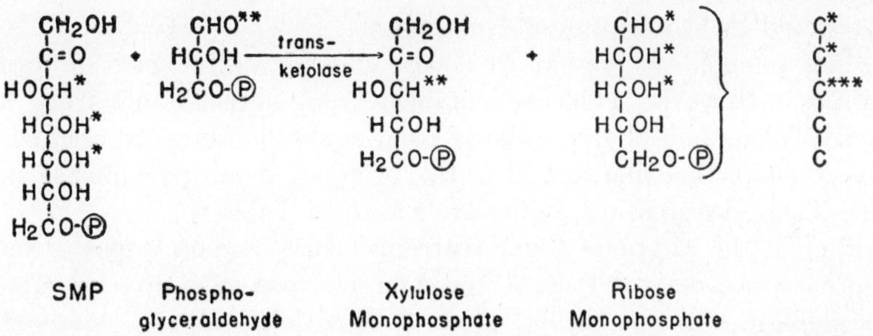

FIG. 6.9. Formation of 5-carbon sugars from sedoheptulose phosphate.

C_6, respectively. Until we realized that the C_5 might have more than one origin we were not able to deduce a possible route for its formation. This route is shown in Fig. 6.9. By taking two carbons off the top of the C_7 and adding them onto a three-carbon piece labeled as is phosphoglyceraldehyde, we would get two five-carbon pieces—one ribulose and one ribose—with their labeling distributed as shown. The average of their labeling would be the actual one found. This evidence, therefore, indicates that the origin of the ribose and ribulose phosphates is in a transketolase reaction of the sedoheptulose phosphate with the triose phosphate to give the two pentose phosphates. These can be interconverted by suitable isomerization. Thus, the pentoses are formed from heptose and triose.

As was shown earlier, the hexose is formed from two trioses. The question then remains, Where does the heptose come from? And here, again, a similar detailed analysis was made of the carbon distribution within the heptose molecule as a function of time. This analysis led to the realization that the heptose must have been made by the combination of a four-carbon with a three-carbon piece. The question arose then: Where do the properly labeled four-carbon and three-carbon pieces come from? The four-carbon piece could only come by splitting the C_6 (hexose) into a C_4 and a C_2.

This is accomplished by the transketolase enzyme which removes the two "top" atoms from the fructose molecule and adds them to a molecule of glyceraldehyde–3-phosphate to produce a molecule of ribulose–5-phosphate (RuMP). The four-carbon piece that remains (erythrose–4-phosphate) has the distribution of radiocarbon that is required by the observed labeling in the four "bottom" carbon atoms of sedoheptulose.

The three-carbon piece required for the three "top" carbons of sedoheptulose might be dihydroxyacetone phosphate. In this case the condensing enzyme would be aldolase and the product would be sedoheptulose diphosphate.

Alternately, the three-carbon piece might be obtained by the splitting of hexose by the enzyme transaldolase, which would transfer the three top atoms of fructose–6-phosphate to the four-carbon piece (erythrose phosphate) formed from the four bottom atoms of another fructose molecule. In this case the product would be sedoheptulose–7-phosphate.

It is not possible at present to choose unequivocally between these two possibilities. However, evidence obtained from degradations from various radioactive sugar phosphates isolated from soybean leaves exposed to $C^{14}O_2$ for a very short time indicate that the proposal requiring aldolase may be correct. These degradation results are shown in Table 6.1.

In either of the alternate sugar rearrangements, carbon atoms 4 and 5 of sedoheptulose are derived from 3 and 4 of fructose, respectively. However, in the aldolase version, carbon atom 3 of sedoheptulose is derived from carbon 1 of dihydroxyacetone phosphate. Alternatively, in the transaldolase version, carbon atom 3 of sedoheptulose is derived from carbon 3 of fructose and therefore should have the same label at all times. Since the latter condition is not experimentally fulfilled, the aldolase reaction appears to be the correct one. However, it must be noted that this argument rests on the assumption that the concentration of the intermediate erythrose phosphate is small compared with that of fructose–6-phosphate. Also it may be noted that a small amount of labeled sedoheptulose has been obtained from hydrolysis of

TABLE 6.1. DISTRIBUTION OF CARBON-14 IN SEDOHEPTULOSE ISOLATED FROM SOYBEAN LEAF

	Time of Exposure to $C^{14}O_2$	
	0.4 Sec	0.8 Sec
$H_2C—OH$	0	2
C=O	0	2
HO—C—H	33	39
HC—OH	8	18
HC—OH	49	38
HC—OH	0	2
$H_2C—OPO_3H^-$	0	2

the sugar diphosphate area, indicating the presence of labeled sedoheptulose diphosphate. The presence of this compound may be accounted for by assuming its formation by aldolase from dihydroxyacetone phosphate and erythrose phosphate. Therefore, this route is tentatively accepted for the formation of sedoheptulose. The transformation described is shown in Fig. 6.10. Here are

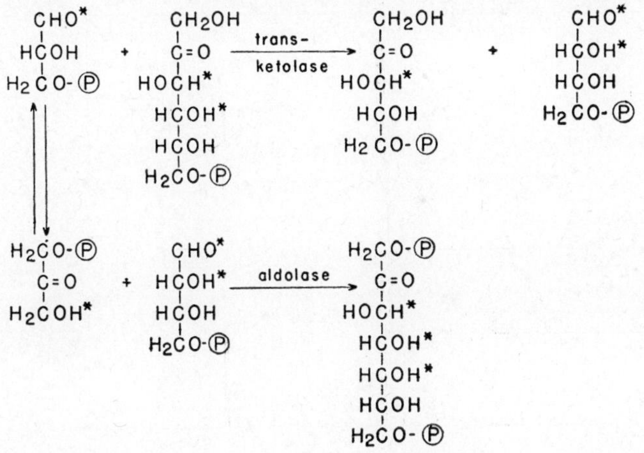

FIG. 6.10. Formation of a heptose from triose and hexose.

shown the two trioses that can make one hexose. One hexose then reacts with another triose to give pentose and tetrose, by means of the action of the enzyme transketolase. Tetrose and triose are then condensed by aldolase to give sedoheptulose. The net result of the reactions shown in Figs. 6.9 and 6.10 is the formation of three molecules of pentose from five molecules of triose.

Identification of the CO_2 acceptor. All the results thus far were obtained with the first type of experiment, in which $C^{14}O_2$ was added to plants for a very short period (1 to 60 sec) before the plants were killed. A second type of experiment was used for the identification of the CO_2 acceptor. In this case once again the starting condition was an actively photosynthesizing algae suspension in "steady-state" condition. In addition, the intermediate compounds were "saturated" with C^{14}. This was accomplished by leaving the plants in contact with an atmosphere of C^{14}-labeled CO_2, maintained at constant specific activity and CO_2 pressure, for more than an hour prior to the start of the experiment. Under this condition, the concentration of each labeled intermediate compound can be determined from the radiocarbon found in that compound on subsequent analysis by chromatography and radioautography.

After this initial C^{14}-saturated steady state was obtained, aliquots of the algal suspension were taken at frequent intervals for analysis. Then some environmental condition such as light was suddenly changed. Aliquots of the algae were taken every 2 or 3 sec for about a minute, and then at less frequent intervals. Analysis of these aliquots showed the way in which the concentrations of the various intermediates varied as a result of the environmental change.

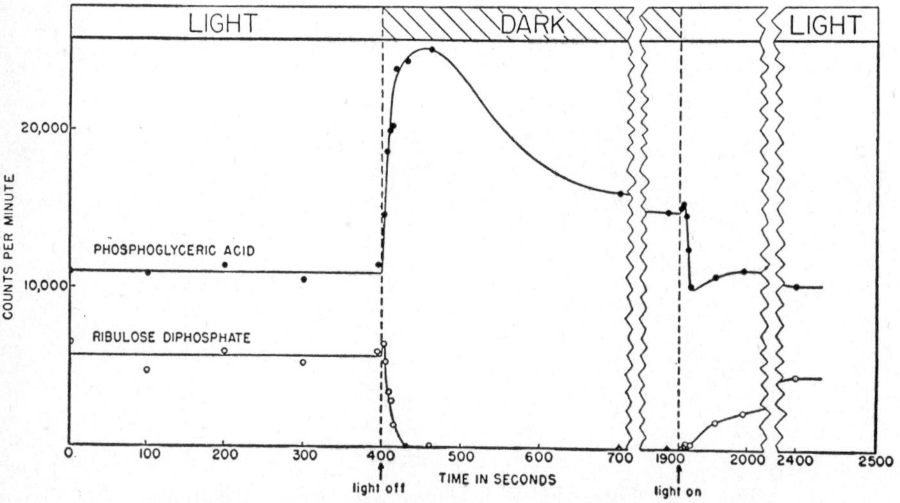

FIG. 6.11. Light-dark transients in PGA and RuDP concentrations.

In the first such study (Ref. 2) the light was turned off. It was found that the concentration of PGA increased very rapidly while that of ribulose diphosphate (RuDP) decreased rapidly. The results of a later, somewhat more refined, experiment are shown in Fig. 6.11 (Ref. 8). Here it is seen that the concentration of RuDP decreases to below a detectable amount ($<1\%$ of its initial value) in about 30 sec. These changes in concentrations can be accounted for if we assume the following: The reduction of PGA to triose and the formation of RuDP are reactions requiring light; RuDP is converted to PGA via a carboxylation reaction that does not require light.

These relations are shown in Fig. 6.12. PGA is reduced to triose phosphate (at the sugar level); the triose phosphate then undergoes a series of rearrange-

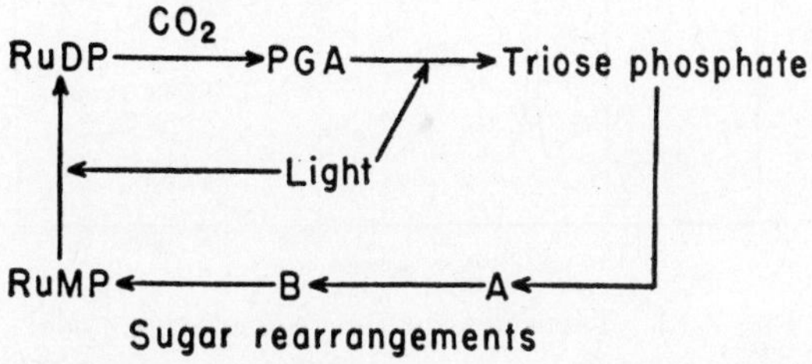

FIG. 6.12. Suggested cyclic scheme for relationships in photosynthesis.

ments, such as the ones described earlier, through the hexose, pentose, and heptose, back again to the ribulose–5-phosphate. This is all at the sugar level of oxidation and requires very little energy for its operation. There is then some light requirement for the formation of RuDP from RuMP. The reduction of PGA requires both reducing power, reduced triphosphopyridine nucleotide (TPNH), and adenosine triphosphate (ATP), while the formation of RuDP from RuMP requires ATP, as will be seen later. Both these cofactors are produced at the required rate only when the light is on. Thus, when the light is turned off, the rate of formation of RuDP and the rate of reduction of PGA decrease, but the rate of carboxylation of RuDP to form PGA continues unaffected except by the concentration of RuDP.

From the above scheme it was possible to predict the result if the light were left on but the CO_2 pressure were suddenly decreased. In that event, the carboxylation of RuDP to form PGA should decrease, but the formation of RuDP and the reduction of PGA should be unaffected. Consequently the concentration of RuDP should rise while that of PGA should fall. This experiment was performed (Ref. 9) and the expected result, shown in Fig. 6.13, was obtained. When the CO_2 pressure is decreased, the first compound to increase in concentration is RuDP and the second is its immediate precursor, RuMP. Last to rise in concentration is triose phosphate which is one of the precursors of RuMP. The first compound to decrease in concentration is PGA

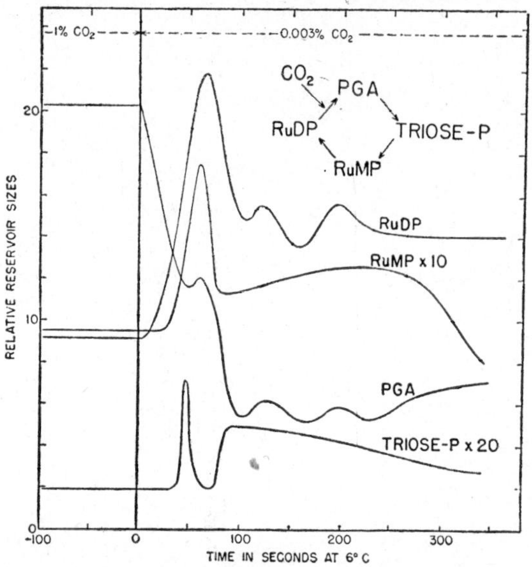

FIG. 6.13. Transients in the regenerative cycle.

and the second is its immediate product, triose phosphate. Next to decrease is RuMP, and last to decrease is RuDP. These changes provide excellent confirmation for the proposal of the cyclic system.

There remained some question whether the carboxylation of a molecule of RuDP produced two molecules of PGA or whether some other reaction might occur *in vitro* in which only one molecule of PGA is produced along with a molecule of triose. In order to test this alternative, a rather careful experiment was performed (Ref. 8) in which the rate of increase of PGA when the light was turned off was compared with the steady-state uptake of CO_2. During the first few seconds after turning off the light, the rate of increase of PGA should approximately equal the rate of its formation during steady-state conditions, provided reduction of PGA could be suddenly halted. The ratio of molecules of PGA increase per second to molecules of CO_2 taken up per second should indicate the number of molecules of PGA actually formed per molecule of CO_2. If this ratio experimentally approached 2 at short times, or even exceeded 1, we would have evidence for the formation of two molecules of PGA for each molecule of RuDP carboxylated. The ratio was calculated from the data shown in Fig. 6.11 and the measured CO_2 entry rate, and was found to be between 1.5 and 2. Thus, kinetic *in vitro* evidence is provided for the carboxylation reaction

$$\text{RuDP} + CO_2 + H_2O \rightarrow \text{2PGA}$$

THE CARBON-REDUCTION CYCLE

The complete carbon-reduction cycle is shown in Fig. 6.14. Here are shown all the details, including the intermediate compounds and enzymes required for the various transformations. The net result of each turn of the complete cycle is the introduction of 3 molecules of CO_2 and the carboxylation of 3 molecules of ribulose diphosphate, leading to the formation of 6 molecules of phosphoglyceric acid. These 6 molecules of PGA are then reduced to provide

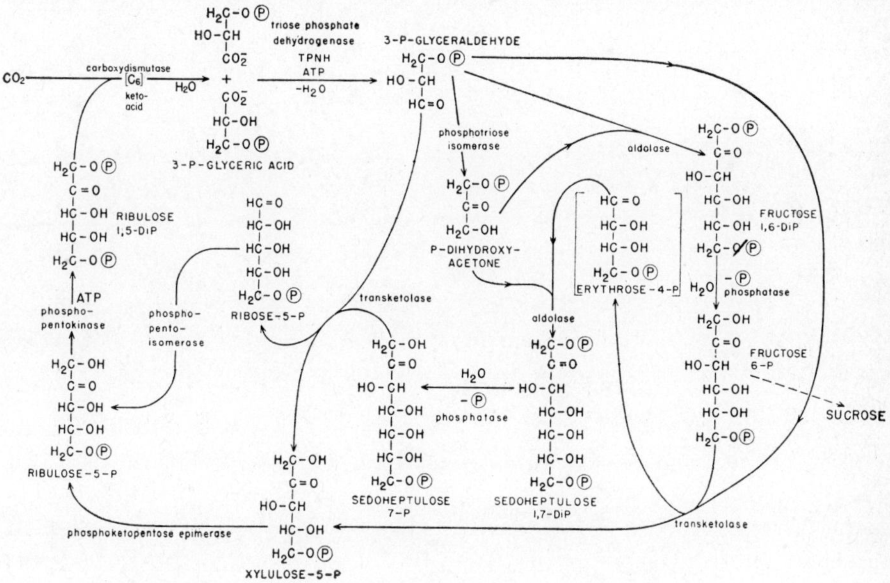

FIG. 6.14. The complete photosynthetic carbon cycle.

6 molecules of triose phosphate. Of these, 5 are eventually converted to ribulose diphosphate, thus completing the cycle, while the sixth finds its way ultimately into sucrose and represents the net gain in reduced carbon per turn of the cycle. All the enzymes shown had been previously isolated separately except for the carboxylation enzyme which converts CO_2 and ribulose diphosphate to PGA.

Carboxydismutase. About a year ago, using tracer studies, we sought and found a cell-free preparation, both from algae and from other green plants, that was capable of catalyzing the production of PGA specifically from ribulose diphosphate (RuDP) and sodium bicarbonate. The RuDP used in these experiments was isolated by chromatography from green-plant extracts. The technique was to expose the RuDP and the enzyme preparation to $NaHC^{14}O_3$

and show that carboxyl-labeled PGA was formed (Fig. 6.15). The traces of malic, citric, and aspartic acids and alanine formed indicate the presence in the preparation of some Krebs-cycle enzymes which could convert some of the PGA initially formed to other compounds. Indeed, upon longer exposure (>3 min) to these crude preparations, much of the PGA was converted. The formation of a little labeled malic acid in the absence of substrate (RuDP) indicates the presence of pyruvic acid and malic enzyme.

Because in this experiment the tracer was in the CO_2 and not in the RuDP, it did not give direct information about the fate of the five carbon atoms of ribulose. It was therefore necessary to do the experiment with labeled

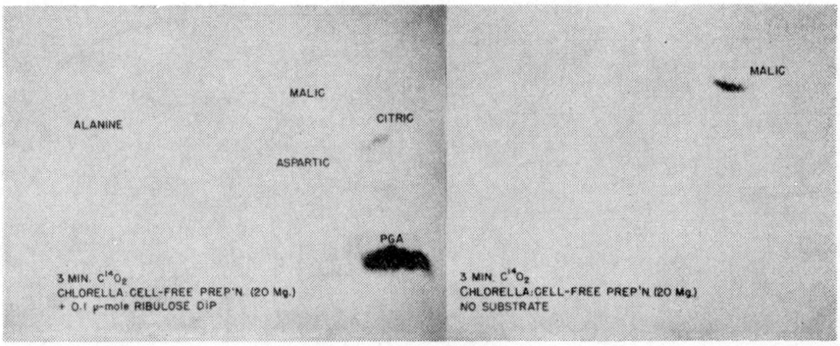

FIG. 6.15. Chromatograms indicating formation of carboxyl-labeled PGA.

RuDP and unlabeled CO_2. This was not very satisfactory in the first instance when the crude preparation was used. Although labeled PGA was formed, a good many other labeled compounds were formed as well, because of the presence in the preparation of enzymes that could act on ribulose diphosphate and compounds formed from it. In particular there was present a phosphatase which permitted the formation of ribulose–5-phosphate. This compound, in the presence of transketolase and aldolase (and possibly transaldolase), would rapidly find its way into hexose, heptose, and triose. The triose may have given rise to some PGA by oxidation. Although attempts to bypass this difficulty by inhibiting the initial phosphatase reaction on RuDP were partially successful, they were not conclusive, because of the insensitivity of the

$$\text{RuDP} \xrightarrow{HCO_3^-} \text{PGA}$$

system to fluoride ion (F^-). It was therefore necessary to proceed with the attempt to free the preparation from any other enzymes capable of acting upon RuDP except the one(s) required for the PGA-forming reaction (from CO_2). This was accomplished first from neutral extracts of New Zealand spinach (*Tetragonia expansa*) and later from extracts of sonically ruptured algae. The enzyme appears in the protein fraction, salted out of neutral extracts, between approximately 0.3 and 0.4 of saturation with $(NH_4)_2SO_4$. The results of an early experiment with such a preparation acting on labeled RuDP are shown in Fig. 6.16 (Ref. 10). Here the fate of the ribulose carbon is clearly its conversion to PGA when both enzyme and $NaHCO_3$ are present. There appears to be some sugar monophosphate present in all the experiments, partly because of its presence in the original RuDP sample and perhaps partly because of the presence of some residual phosphatase in the enzyme prepara-

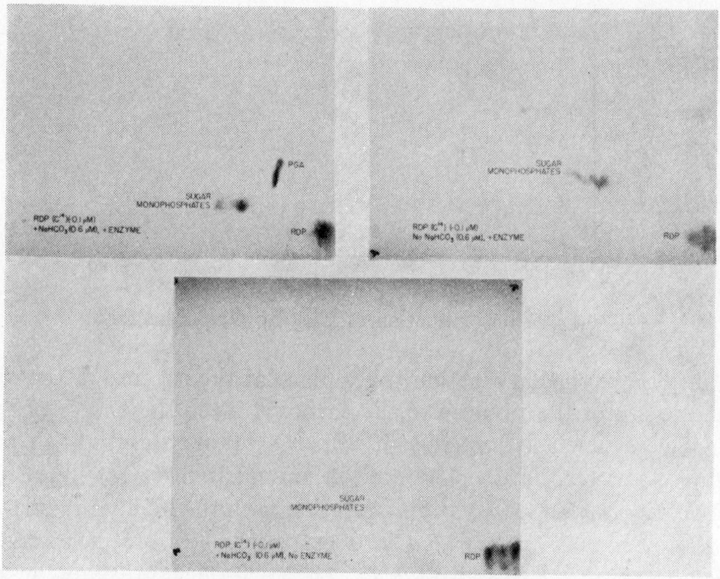

FIG. 6.16. Chromatograms showing effect of enzyme action on ribulose diphosphate.

tion. Later experiments have given preparations that convert essentially *all* of the ribulose carbon into PGA and nothing else.

It thus appears that the original formulation of the reaction is at least a likely one.

$$\begin{array}{c} O \quad CH_2\text{—}O\circledP \\ \| \quad | \\ HO\text{—}C^+ \quad C\text{—}OH \\ | \quad \| \\ ^-O \quad C\text{—}OH^+ \\ | \\ CHOH \\ | \\ CH_2O\circledP \end{array} \rightarrow \left[\begin{array}{c} O \quad CH_2O\circledP \\ \| \quad | \\ ^-O\text{—}C\text{—}C\text{—}OH \\ | \quad H^+ \\ ^-OH \quad C\text{=}O \\ | \\ CHOH \\ | \\ CH_2O\circledP \end{array} \right]$$

$$\rightarrow 2CH_2\text{—}CHOH\text{—}CO_2^- + H^+$$
$$\quad\quad\quad | $$
$$\quad\quad\quad O\circledP$$

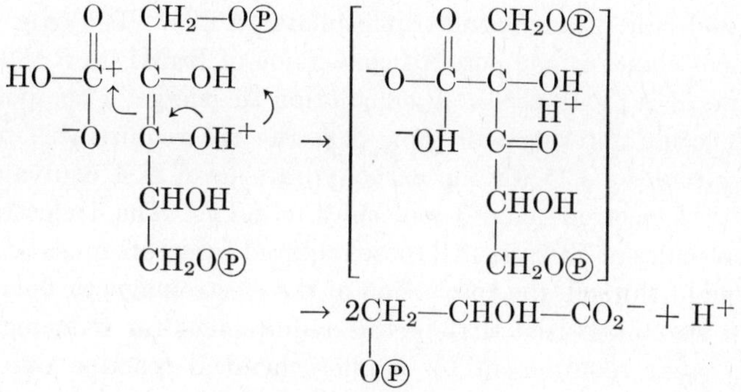

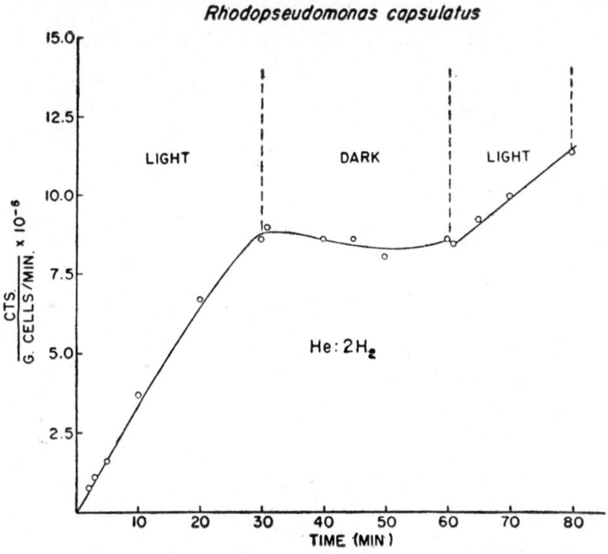

FIG. 6.17. Photoreduction of CO_2 by purple bacteria.

Because the carboxylation reaction takes place at the expense of the oxidation of carbon atom 3 of the ribulose to the carboxyl level, the name "carboxydismutase" suggests itself as uniquely descriptive. It is interesting to note that the enzyme is not readily demonstrated in animal tissues (rat liver) and that it can be obtained from spinach in association with the highly organized intact chloroplasts (Ref. 11) from which it is extremely easily separated. It does not appear to be especially sensitive to versene, *o*-phenanthroline, or cyanide, but it is sensitive to *p*-chloromercuribenzoate, an inhibition that is reversed by cysteine.

Chemical requirements to run the cycle. We now have the cycle in its details (Fig. 6.14), and we now know precisely what reagents are required to make the cycle turn. It can be seen that the requirement for the reduction of a PGA molecule to a triose is 1 molecule of triphosphopyridine nucleotide (TPNH) and 1 molecule of adenosine triphosphate (ATP). The only other energy requirement comes at the point of conversion of RuMP to RuDP, where another molecule of ATP is used. A calculation of energetic compounds needed per CO_2 molecule entering will show that the net requirement for the reduction of 1 molecule of CO_2 to the carbohydrate level is 4 equivalents of reducing agent, or 4 electrons, and 3 molecules of ATP. The 4 electrons are supplied by 2 molecules of TPNH. All these required cofactors must be made ultimately by the light through the conversion of the electromagnetic energy in some way. It must be emphasized that in this requirement for reducing carbon there is no particular requirement for a photochemical reaction other than the production of the two reagents. If we could supply those two things from some other source than the photochemical reaction, we should be able to make this whole sequence of operations function. We have reason to believe that this is indeed being done by the use of the required collection of enzymes. But a suitable situation exists in nature also. The situation is such that we must have simultaneously a high level of this particular reducing agent—which we now know can be triphosphopyridine nucleotide (TPN) and ATP at the same time and the same place.

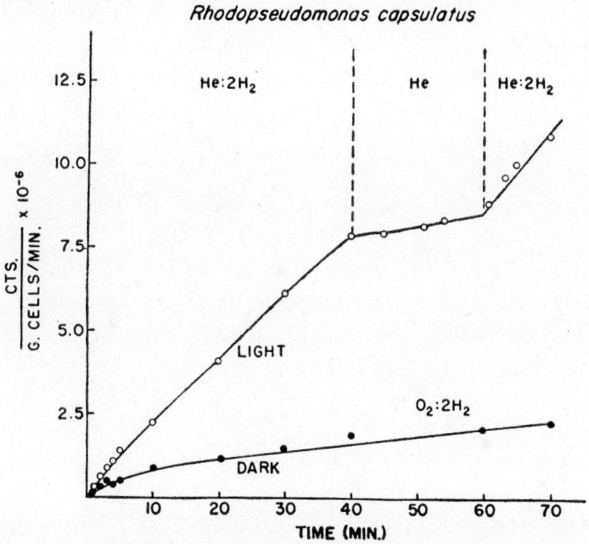

FIG. 6.18. Chemical reduction of CO_2 by purple bacteria.

Running the cycle without light. There is one known system in nature, aside from the green plants, in which that situation occurs. This situation exists in one of the photosynthetic purple bacteria that does not make oxygen, but does reduce carbon dioxide with molecular hydrogen. Figs. 6.17 and 6.18 (Ref. 12) show that it is possible to have the reduction of CO_2 take place either through the agency of light or through the agency of a chemical oxidation system. The organism is the purple bacterium, *Rhodopseudomonas capsulatus*. The initial slope corresponds to the reduction of carbon dioxide in the light. In this case both hydrogen—as the reducing agent—and light are required. As soon as the light is turned off, the reduction of carbon dioxide stops. Fig. 6.18 shows the same organism. This is a dark fixation. Here it is exposed only to helium and hydrogen, and there is an initial fixation which immediately saturates and stops. When oxygen is then admitted to the system, the fixation again continues in the same way as it does with light. The intermediates in the dark are very much the same as in the light. The hydrogen presumably provides the reducing power that is needed. The oxygen is required to oxidize some of that hydrogen to make ATP, and the two together can make the carbon dioxide cycle function. This suggests that a prime function of the light, in this case where hydrogen is the reducing agent, is to supply the oxidizing agent necessary for the production of the required ATP.

Quantum requirements. In order to estimate what a minimum quantum requirement for photosynthesis may be, on the basis of the information we have so far accumulated about the detailed chemistry of the process, at least one assumption is necessary. This is related to the mode of interaction of electromagnetic radiation and matter. It is that a single quantum can excite not more than a single electron. Another assumption about the behavior of the excited electron is required, namely, that it does not by some chemical (or physical) dismutation process give rise to more than one equivalent of reducing power at the potential of TPNH. And if that is the case, inspection of the requirements mentioned above allows one to predict what the minimum quantum requirement for such an operation would be. Four electrons are needed for the reduction, and three molecules of ATP.

Something about the various ways in which ATP can be produced is already known. For example, during the transfer of 2 electrons from DPNH to an atom of oxygen, 2 or 3 molecules of ATP can be produced. Therefore, one can suppose that when all the energy for the operation of this cycle comes from light, the minimum quantum requirement must be 6 or 7. That is, 4 electrons are needed for the reduction and 2 or 3 more for the 3 molecules of ATP that are required. However, it should be possible to find conditions under which the quantum requirement for the reduction of CO_2 and the evolution of oxygen would be as little as 4, provided there were some other source besides the light for the 3 molecules of ATP. These conditions have been realized (Ref. 13). The quantum-requirement determination was carried out by use of an apparatus in which one could measure directly, without any ambiguity, the production of oxygen by a direct measurement of a unique quality of the oxygen, paramagnetism, rather than merely by a gas pressure. Also it was possible to measure directly the amount of carbon dioxide absorbed by measuring a property of the CO_2 in the gas phase, in this case its infrared spectrum.

As a result of these measurements, it was found that the quantum requirement ranged experimentally from 7.4 at high light intensities, where photosynthesis exceeded respiration by a factor of 12, to 4.9 at low light intensities, where photosynthesis and respiration were nearly equal. At zero light intensity the value of the quantum requirement extrapolated to 4.* This result indicates that some of the ATP requirement of photosynthesis can be met by reactions of respiration which produce ATP, but that the 4 electrons of re-

* This value of 4 as the quantum requirement at low photosynthetic rates is in no way comparable to the values between 3 and 4 reported by Warburg and his associates at very high P/R ratios (>20). See Ref. 14.

ducing agent must be supplied by the light reaction, or with special organisms, by externally supplied reducing agents.

Quantum Conversion

So far only the reduction of carbon has been considered. Since this seems to be quite a separate system from the oxygen-evolution reaction, it might appear that one should not expect to learn much about the photoproduction of the

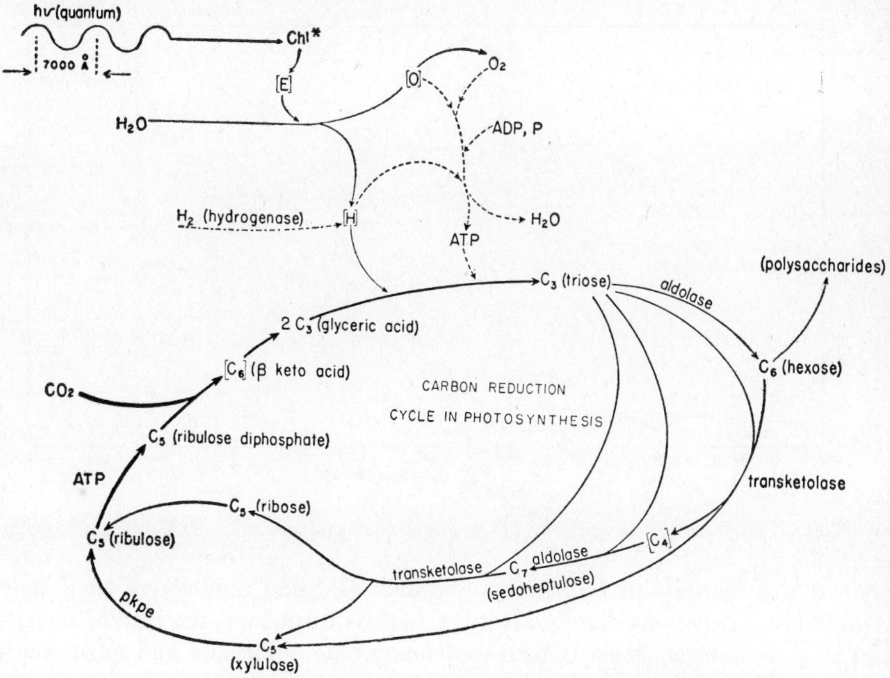

FIG. 6.19. Proposed cycle for carbon reduction in photosynthesis.

electrons and the ATP from studying the carbon reduction. But there must be a connection between the two. By suitable observations it is possible to see at least one point at which the carbon-reduction cycle makes contact directly with the photochemical apparatus. This is shown in Fig. 6.19. Here the cycle is shown again. The quantum is first absorbed by chlorophyll and converts water into something that makes a reducing agent [H] and some oxidizing agent [O]. The reducing agent can reduce the glyceric acid to triose. Some of the reducing agent must be used to make ATP, with oxygen or the intermediates on the way to oxygen, because that is necessary for the cycle to run. What we wish to consider now is this point of contact, [H], between the photochemical apparatus and the carbon cycle and what information about the quantum conversion we can gain from this study.

Light inhibition of TCA-cycle incorporation. An experiment was carried out in which a steady state was examined and the changes induced by a sudden change of conditions were observed. Fig. 6.20 shows the result of this experiment. Here is the same type of experiment as before, but with the examination directed toward different substances. Attention is focused on glutamic acid and citric acid, and it will be seen that while the light is on,

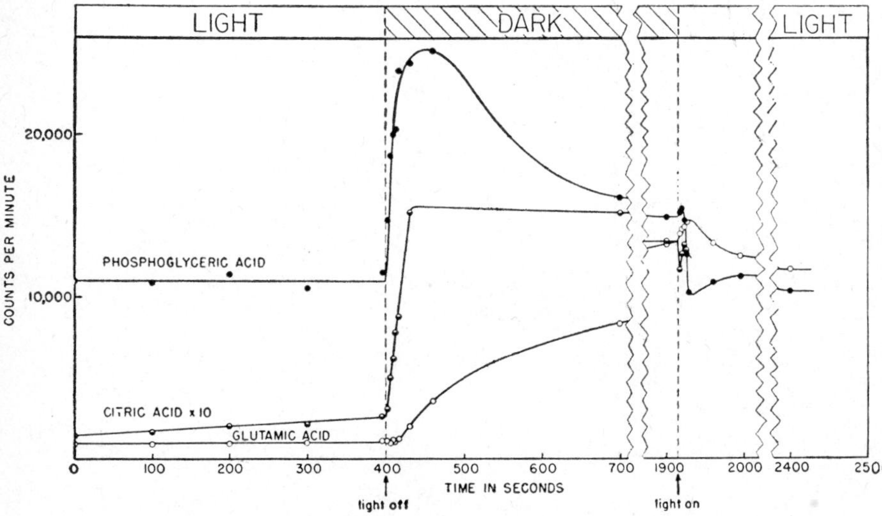

FIG. 6.20. Light-dark transients in PGA, citric acid and glutamic acid concentrations.

the rate of formation of radioactive glutamic acid and radioactive citric acid is quite low. But immediately after the light is turned off, the rate of formation of these labeled acids is increased manyfold. Glutamic and citric acids are two compounds very closely related to the respiratory cycle known as the Krebs cycle, and Fig. 6.21 describes in schematic terms the metabolic relationships leading to the experimental facts we have just seen. Here is shown the photosynthetic cycle and the Krebs (tricarboxylic acid) cycle. The glutamic acid and citric acid are in or related to the Krebs cycle. The photosynthetic cycle does not contain either glutamic or citric acid but does form PGA and sugars. Eventually these direct products of the photosynthetic cycle have to become carbohydrates, proteins, and fats, and ultimately they will get back into the tricarboxylic acid cycle. That is the major route in the light. But immediately after the light is turned off a direct connection between the two cycles is apparently made which allows the PGA to be transformed directly into the compounds of the tricarboxylic acid cycle. Fig. 6.22 shows the details of that mechanism. Carbon can enter the tricarboxylic acid cycle via acetyl

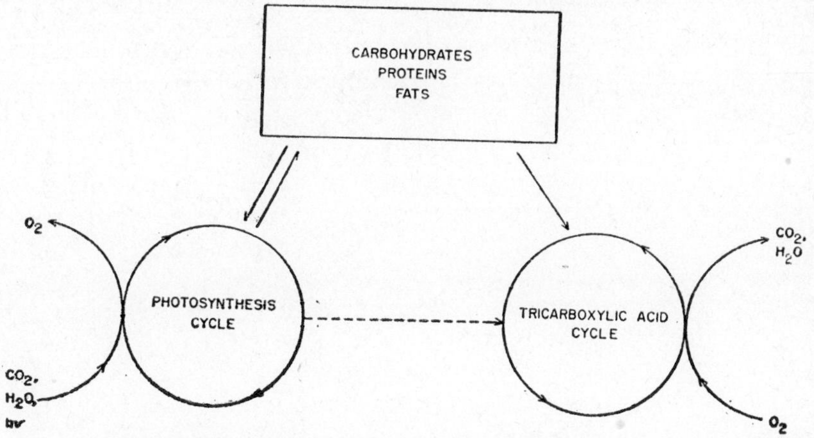

FIG. 6.21. Schematic relationships between the photosynthetic cycle, the tricarboxylic acid cycle, and storage products in the plant.

coenzyme A, condensing with oxalacetic acid to give citric acid, thence continuing around this cycle and via a side reaction to glutamic acid. The question is, How is glyceric acid converted to acetyl coenzyme A? This must happen rapidly in the dark, but not very rapidly in the light. Fortunately we

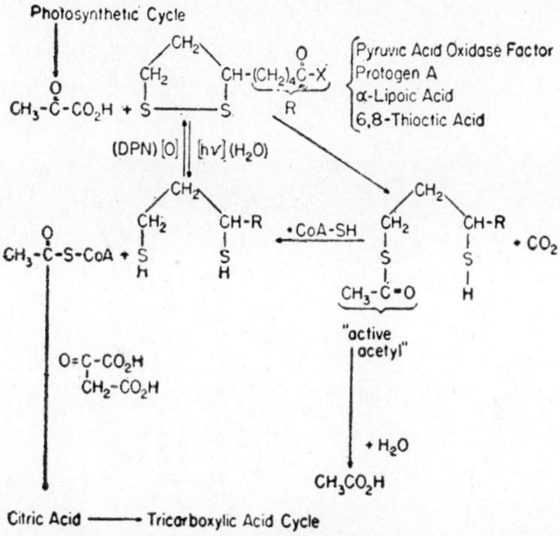

FIG. 6.22. Mechanism of photochemical control of the relationships between the photosynthesis cycle and the tricarboxylic acid cycle.

have some idea how acetyl-CoA may be formed from glyceric acid, and Fig. 6.23 shows this. The glyceric acid is dephosphorylated to form pyruvic acid; the pyruvic acid then reacts with an enzyme system, of which thioctic acid is a coenzyme, to form acetyl-thioctic acid and carbon dioxide. The acetyl-thioctic acid then undergoes a thiol ester interchange with CoA to form re-

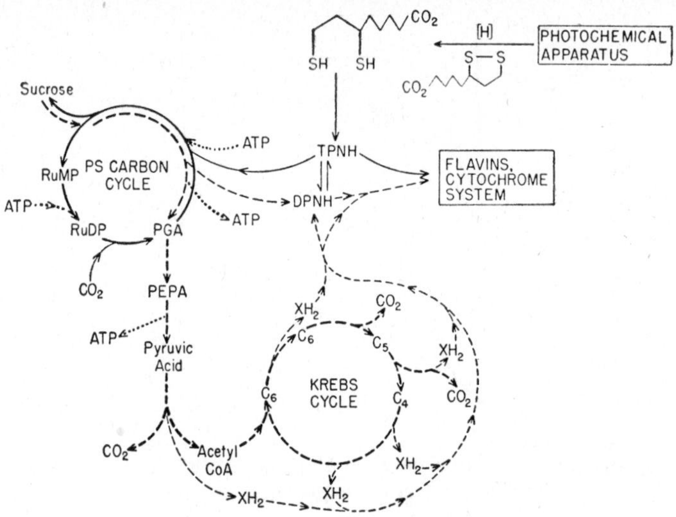

FIG. 6.23. Diagram of the suggested nature of the photochemical apparatus and its relationship to other functions. ------ Oxidative, or respiratory, pathways; ——— Reductive, or photosynthetic, pathways.

duced thioctic acid and acetyl-CoA, which then goes on into the citric acid cycle, Fig. 6.23 (Ref. 15).

How does light affect these reactions? The conversion of PGA to citric acid provides for the entrance of carbon into the tricarboxylic acid cycle, and if somehow this pathway is closed by reduction of the level of the disulfide, the rate of transfer of radioactive carbon from the photosynthetic cycle to the citric acid cycle will be reduced. This suggests that the light shifts the equilibrium from the disulfide to the dithiol form of thioctic acid by inducing reaction with something other than pyruvic acid, perhaps ultimately water. In the dark, oxidation converts the dithiol form to the disulfide, which can again catalyze the oxidation of pyruvic acid to CO_2 and acetyl-CoA. This system is like a valve that is closed by light and that controls the flow of carbon from the photosynthetic cycle directly into the tricarboxylic acid cycle. It suggests further that the disulfide may be closely allied to, if not identical with, the electron acceptor from the photochemical act. Actually a number of experiments have been performed that indicate that this may be so.

The proposed relations between the photosynthetic carbon reduction cycle, the photochemical reactions, and the Krebs cycle are shown in Fig. 6.23. It is suggested that the required ATP is generated by reactions coupled with the oxidation of TPNH or DPNH through the cytochrome system.

It can be seen that the use of radioactive elements, employed as tracers, has made possible the elucidation of the path of carbon reduction in photosynthesis. In addition, information gained from the study of the path of carbon in photosynthesis and its relation to reactions of respiration has provided the basis for proposals regarding the energy transport from the primary photochemical act.

References for Chapter 6

1. A. A. Benson, J. A. Bassham, M. Calvin, T. C. Goodale, V. A. Haas, and W. Stepka, "The path of carbon in photosynthesis: V. Paper chromatography and radioautography of the products," *J. Am. Chem. Soc.* 72:1710–18 (1950).
2. M. Calvin and Peter Massini, "The path of carbon in photosynthesis: XX. Steady state," *Exper.* 8:445–7 (1952).
3. J. A. Bassham, A. A. Benson, L. D. Kay, A. Z. Harris, A. T. Wilson, and M. Calvin, "The path of carbon in photosynthesis: XXI. The cyclic regeneration of carbon dioxide acceptor," *J. Am. Chem. Soc.* 76:1760–70 (1954).
4. W. Stepka, A. A. Benson, and M. Calvin, "The path of carbon in photosynthesis: II. Amino acids," *Science* 107:304–6 (1948).
5. A. A. Benson, J. A. Bassham, M. Calvin, A. G. Hall, H. Hirsch, S. Kawaguchi, V. Lynch, and N. E. Tolbert, "The path of carbon in photosynthesis: XV. Ribulose and sedoheptulose," *J. Biol. Chem.* 196:703–16 (1952).
6. J. G. Buchanan, J. A. Bassham, A. A. Benson, D. F. Bradley, M. Calvin, L. L. Daus, M. Goodman, P. M. Hayes, V. H. Lynch, L. T. Norris, and A. T. Wilson, "The path of carbon in photosynthesis: XVII. Phosphorus compounds as intermediates in photosynthesis," *Phosphorus Metabolism,* Johns Hopkins Press, Baltimore, 1952, Vol. 2, pp. 440–59.
7. J. G. Buchanan, "The path of carbon in photosynthesis: XIX. The identification of sucrose phosphate in sugar beet leaves," *Arch. Biochem. Biophys.* 44:140–49 (1953).
8. K. Shibata, J. A. Bassham, and M. Calvin: Unpublished results.
9. A. T. Wilson, *A Quantitative Study of Photosynthesis on a Molecular Level.* Thesis, University of California, Berkeley, 1954.
10. J. Mayaudon: Unpublished results in this laboratory.
11. R. C. Fuller: Unpublished observations in this laboratory.
12. A. O. M. Stoppani, R. C. Fuller, and M. Calvin, "Carbon Dioxide Fixation by *Rhodopseudomonas capsulatus,*" *J. Bact.* 69:491–501, May 1955.
13. J. A. Bassham, K. Shibata, and M. Calvin, "The relation of quantum requirement in photosynthesis to respiration," *Biochem. Biophys. Acta* 17:332–340 (1955).
14. O. Warburg, G. Krippahl, W. Buchholz, and W. Schroder, "Weiterentwicklung der Methoden zur Messung der Photosynthese," *Z. Naturf.* 86:675–86 (1953).
15. J. A. Bassham and M. Calvin, *Photosynthesis: Currents in Biochemical Research,* Interscience Publishers Inc., New York, 1956, pp. 29–69. (University of California Radiation Laboratory Report No. 2853.)

Chapter 7

INFLUENCE OF ENVIRONMENT ON THE PRODUCTS OF PHOTOSYNTHESIS *

The plants of various systematic groups are distinguished by the enormous diversity of their biochemical characteristics. There is marked regularity and distinctness in the changes effected in plant metabolism in ontogenesis, or when the plants are grown in different geographic zones, under different conditions of nutrition, light, watering, temperature, etc.

It was and still is supposed that biochemical variety of plants is an outcome of differences in the secondary transformations of the direct products of photosynthesis. On the other hand, the latter are supposed to be alike in all plants or at least to present various carbohydrates (Ref. 1).

On the other hand, there have long been surmises that the variety of the biochemical peculiarities of plants finds its origin in the variety of the direct products of photosynthesis. Thus, Schimper (Ref. 2), Chrapovicki (Ref. 3) and others believed that besides carbohydrates, the direct products of photosynthesis include proteins and organic acids.

One of the first attempts at experimental demonstration (Refs. 4, 5) of the variety of the direct products of photosynthesis exposed the leaves of plants in an atmosphere of CO_2, illuminated them from 12 to 24 hr, and by means of chemical analysis established an increase not only of carbohydrates, but of proteins as well. The quantitative ratio between the produced carbohydrates and proteins altered depending on the age of the leaves, the supply of nitrogen-bearing salts, the light intensity, and the type of plant. Evidence in favor of the possible photosynthetic formation of amino acids and proteins was likewise obtained by a number of other authors (Refs. 6 to 18).

Thus there have long been appearing facts which raised the question of where, at what stages of the transformation of carbon assimilated in the course of photosynthesis, begins the biochemical specificity connected with the characteristics of different plants existing in various physiological conditions

* This chapter is taken from Geneva Conference Paper 697, "Tracer Atoms Used to Study the Products of Photosynthesis as Depending on the Conditions in Which the Process Takes Place" by A. A. Nichiporovich of the U.S.S.R.

and environmental circumstances. One may inquire: Is this specificity due to the variety of transformations of some first direct products of photosynthesis common to all plants and universal in their physiological significance? Or to diversity of the very earliest stages of the process of photosynthesis and the variety of its direct products? (Refs. 19, 20.)

The large theoretical and practical importance of this question cannot be denied. Its solution must determine our concepts of the very content and volume of the process of photosynthesis, of its mechanism and the systems partaking in its operation, the part it plays in the vital activity of the plant, the possible methods and outcomes of influencing plants by regulating the course of their carbon nutrition, etc. But while such problems were being resolved by means of drawing up balances of substances with the aid of conventional chemical analysis, any reliable solution of the question of what the direct products of photosynthesis are was practically impossible. Exposures of the plant leaves were necessarily long, and therefore it was not possible to provide solid proof that the observed changes in the chemical composition of the leaves are an outcome of primary syntheses and not of secondary transformations.

Obviously, a decisive role in the solution of the above-stated problem could be played by the method of labeled atoms and the use of labeled carbon in particular. Investigations in this direction had already begun in the U.S.S.R. in the Plant Physiology Institute of the U.S.S.R., Academy of Sciences, in 1950.

Distribution of Carbon-14 in $C^{14}O_2$ Absorption by Plants

The first experiments were directed at elucidating the possible variety of ways of inclusion of C^{14} into different groups of substances.

In one of the experimental series (Ref. 21) leaves were exposed to light in an atmosphere of $C^{14}O_2$ for 20 min. After exposition the leaves were fixed with 85% ethyl alcohol and successively submitted to extraction by 85% ethyl alcohol, petroleum, ether, and boiling water.

The petroleum and ether fractions contained only traces of activity and were not included in subsequent reckoning. The alcoholic extract, which contained up to 90% of the total activity, was precipitated first by lead acetate and afterwards by copper sulfate. Each of the fractions mentioned contained a mixture of substances, but it may be considered that the alcohol-soluble proteins were in the precipitate with lead acetate, the alcohol-soluble carbohydrates were mainly concentrated in the precipitate obtained from the alcohol fraction with copper sulfate, the water fraction mainly held starch, dextrines, and water-soluble proteins, and the insoluble residue contained cellulose, proteins insoluble in water and alcohol, and other substances.

The distribution of the radioactivity for various plants is given in Table 7.1. The data in the table show that the transformations of the carbon assimilated by plants in the process of photosynthesis are very diverse. This is

TABLE 7.1. DISTRIBUTION OF CARBON-14 IN FRACTIONS FROM DIFFERENT PLANTS

(Leaves exposed for 20 min to the light in an atmosphere containing $C^{14}O_2$)

Plants	Activity in Fractions, Percentage of Total				
	Insoluble	In Water	In Alcohol		
			Precip. with Lead Acetate	Precip. with $CuSO_4$	Others
Phaseolus vulgaris	3	36	51	5	5
Asclepias cornuti	4	20	73	2	1
Brassica napus	5	19	70	4	2
Nicotiana rustica	5	10	82	2	1
Ricinus communis	8	8	76	2	6
Vicia faba	5	5	77	4	9
Taraxacum kok-saghyz	1	5	93	0	1
Beta vulgaris	3	1	92	2	2

obvious, for instance, from the sharp differences in C^{14} distribution in the water fraction and the precipitate obtained by using lead acetate.

In 1955 (Ref. 22) chromatographic analysis of photosynthesis products, obtained after 5-min exposure of the photosynthesis organs of many plants of various systematic groups, likewise established considerable differences in the nomenclature of their products.

Young and adult leaves. Our laboratory has found differences in the distribution of C^{14} calculated for substances of the alcohol fraction when the leaves exposed were young and intensively developing as well as in mature leaves (see Table 7.2 and Ref. 21). The principal concentration of activity in the case of young leaves in the precipitate caused by lead acetate—and in adult leaves in the precipitate from $CuSO_4$—shows more intensive inclusion of C^{14} into proteins in the first case, and into the carbohydrates in the second.

Effect of spectral composition of light. Other experiments with sunflower leaves (Ref. 15) furnished data pointing at differences in transformations involving the carbon assimilated by photosynthesis under light of various spectral composition—in particular, in the filtered light of an electric lamp (580–720 millimicrons, red light) and mercury-quartz lamps (400–575 millimicrons, blue light).

TABLE 7.2. DISTRIBUTION OF CARBON-14 FROM LEAVES EXPOSED TO LIGHT FOR 20 MIN

Plants	Per Cent Activity in Fractions of Alcohol Extract					
	Precip. with Lead Acetate		Precip. from $CuSO_4$		Residue After Precipitation	
	Young Leaf	Adult Leaf	Young Leaf	Adult Leaf	Young Leaf	Adult Leaf
Nicotiana rustica	87	56	10	39	3	5
Zea mays	83	53	3	40	14	7
Asclepias cornuti	43	34	51	59	6	7
Sorghum vulgare	73	72	22	24	5	4

In the given case the substances taken into account were those extracted with water. One of the fractions included substances precipitated from the water extract with lead acetate, the latter being removed with hydrogen sulfide, while the other comprised neutral compounds (carbohydrates), the substances with alkaline and acid properties here being removed by passing them through ion-exchange resins.

The results are presented in Table 7.3, which shows that after as little as

TABLE 7.3. CARBON-14 IN FRACTIONS OF SUBSTANCES FROM SUNFLOWER LEAVES EXPOSED TO RED AND BLUE LIGHT

Fraction Activity	Time of Exposure, sec							
	Experiment I				Experiment II			
	5	30	90	180	5	30	90	180
Precipitated with lead acetate:								
Red light, %	46	43	12	15	69	72	33	15
Blue light, %	47	41	18	23	40	52	24	22
Neutral:								
Red light, %	7	7	11	50	2	7	14	50
Blue light, %	6	9	11	16	0	4	11	20
Absorbed in ion-exchange resins:								
Red light, %	47	50	77	35	28	21	53	40
Blue light, %	46	50	71	61	60	44	65	58

3 min of exposure considerable differences begin to appear in the distribution of C^{14} in the substances of various fractions. The data of the table give support to the conclusion that the spectral composition of the light has significant influence on the course of the transformations of carbon assimilated in photosynthesis. Short-wave rays, in particular, by the third minute of exposure considerably diminish its inclusion into the fraction in which carbohydrates are the main component and increase it in the water-soluble protein fraction (precipitated with lead acetate).

Further investigations of the problem (Refs. 15, 16) have shown that with the exposure (5 min) of leaves to red or blue light there is a difference in the distribution of C^{14} among the amino and organic acids. In blue light the fraction incorporated in amino acids is considerably higher than in red

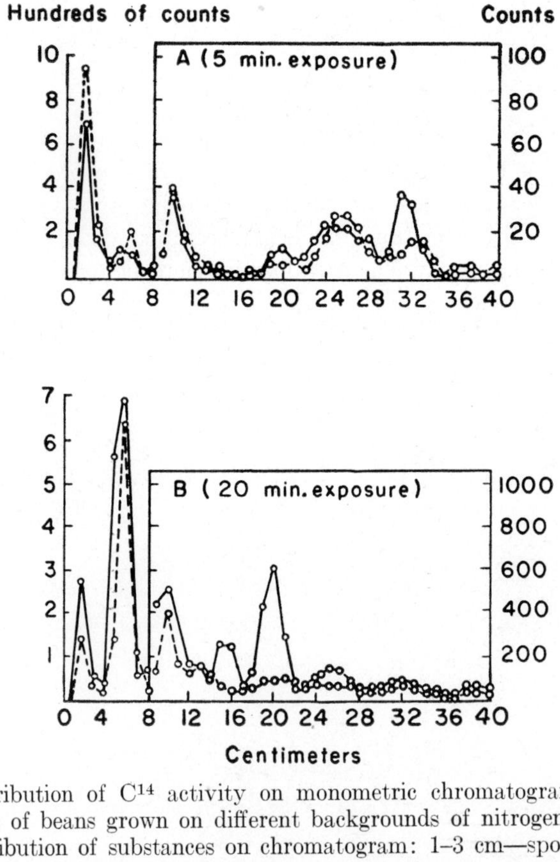

FIG. 7.1. Distribution of C^{14} activity on monometric chromatogram of an extract from the leaves of beans grown on different backgrounds of nitrogen nutrition. Approximate distribution of substances on chromatogram: 1–3 cm—spot of application; 5–7 cm—saccharose, arginine, asparagine; 9–11 cm—glucose, asparagic acid, glycine. Broken line is for insufficient nitrogen nutrition; solid line for normal nitrogen nutrition. 14–17 cm—glutaminic acid, treonine; 18–22 cm—alamine, proline; 24–28 cm—tryptophan; 30–34 cm malic acid.

light, while the fraction incorporated in organic acids is lower. This is explained by the fact that blue light causes a more intensive synthesis of amino acids (alanine in particular) at the expense of the organic acids.

Mineral and nitrogen nutrition effects. A number of experiments vividly show that the course of the transformations of carbon assimilated in the process of photosynthesis is significantly influenced by the conditions of the plant's mineral nutrition, and particularly of its nitrogen nutrition. Fig. 7.1 presents curves characterizing the results of measurement of the counting rate due to radioactive carbon on one-dimensional paper chromatograms containing substances from an alcohol extract obtained from the leaves of beans cultivated with insufficient and normal nitrogen nutrition and exposed to light in an atmosphere with $C^{14}O_2$ for 5 or 20 min. The data bear witness that intensified nitrogen nutrition causes a number of characteristic alterations in the transformations of photosynthetically assimilated carbon.

Thus, we see that the trend of the aforementioned transformations is rapidly and significantly altered, depending on a multitude of conditions—i.e., both on the type, nature, and condition of the plant itself and on environmental conditions.

Primary and Secondary Products of Photosynthesis

Even the experiments described above, in which the plants were exposed from 30 sec to 20 min, do not completely solve the question as to which of the observed phenomena must be referred to photosynthetic carbon transformations and which to secondary, nonphotosynthetic transformations. Certain clarity may be brought into the question by the data of Ref. 23, which show marked differences in the distribution of C^{14} in substances of various fractions from the leaves of different plants exposed to the light for 1 sec. Even in this short period, the distribution of C^{14} in the substances of various fractions from various plants is different. Likewise varying was the redistribution of the assimilated C^{14} in fractions during subsequent exposure to light in an atmosphere with $C^{12}O_2$ for 4, 14, 29, 59, 299 sec. The diversity of products formed in the leaves of various plants exposed for 1 sec is also revealed by ionophoretic separation of substances on paper, with the subsequent production of radioautographs.

Thus we have cause to think that the quantitative relations of products formed as a result of photosynthetic carbon assimilation may vary, beginning with the earliest stages of its transformation; and this testifies to the possible multitude of the direct products of photosynthesis.

In order to make a more definite statement on this subject, it is necessary to give a precise definition of what may be considered the direct products of photosynthesis—all the more necessary since there may be different interpretations of this problem.

From the purely photochemical or energetic viewpoint, the first products of the reduction of CO_2 or the carboxyl group which, as a result of photochemical electron or hydrogen transference, accumulate the energy of the light absorbed by the chlorophyll are those that could be considered the direct products of photosynthesis. Phosphoglyceric aldehyde, for example, might be considered one of these direct products of photosynthesis and its formation might be assumed as the end of the process of photosynthetic transformations, the reduction of carbon and binding of energy having been accomplished. Ensuing transformations of carbon must proceed at the cost of the plant's usual enzyme systems, so that these transformations seem to give no cause for relating them to photosynthesis proper.

The above definition of the boundaries of photosynthesis seems to be logical. Yet it is purely formal and based only on physicochemical criteria, whereas the process of photosynthesis is first of all physiological and it is this criterion that should prevail in determining the content and the boundaries of photosynthesis (Ref. 20). From the physiological viewpoint photosynthesis is the process of carbon nutrition of plants, including the primary formation of organic substances from CO_2 and H_2O and the accumulation in them of the energy of sunlight.

Role of the chloroplasts. The direct photosynthetic apparatus of the plant, i.e., the place where the main reactions of photosynthesis go on, is the chloroplast. In the chloroplasts or on their surface during photosynthesis, such substances as sugar and starch are primarily formed, which according to classical concepts are the direct products of photosynthesis. The chloroplast subsequently supplies and nourishes the cell with these substances, which are then used for different purposes and in different ways. Between the main photochemical reaction of photosynthesis and the formation of the first product of the reduction of CO_2 (for instance, glyceric aldehyde) and particularly of carbohydrates, there is a more or less complex chain of enzymatic reactions which may likewise take place in the plant in the course of secondary transformations of carbon without the participation of the photosynthetic apparatus.

However, this does not mean that sugar and starch are not direct products when they are primarily formed in the chloroplasts during the photosynthetic assimilation of carbon. In this case they are recognized as direct products of the complicated work of the photosynthetic apparatus of plants, the beginning of which is primary fixation, photochemical reduction of carbon, and whose completion is the formation of stable products with definite physiological characteristics, with which the chloroplast supplies the cell.

Therefore, if we could demonstrate that during photosynthesis in the chloroplast other substances are also directly formed (for instance, amino acids, proteins) which pass into the protoplasm as nutritive or physiologically active substances, we could speak of the diversity of the direct products of the work of the photosynthetic apparatus, as the apparatus of the plant's carbon nutrition.

Carbon and nitrogen in chloroplast protein photosynthesis. It was this problem that our laboratory sought to solve (Refs. 12, 13) in its work with the combined use of the radioactive isotope of carbon (C^{14}) and the heavy isotope of nitrogen (N^{15}). In these experiments the leaves were supplied with ammonium sulfate labeled with N^{15} (through the stalks, by means of infiltration or floating on a solution). Afterward the leaves were exposed to light for 1 to 2 hr in an atmosphere with usual $C^{12}O_2$ or radioactive carbon dioxide, $C^{14}O_2$.

After exposure the chloroplasts were extracted from the leaves by means of a centrifuge. The proteins of the chloroplasts were carefully purified by successive extracting with alcohol and ether and then treated with diastase to remove starch (Ref. 17). Besides this, after extraction of chloroplasts from the leaves, plasmic proteins were extracted by heating the centrifugate.

The final stage of each experiment consisted in the determination of the presence of N^{15} (excess atom per cent of N^{15}) and C^{14} (counts per mg per min) in the chloroplast and plasmic proteins. The experiments showed that during photosynthesis the proteins of chloroplasts are considerably enriched with both C^{14} and N^{15}, which testifies to the intensive new formation of amino acids and proteins in the course of photosynthesis.

It is important to note that the inclusion of C^{14} in proteins is intensified only if it is supplied to the plant in the form of $C^{14}O_2$ in the course of photosynthesis. But if radioactive carbon is supplied to the plant as part of a carbohydrate, in spite of the intensive formation of proteins in the light (which is attested by the increase in the proteins of a surplus atom per cent of N^{15}) there is no intensified uptake of C^{14} into protein. This indicates that the proteins are formed at the expense of the intermediate products of photosynthesis, and not of the ready sugars (see Table 7.4).

In other experiments gladiolus leaves previously kept for 22 hr in darkness, were illuminated for 2 hr in an atmosphere with $C^{14}O_2$. In this manner the leaves were by natural means enriched with radioactive carbohydrates and other still earlier products of photosynthesis. Then the halves of the leaves were deposited on a solution with N^{15} in an atmosphere with $C^{12}O_2$ under light and in darkness. After 1½ or 2 hr the specific activity of osazones, sugars, and proteins obtained from the leaves, and likewise the amount of N^{15} in the proteins of the chloroplasts, were determined. The results of the experiments are given in Table 7.5.

The table indicates intensive photosynthesis in the leaves. This is testified by intensive protein synthesis (increase of N^{15} in proteins under light) and the sharp decrease in the specific activity of osazones (evidently due to the "dilu-

TABLE 7.4. CARBON-14 AND NITROGEN-15 CONTENT OF CHLOROPLAST PROTEINS

(Leaves exposed to light and fed C^{14}-labeled carbohydrates or $C^{14}O_2$)

Treatment of Chloroplast Proteins	Original Condition	Exposure in Dark	Exposure in Light
Exposure in $C^{14}O_2$:			
C^{14} content	0	0	414
N^{15} content	0.1292	0.1292	0.4940
Exposure in $C^{12}O_2$, leaves enriched with labeled invert:			
Experiment I:			
C^{14} content	60	63	60
N^{15} content	0.1140	0.1026	0.1862
Experiment II:			
C^{14} content	16	14	14
N^{15} content	0.0874	0.0950	0.2546

tion" of labeled carbohydrates with common ones forming in the course of photosynthesis).

However, the chloroplast protein activity did not increase under light as compared with the protein radioactivity in darkness. Therefore, the synthesis of chloroplast proteins proceeds mainly at the expense of the direct intermediary products of photosynthesis. This supplies new proof for the fact

TABLE 7.5. INCORPORATION OF CARBON-14 TO CHLOROPLAST PROTEINS FROM PRODUCTS OF PRECEDING PHOTOSYNTHESIS

	Initial Test	Exposure 1½ Hr	Exposure 3 Hr
Activity of osazones (counts per mg) at exposure of leaves	2008		
in the dark		1929	1984
in the light		656	493
Chloroplast protein activity (counts per mg) at exposure	25		
in the dark		71	129
in the light		91	120
N^{15} surplus (atomic per cent) in chloroplast proteins at exposure	—		
in the dark		0.0000	0.1292
in the light		0.0180	0.5890

that chloroplast amino acids and proteins are synthesized in the course of photosynthesis.

Inhibition of photosynthesis by phenylurethane. Another confirmation of this is the sharp suppression of protein synthesis in chloroplasts when photosynthesis is inhibited by phenylurethane, as is seen from the data of Table 7.6.

TABLE 7.6. INCORPORATION OF CARBON-14 AND NITROGEN-15 IN CHLOROPLAST PROTEINS UNDER INHIBITION OF PHOTOSYNTHESIS BY PHENYLURETHANE

Experimental Variants	Content in Chloroplast Proteins	
	N^{15} Surplus (Atomic %)	C^{14} Counts per Mg per Min
Exposure of leaves to light in atmosphere with $C^{14}O_2$	0.0407	471
The same, with addition of phenylurethane	0.0074	275

The summary of such data, which are not exhausted by those cited above, supports the conclusion that apart from carbohydrates, amino acids and proteins are likewise formed in plant chloroplasts. This formation goes on directly, independent of carbohydrates, possibly owing to divergence of the course of the transformations of intermediary products of photosynthesis. One path in this case leads to the formation of carbohydrates, while the other or others lead to the formation of other compounds, and proteins in particular.

Influence of nitrogen supply. Alteration in conditions of environment, nutrition and the state of the plant itself may change the intensity of the course of photosynthetical transformations of carbon in this or that direction and lead to differences in the qualitative relationships between the direct products of photosynthesis. Thus, for instance, variations in the plant's nitrogen supply have considerable influence on the rate of the photosynthetic formation of proteins in the chloroplasts, as is seen from the rate of the incorporation of C^{14} in Table 7.7.

In the experiment described, the plants of *Nicotiana rustica* were cultivated with both normal and lowered nitrogen supply. The leaves of the control specimens were exposed to light in an atmosphere with $C^{14}O_2$. The leaves of the experimental variant prior to a similar exposure were kept with their stems submerged in $(NH_4)_2SO_4$ for 14 hr.

TABLE 7.7. INCORPORATION OF CARBON-14 IN LEAF CHLOROPLAST PROTEINS OF *Nicotiana rustica* CULTIVATED WITH VARYING NITROGEN NUTRITION

	Rate of Application of Nitrogen in Cultivation		
	Full	One-half	One-fourth
Specific activity in chloroplast proteins:			
Control plants:			
Counts per mg per min	173	212	105
Per cent	100	100	100
Plants with leaves saturated in $(NH_4)_2SO_4$:			
Counts per mg per min	182	466	408
Per cent	105	220	145

Formation of Ascorbic Acid During Photosynthesis

If at earlier stages of the photosynthetic transformations of carbon—at stages preceding the formation of carbohydrates—the transformation of intermediary products of photosynthesis diverges and leads to the production of a number of stable compounds having independent physiological significance, then we may suppose that apart from carbohydrates and proteins there may be other substances among them as well.

In confirmation of the latter supposition are the results of an experimental work with ascorbic acid, where an attempt was made to ascertain the possibility of direct formation of this acid in the course of photosynthesis (I. V. Ogolevetz). The reasons for a special concern with ascorbic acid are as follows:

1. Its close genetic affinity to carbohydrates (Refs. 24, 25).
2. The availability of numerous data testifying to its principal formation in illuminated leaves and localization in the chloroplasts (Ref. 26).
3. Its possible active role in the process of photosynthesis as one of the components of oxidation-reduction systems participating in the transfer of hydrogen (Refs. 27, 28, 29).

The procedure was as follows: The leaves were exposed for different lengths of time to light in an atmosphere with $C^{14}O_2$. Later, when the ascorbic acid was identified on paper chromatograms, C^{14} incorporation was studied. Curves were drawn based on three points each and were further extrapolated to the zero ordinate on the assumption that the passage of the extrapolated part of the curve through the zero point should testify to the direct photosynthetic formation of the ascorbic acid, while its passage through a point not corresponding to the beginning of exposure—as in particular, through the point where the formation of carbohydrates may be stated to have begun—should be indicative of its secondary origin—for instance, from carbohydrates.

The data in Fig. 7.2 show that some of the experiments gave kinetic curves testifying to the direct photosynthetic formation of ascorbic acid, as the incorporation of C^{14} in it simultaneously began with initiation of photosynthesis. At the same time, when leaves exposed in the described manner were kept in the dark for 4 hr a marked shift to the right in the dynamic curves (Fig. 7.3) was observed, revealing intensive subsequent transformations of the ascorbic acid.

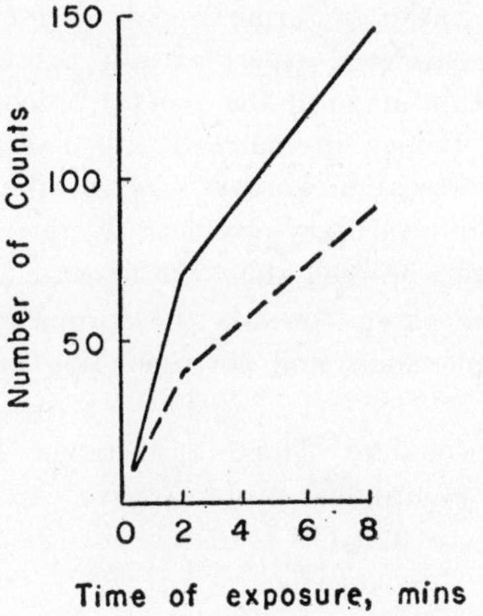

FIG. 7.2. Penetration of C^{14} into asparagic acid in the leaves of gladiolus exposed to light in an atmosphere with $C^{14}O_2$ for various lengths of time.

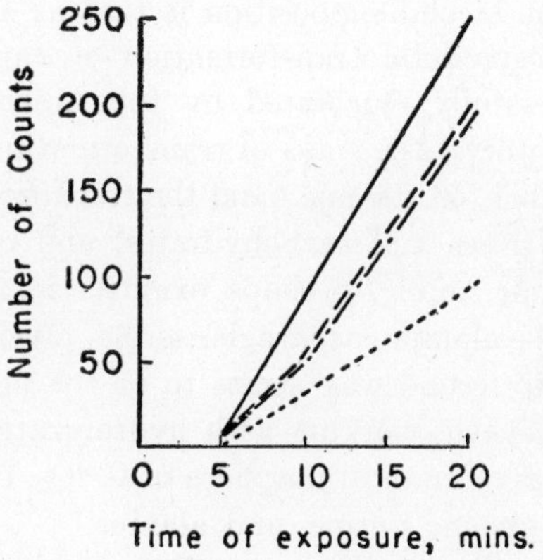

FIG. 7.3. Presence of C^{14} in ascorbic acid from gladiolus leaves exposed to light in an atmosphere of $C^{14}O_2$ for different periods and then held in the dark for four hours.

Thus ascorbic acid may be considered one of the direct products of photosynthesis, its formation possibly being an important prerequisite for the normal course of that process itself.

ORIGIN OF VARIETY IN PHOTOSYNTHETIC PRODUCTS

At present we do not yet know the details of the origin of the variety of photosynthesis products. Only a number of suppositions may as yet be made to this effect. One of them is that CO_2 is primarily fixed not by one special receptor which it carboxylizes, but by several.

The further reduction of the carboxyl groups which may be in several substances may be one of the starting points of the multiplicity of the direct products of photosynthesis. A certain confirmation to this viewpoint is furnished by the presence of C^{14} in various substances even after fixation in darkness. It may be supposed that the comparatively simple substances found among the intermediary compounds may present mere fragments of some more complex compounds which are, strictly speaking, the real objects of photosynthesis, but which quickly decompose during usual preparatory processing of materials. This supposition, though it has been formulated by a number of authors (Refs. 21, 23, 30) may not be considered experimentally proved.

The second supposition is that at a certain stage of the general process of photosynthetic transformation of carbon (which in many details has been successfully elucidated by the American research workers Calvin, Benson, and others) the ways of transformation of intermediary products diverge. For instance, on the one hand the transformations proceed along the lines of forming trioses and carbohydrates, and on the other, towards the formation of pyruvic, malic, perhaps oxaloacetic, glycole acids and corresponding amino acids—alanine, asparagic, serine, glycine.

The second way seems to be the most probable. Thus, for instance, T. F. Andreyeva, working with hydrolyzates of chloroplast proteins after exposure of leaves in an atmosphere of $C^{14}O_2$, observed that C^{14} is mainly incorporated into glycine, serine, and alanine.

Some interesting data (Ref. 31) show that light promotes the formation of

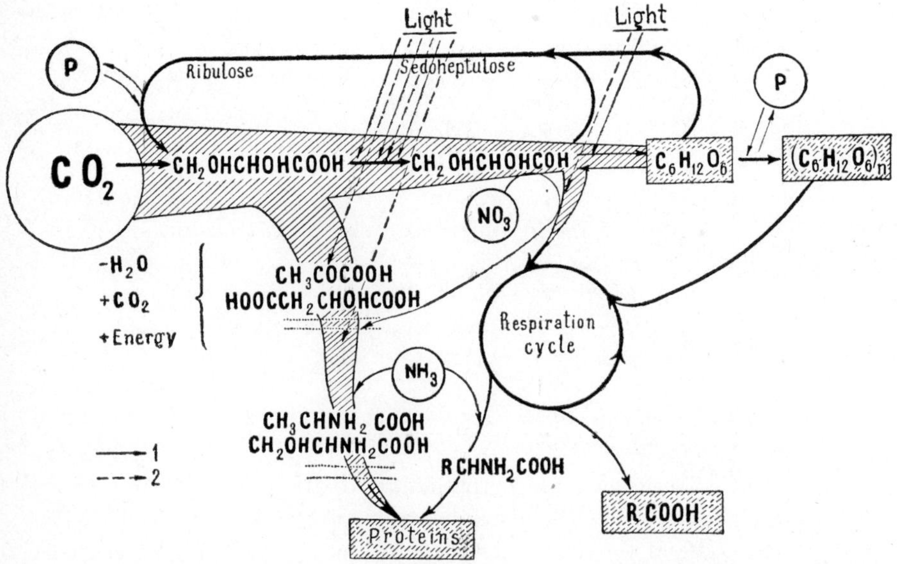

FIG. 7.4. Diagram showing products of photosynthesis.

amino acids with branching chains and aromatic rings. Light also had strong influence on the incorporation of amino acids in protein. The relative amounts of various acids entering proteins in the light differs from that of amino acids entering in the dark.

From the foregoing one may conclude not only that the process of photosynthesis is accompanied by the formation of several direct products, but that some of them (for instance proteins) possess specific properties. This problem merits the most serious attention. (See Fig. 7.4.)

Summary

A large amount of available data shows that the direct products produced by the photosynthetic apparatus may include not only carbohydrates, but also a number of other substances formed in the chloroplasts and translocated from them to cytoplasm.

The formation of various products in the photosynthetic organs—the chloroplast—omitting the stage of carbohydrates, seems to be reasonable biologically. Such synthesis may proceed in energetically extremely favorable conditions, because not only enzymes but photochemical reactions may then participate.

Owing to the pronounced influence of rays of different wavelength on the assortment of products of photosynthesis, it may be supposed that the complex cycles of the transformation of carbon take place with the participation not only of the main photochemical reaction of photosynthesis leading to the accumulation of energy in products, but also of secondary photochemical reactions not always leading to the accumulation of energy, but directing some of the links of the chain of carbon conversions along specific lines.

As the chloroplasts have different structure and different enzyme, and as their condition may alter with age of the plants and the conditions of their cultivation (Refs. 18, 32, 33, 34, 35) the character and direction of their work as of photosynthetic organs may considerably vary, and the cells may thus be supplied with materials varying in both composition and quality.

This view of the peculiarities of the plant's photosynthetic apparatus does not alter our concepts on the important and in many cases decisive role of the diverse secondary syntheses, but broadens our information on the nature and procedure of the initial stages of the carbon nutrition of plants and our knowledge about photosynthesis.

Thus the photosynthetic apparatus does more than supply plants with universal and uniform nutritive material—carbohydrates. It forms other important products as well, altering their composition and quality depending on the kind of plants, their state, and environmental conditions. Thereby the photosynthetic apparatus defines the biochemical nature of plant metabolism and determines their complex and diverse relationships with the environmental conditions and especially their reactions to conditions of illumination (intensity, spectral composition, frequency of illumination).

The further development of this important and complicated problem must be conducted with extensive use of the method of labeled atoms.

References for Chapter 7

1. I. Smith, "Products of photosynthesis." In Franck and Loomis, eds., *Photosynthesis in Plants,* Iowa, 1949.
2. Schimper, *Botanische Zeitung* 5:(1888).
3. Chrapowicki, *Botanisches Centralbl.* 38:352 (1889).
4. V. V. Sapozhnikov, *Formation of Carbohydrates in Leaves and Their Movement Through the Plant,* Moscow, 1890.
5. V. V. Sapozhnikov, *Proteins and Carbohydrates of Green Leaves as Products of Assimilation,* Tomsk, 1894.
6. A. M. Lyovshin, *Experimental Cytological Study of Adult Leaves of Autotrophic Plants with Regard to the Problem of the Nature of Chondriosomes,* Saratov, 1917.
7. A. M. Lyovshin, *Scientific Proc. of Kiev Univ.* I, 6:49 (1935).
8. W. J. Kabos, *Réc. Trav. Bot. Néerl.* 133:446. Mulder, Amsterdam, 1936.
9. H. Burström, *Die Naturwissenschaften II,* 41/42 (1942).
10. H. Burström, *Arkiv f. Botanik.* 30,8:1 (1943).
11. J. Myers, "The pattern of photosynthesis in *Chlorella,*" in Franck and Loomis, eds., *Photosynthesis in Plants,* Iowa, 1949.
12. T. F. Andreyeva, *Proc. AS USSR* 78,5:1033 (1951).
13. T. F. Andreyeva and E. G. Plyshevskaya, *Proc. AS USSR* 87,2:301 (1952).
14. A. N. Pechenitsina, *Proc. Timiryazev Plant Physiol. Inst.* 8,2:212 (1951).
15. N. P. Voskresenskaya, *Proc. AS USSR* 93,6:911 (1953).
16. N. P. Voskresenskaya, *Proc. Timiryazev Plant Physiol. Inst.* 8,1:42 (1953).
17. S. P. Osipova and I. V. Timofeyeva, *Proc. AS USSR* 80,3:449 (1951).
18. O. P. Osipova, *Izv. AS USSR Biol. Ser.* 1:96 (1953).
19. V. O. Tauson, *Izv. AS USSR Biol. Ser.* 3:423 (1947).
20. A. A. Nichiporovich, *Proc. Plant Physiol. Inst. AS USSR* 1:3 (1953).
21. L. A. Nezgovorova, *Proc. AS USSR* 79,3:537 (1951).
22. L. Norris, R. E. Norris, and M. Calvin, *J. Experim. Botany* 6,16:64 (1955).
23. N. G. Doman, A. M. Kuzin, Y. V. Mamul, and R. I. Khudyakov, *Proc. AS USSR* 86,2:369 (1952).
24. V. A. Devyatin, *Biochemistry* 15,4:3 (1950).
25. F. A. Isherwood, J. T. Chen, and L. Mapson, *Biochem. J.* 56,1:1 (1954).
26. H. Metzner, *Protoplasm* 41:129 (1952).
27. A. A. Krasnovsky and G. P. Brin, *Proc. AS USSR* 73,6:1239 (1950).
28. D. J. Arnon, M. B. Allen, and F. R. Whatley, *Nature* 174,4426:394 (1954).
29. H. Luger, *Protoplasma* 44,2:212 (1954).
30. L. A. Nezgovorova, *Proc. AS USSR* 85,6:1387 (1952).
31. D. M. Racusen and S. Aronoff, *Archiv. Biochem. & Biophys.* 51,1:68 (1954).
32. A. A. Tabentzky, *Izv. AS USSR Biol. Ser.* 5:611 (1947).
33. A. A. Tabentzky, *Belozerkovsky Agr. Inst., Sci. Series* 3,1(4):39 (1952).
34. A. A. Tabentsky, *Izv. AS USSR Biol. Ser.* 1:72 (1953).
35. N. M. Sisakian, *The Fermentative Activity of Protoplasmic Structures.* Paper by V. Bach, Moscow, 1951.

Part III

PLANT PHYSIOLOGY, PATHOLOGY, AND CYTOLOGY

Chapter 8

STIMULATION OF PLANT GROWTH WITH IONIZING RADIATIONS *

The ever-growing availability of the various sources of ionizing radiation makes it desirable to look for means of effective application of this type of radiation to food industry and agriculture. Research conducted in the Soviet Union in the sphere of application of ionizing radiation embraces (*a*) sterilization and conserving of foods and extension of the period of storage of potatoes and vegetables, and (*b*) acceleration of seed germination and of the initial stages of development of agricultural crops, as well as raising their yields.

The present paper brings up for discussion the future prospects of utilizing the stimulating action of ionizing radiation for the growth and development of agricultural plants.

The possibility of using ionizing radiation for stimulation of the initial stages of plant growth was proved by a number of workers (Refs. 1–18). In a series of studies (Refs. 19–22) the stimulating effect was followed up to the ripening period, and higher crop yields were registered.

However, since in spite of similar treatment contradictory results were eventually obtained, the stimulating effect on the quality and the amount of yield still remains disputable. Thus, Johnson (Ref. 23) obtained no stimulating effect, probably owing to low-precision dosimetry; besides, his conclusions are based on too scanty material. The experiments by the same author (1948) on a decorative plant (*Calanhoe tubiflora*) showed accelerated florescence and ripening of plants under the influence of radiation.

On the basis of experimental evidence obtained in twelve different species of plants (wheat, barley, oats, peas, beans, horse beans, mustard, lettuce, esparcet, and three meadow cereals) Schwarz, Szepa, and Schindler (Ref. 24) denied the stimulating action of X rays. In treating dry and sprouting seeds of wheat and peas with various doses of X rays, A. V. Koltsov and A. I. Koltsov (Ref. 25) did not observe any stable stimulating effect. Patten and

* This chapter is taken from Geneva Conference Paper 699, "The Utilization of Ionizing Radiation in Agriculture" by A. M. Kuzin of the U.S.S.R.

Wigoder (Ref. 26) subjected to radiation the seeds of beans, mustard, and barley, but also failed to obtain any marked stimulation.

The contradictory results of a number of studies are largely due to a difference in the methods of treatment, to insufficiency of dosimetry in the past, and to underestimation of radiosensitivity of different plant species, as well as to the conditions of the environment in which the treated plants developed.

The possibility of obtaining a positive effect makes it expedient to continue research in this field. Within recent years Breslavets, Berezina, Butenko, Vlasyuk, Drobkov, Zhezhel, Kuzin, Kuznetsov, Nichiporovich, Shirshov, Engel and many other Soviet scientists have studied the action of ionizing radiation on growth and development of plants. It should be emphasized that some data obtained by Soviet scientists under practically similar experimental conditions are indicative both of a stimulating effect and the absence of a stable positive action.

In the present paper we shall confine ourselves to those studies which conclusively enough demonstrate the presence of the stimulating effect and which seem to be of interest for the problem at issue. The following methods of treatment were applied:

Irradiation of seeds before sowing.

Soaking of seeds before sowing in solutions containing natural and artificial radioactive substances.

Treatment of the soil with radioactive substances serving as microfertilizers.

Continuous irradiation of growing crops with Co^{60} gamma rays.

The results are briefly discussed below.

Irradiation of Seeds Before Sowing

Irradiation of seeds before sowing has great advantages as compared with other methods of treatment. Its chief merits are:

1. Organization of irradiation in specially equipped places with subsequent transportation of seeds.
2. Irradiation of material in due time.
3. Complete absence of radioactivity both in the sowing material and in the yield.

In a series of experiments conducted in 1953–1954 the difference in radiosensitivity of several species of agricultural plants was tested.

Radiation of dry seeds with X rays was carried out at 170 kilovolts, 6 milliamperes with a cardboard filter at a distance of 60 cm from the source of radiation (43 roentgens per min). The subsequent procedure consisted in germinating irradiated seeds on filter paper in Petri dishes in duplicates or triplicates and measuring the length of the shoot roots in the course of growth. One hundred irradiated seeds were sown in each Petri dish.

For irradiation of seeds of a number of plants with various doses of X rays, the following stimulating doses were used for short-time exposures:

For rye seeds	750–1000 r
For pea seeds	350–500 r
For radish seeds	500–1000 r
For cabbage seeds	1000–2000 r
For cucumber seeds	100–300 r

Table 8.1 illustrates the results of measurement of root length of rye shoots for various radiation doses.

TABLE 8.1. LENGTH OF ROOTS OF RYE SHOOTS AFTER IRRADIATION WITH X RAYS

Doses, r	Third Day of Development, mm	Fourth Day of Development, mm
Control	17.8	43.5
250	18.9	49.8
500	20.7	52.2
750	18.9	49.7
1000	24.1	55.6

Thus, the maximum length of the roots was registered in seeds irradiated with 1000.

In one of the experiments not only the length of the shoot roots was measured, but also their diameter at the site of tissue differentiation (Table 8.2).

TABLE 8.2. ROOT DIAMETER OF RYE SHOOTS AFTER IRRADIATION WITH X RAYS

Doses, r	Third Day of Development, microns
Control	304
250	363
500	387
750	393
1000	359
2000	326
4000	317
8000	305

It will be seen from the data of Table 8.2, that the maximum root diameter corresponds to the radiation dose of 750 r.

It is theoretically important that the increase in the root diameter is due not to the growing size of the cells, but to the increase in their number (Table 8.3). The increase in the number of cells is indicative of accelerated cell division as caused by irradiation.

TABLE 8.3. NUMBER OF CELLS IN THE SUBEPIDERMAL LAYER OF SHOOT ROOTS AFTER IRRADIATION WITH X RAYS

Doses, r	Number of Cells
Control	40
250	56
500	61
750	68
1000	57
2000	39
4000	40
8000	36

Another object of study was radish seeds. The conditions of irradiation were the same as for rye seeds. The stimulation of development is shown in Table 8.4.

TABLE 8.4. ROOT LENGTH OF RADISH SHOOTS AFTER IRRADIATION WITH X RAYS (IN MM)

Doses, r	Day of Development		
	3rd	4th	5th
Control	27.2	37.6	50.0
250	27.4	50.7	63.9
500	32.6	64.0	72.2
1000	28.2	43.6	53.0
2000	17.2	34.0	48.0

Similar experiments with pea seeds gave the following results. It will be seen that the dose of 500 exercises a distinct stimulating effect (Table 8.5).

The effect of X-ray irradiation of dry cucumber seeds was rather peculiar: while no clearcut results were obtained in measuring the root length of the shoots of irradiated seeds (Table 8.6), an optimal dose of 300 r accelerated the growth of true leaves (Table 8.7).

It will be seen that radiation doses of 100–300 r stimulate the growth of the

TABLE 8.5. ROOT LENGTH OF PEA SHOOTS AFTER IRRADIATION WITH X RAYS (IN MM)

Doses, r	Day of Development			
	3rd	4th	5th	6th
Control	15.5	21.6	53.6	65.4
500	21.3	32.5	61.2	70.4
1000	22.4	28.6	49.7	57.1
2000	22.7	26.7	38.4	41.6

TABLE 8.6. ROOT LENGTH OF CUCUMBER SHOOTS IRRADIATED WITH X RAYS (IN MM)

Doses, r	Day of Development		
	3rd	4th	5th
Control	21.2	41.4	82.6
100	18.6	40.6	97.0
200	16.6	21.9	67.4
300	21.6	39.2	92.0
400	16.1	21.6	35.0
500	16.6	33.3	89.5
1000	13.2	15.7	49.9
2000	14.1	18.6	43.7
4000	17.1	29.4	59.1
8000	12.1	18.9	34.5
16000	11.1	19.2	26.6

TABLE 8.7. LENGTH AND WIDTH OF FIRST TRUE LEAVES AFTER IRRADIATION WITH X RAYS (IN MM)

Doses, r	1st True Leaf		2nd True Leaf		3rd True Leaf		4th True Leaf	
	Length	Width	Length	Width	Length	Width	Length	Width
Control	31.9	—	86.6	76.0	59.0	62.3	No 4th leaf yet	
100	36.8	—	74.3	76.0	57.0	71.3	No 4th leaf yet	
200	30.3	—	79.3	73.3	42.6	40.3	No 4th leaf yet	
300	34.2	—	97.0	91.6	74.3	74.0	22.0	20.3
400	23.2	—	86.3	82.6	45.3	44.3	No 4th leaf yet	
500	30.4	—	86.6	76.0	44.3	37.3	No 4th leaf yet	

first true cucumber leaves. However, the stimulating effect was not universal. Thus, irradiation of dry soaked seeds and shoots of soft wheat did not call forth any stimulation. The wheat did not react to small doses of radiation, while doses of 1000 r or more produced an inhibitory effect. Frolov's experiments (Ref. 12) with another variety of wheat showed a stimulating action of X rays and a higher yield. This fact points to the necessity of taking into account the radiosensitivity not only of individual species but of different plant varieties as well.

The acceleration of the initial stages of plant development is of primary importance since it may essentially influence the yield in the arid districts as well as in those where the sowing period is limited. The acceleration of growth at the initial stage of plant development results (*a*) in earlier ripening and (*b*) in a higher yield. This will be illustrated by a few examples.

Experiments with radish. On January 1, 1954, in the Marfino State Farm, seeds of radish of the Saks variety were sown in the hothouse on an area of 12.5 m^2. Part of the dry seeds were irradiated with different doses of X rays; another part were left for control. The results are summarized in Table 8.8.

TABLE 8.8. RADISH YIELD AS INFLUENCED BY X RAYS

Doses, r	Total Weight of Root Yield from the Test Area, kg	Yield, %
Control	3.74	100
500	4.48	119
750	4.47	119
1000	4.98	133

This little experiment, in which the plants were grown to commercial ripeness, illustrates the significance that may be acquired by this kind of radiation.

In another experiment carried out in the same State Farm on March 18, irradiated seeds of the "Moscow hothouse" variety were sown. The same radiation doses were used as in the first experiment. When treated with 1000 r, the radish ripened several days earlier than control specimens. The grown roots were bigger in size and of prime quality, as regards both taste and succulence.

At the Vegetable Station of the Timiryazev Agricultural Academy in Moscow, irradiated radish seeds of the "white-pinkish" variety were sown in the open ground on 6-m^2 plots over an area totaling 72 m^2 (Table 8.9). The roots grown from irradiated seeds were of prime quality, both as regards their taste and succulence.

TABLE 8.9. RADISH ROOT YIELD IN THE OPEN GROUND AFTER TREATMENT WITH X RAYS

Doses, r	Total Yield from 6-m^2 plot, kg	Yield, %
Control	9.42	100
500	11.30	119
1000	12.27	140

Experiment with cabbage. An experiment with cabbage of the Kolkhoznitsa variety was made at the Gribovo Vegetable Station. Dry seeds were subjected to irradiation with X rays. Radiation doses of 1000 and 2000 r were tested, which caused an acceleration of ripening and a certain increase in the yield (Table 8.10).

TABLE 8.10. YIELD OF CABBAGE OF THE KOLKHOGNITSA VARIETY TREATED WITH X RAYS

Doses, r	Yield per Hectare, centners	Yield, %
Control	222.4	100
1000	265.6	119
2000	266.3	119

Experiment with peas. This experiment was conducted at a Vegetable Station of the Timiryazev Agricultural Academy in Moscow. It consisted in planting in triplicate a split variety of peas on 6-m^2 plots. To estimate the yield, 50 plants were taken from each plot and the number of beans contained in each plant was determined (Tables 8.11 and 8.12). Any increase in radiation doses resulted in reducing the yield.

TABLE 8.11. HARVEST YIELDED BY SPLIT PEAS TREATED WITH X RAYS

Doses, r	Number of Beans	Per Cent to Control
Control	73	100
350	80	110
500	88	121

TABLE 8.12. WEIGHT OF PEA BEANS AND GRAINS AFTER TREATMENT WITH X RAYS

Doses, r	Weight of 1000 Beans, gm	Weight of 1000 Grains, gm
Control	3700	338
350	4290	407

Experiment with rye. Irradiation of spring rye seeds with X rays (doses of 750–1000 r) resulted in a 21–22% increase in the weight of 1000 grains. Irradiation of soaked pea seeds with Co^{60} gamma rays (dose 250 r) likewise resulted in increasing the yield (Table 8.13).

TABLE 8.13. THE EFFECT OF IRRADIATION WITH Co^{60} γ-RAYS ON PEA YIELD

Exposure Time	Weight of One Plant			
	Grains		Vegetation Mass	
	Gm	Per Cent	Gm	Per Cent
Control	2.64	100	5.48	100
24 hr	3.87	150	6.80	124
6 hr	3.47	131	6.14	112
6 hr	3.68	139	6.40	117
30 min	3.24	123	6.18	113
30 min	3.60	136	6.68	121

The above experiments prove that exposure of seeds to ionizing radiation before sowing may result in higher yields and accelerate the ripening of some crops.

The use of Roentgen apparatus in rural conditions is not profitable. The most convenient source of radiation for these purposes is Co^{60}. It has been experimentally proved that there is no essential difference between the effect produced upon seeds by X rays and by Co^{60} gamma rays, while the use of cobalt as a source of radiation is more expedient by far. A preliminary comparative study of short exposures to large doses of X rays and of long exposures to small doses of Co^{60} gamma rays shows that by increasing the exposure time it is possible to reduce considerably the effective dose.

Soaking of Seeds in Solutions Containing Radioactive Substances Before Planting

Extensive work in this direction was conducted in various research institutes of the Soviet Union. The object of study was the influence exerted by a number of radioactive isotopes upon the growth and development of plants, particular attention being paid to the use of non-separated mixtures of the fission products of uranium. These studies demonstrated that prolonged action of diluted solutions of beta and gamma radiators is more effective than that of concentrated solutions with a lesser exposure time.

The best results were obtained with solutions of a mixture of beta and gamma radiators with an activity ranging from 0.2 to 0.5 millicuries per liter, the seeds being soaked for 24 hr.

The experiments were made in a laboratory and verified in field conditions. The following cultures were studied: peas, vetch, haricot, beans, soya, lucerne, wheat, barley, oats, millet, buckwheat, sugar beet, flax, maize, and tomatoes. A few examples are cited below.

The experiment with peas * was conducted on 5-m^2 plots and repeated six times. The seeds were soaked for 24 hr in a solution of non-separated fission products of uranium with activity of 0.5 millicuries per liter. Control seeds were soaked in water. An analysis of the yield gave the following results (Table 8.14). As a rule, in the experiments with peas, vetch, beans, and soya conducted on plots, earlier (for 2–5 days) florescence was observed.

TABLE 8.14. EFFECT OF RADIOACTIVE FRAGMENTS ON PEA YIELD

Yield Indices	Excess over Control, %
Weight of aerial organs	40
Number of beans	60
Number of peas	100
Weight of beans	55
Weight of peas	100
Yield of seeds, in per cent of yield of aerial mass	40

The average data computed from a number of experiments show that the weight of seeds of one plant exceeded that of control seeds: peas—by 34%, vetch—by 57%; haricot—by 98%, beans—by 26%, and soya—by 53%.

In field experiments with leguminous crops the seeds were soaked for 24 hr in a solution of non-separated fission products of uranium with a concentration

* Similar experiments with wheat gave a 16–20% increase in the absolute weight of the seeds, the pre-sowing material being treated with a solution of radioactive fragments.

TABLE 8.15. EFFECT OF RADIOACTIVE FRAGMENTS ON THE YIELD OF LEGUMINOUS CROPS

Crop	Area, m^2	Number of Replications	Yield				
			Kg		Centners per Hectare		Ratio to Control, %
			Control	Experiment	Control	Experiment	
Peas	3	8	4.179	4.626	17.4	19.3	110
Peas	10	5	4.620	5.048	9.2	10.1	109
Peas	90	6	25.450	31.300	4.7	5.8	123
Peas	100	5	56.500	66.100	11.30	13.22	117
Peas	300	5	156.900	165.000	10.46	11.00	106
Beans	2.5	4	2.450	3.290	24.50	32.90	134
Beans	2.5	5	2.69	3.73	23.74	29.84	125
Beans	15	5	12.9	16.7	17.20	22.27	130
Vetch	3	7	4.28	4.80	20.40	22.89	112
Vetch	10	5	3.97	5.09	7.95	10.19	128
Vetch	100	5	25.6	27.8	5.12	5.56	109
Lucerne	50	4	140.6	163.5	70.3	81.7	116
Haricot	3	6	2.06	2.70	11.4	15.0	130

of 0.5 millicuries per liter. Numerical differences in the results obtained in repetitive experiments with leguminous crops are shown in Table 8.16

Wheat seeds were soaked in solutions containing artificial radioactive isotopes of phosphorus and calcium. The experiments were conducted on 125-m^2 plots in quadruplicates (Table 8.17).

The results of 32 field experiments, carried out over a period of four years with seeds soaked in a non-separated mixture of radioactive fragments (0.5 millicurie) are presented in Fig. 8.1. The experiments with different crops were conducted on plots with an area ranging from 2.5 to 300 m^2.

The experiments with soaking seeds were repeated on a production scale on an area of 290 hectares. The results obtained are shown in Table 8.18.

The results obtained in 32 field experiments and 42 production experiments with seeds soaked in solutions of radioactive fragments, conducted in the course of four years, confirmed the validity of the observed positive effects. The real positive deviation amounted, on the average, to about +16% for leguminous crops and +10% for cereals. The above method of pretreatment of seeds is of interest from the point of view of utilization of the radioactive waste of atomic production.

TABLE 8.16. EFFECT OF RADIOACTIVE FRAGMENTS ON REPLICATED YIELD OF LEGUMINOUS CROPS

Crop	Repli-cations	Yield, kg per hectare	
		Control	Experiment
Russian black beans	1	22100	26600
	2	24000	28400
	3	23000	27000
	4	22900	26700
White beans	1	16400	16400
	2	24300	28400
	3	25800	24900
	4	14800	30200
	5	17550	24400
Beans (low-grade)	1	1100	1740
	2	1975	2380
	3	1910	2420
	4	1630	2475
	5	1990	2240
Haricot	1	1110	1725
	2	1535	1608
	3	1440	1660
	4	1120	1293
	5	658	1120
	6	999	1610

TABLE 8.17. EFFECT OF P^{32} AND Ca^{45} ON WHEAT YIELD

Variants	Grain Yield, centners per hectare	Ratio to Control, %
Control P^{31}	29.26	
Experiment—Seeds treated with P^{32}, 50 microcuries per kg of seeds	32.45	110.9
Control (Ca^{40})	29.16	
Experiment—Seeds treated with Ca^{45}, 50 microcuries per kg of seeds	32.23	110

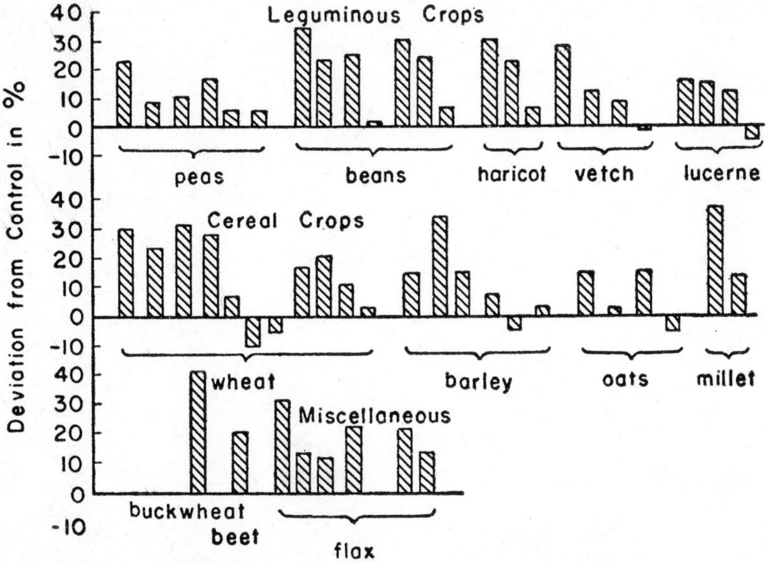

FIG. 8.1. Effect produced by soaking seeds of various agricultural crops in solutions of radioactive fragments.

TABLE 8.18. EFFECT OF RADIOACTIVE FRAGMENTS ON CROP YIELDS

Crops	Ratio of Yield to Control, % Replications in Different Years
Peas	110, 108
Vetch (for grain)	116, 110, 86, 105
(for hay)	103, 124, 122, 120, 115, 111
Lucerne	143, 123
Grass mixture (clover, lucerne, timothy)	108
Turnip	107, 108
Wheat	110, 132, 110, 105, 109, 99, 110, 118, 91
Oats	119, 102, 91, 102, 120

The difficulties of its large-scale application involve the organization of adequate health protection against radioactive substances. As to the traces of radioactivity in the harvest and the radioactive contamination of the soil, this question requires further inquiry. Analyses showed that the activity of the agricultural production amounts to less than 10^{-9} curie per kg.

TREATMENT OF SOIL WITH RADIOACTIVE SUBSTANCES AS MICROFERTILIZERS

The application of radioactive fertilizers has been often discussed in the literature. Some researchers (Refs. 6, 27, 28, 29, 30) point to the positive effect produced by small concentrations of natural radioactive elements of radium, uranium, and thorium in doses varying from 10^{-12} to 10^{-9} curie per liter or curie per kg of soil. No inhibitory effect was produced by these doses. Within the range of doses varying from 10^{-9} to 10^{-5} curie per liter, both a stimulating effect (Refs. 28, 30, 31, 32, 33) and absence of such an effect (Refs, 31, 34, 35, 36) were observed. The use as alpha radiators of uranium and radium produced an inhibitory effect (Ref. 36). Higher doses cause, as a rule, an inhibition of growth and development of plants.

According to a report of the U.S. Atomic Energy Commission (Ref. 37), the use of radium, uranium, and an actinium-containing preparation as fertilizers produces neither a positive nor a negative effect.

Within the past few years extensive research has been carried on again in the Soviet Union on natural radioactive substances (radium, uranium, thorium), radioactive fragments, and artificial radioactive isotopes, as well as minerals and rocks with heightened radioactivity. To avoid contamination of soil, particular attention has been paid to the amount of fertilizers producing a slight increase in its natural radioactivity. No negative effect was observed in treating a number of crops with small doses of radioactive substances (10^{-11} to 10^{-6} curie per kg). The positive effect proved to be more stable in vegetation experiments, while in field experiments it varied greatly according to the soil conditions and cultivation methods.

Thus, in water cultures of tomatoes, an admixture of radioactive phosphorus caused a 25–38% increase in the yield (Table 8.19).

TABLE 8.19. EFFECT OF P^{32} ON THE YIELD OF TOMATO FRUITS

Variant	Total Weight, gm	Ratio to Control, %
Experiment I:		
Control	265	100
2.3×10^{-5} curie per liter	366	138
Experiment II:		
Control	247	100
2.3×10^{-5} curie per liter	311	126
5.3×10^{-5} curie per liter	310	125

In a series of eight-year experiments of cultivating potatoes, cabbage, oats, clover with timothy, spring wheat, sugar beet, and sunflower, the annual yield increases amounted to nearly 27% (10–68%). A single dose of 1×10^{-9} curie per kg of radium and 1×10^{-4} curie per kg of uranium was introduced into the soil. The results of the experiments are summarized in Table 8.20.

TABLE 8.20. EFFECT OF NATURAL RADIOACTIVE SUBSTANCES ON CROP YIELDS

Year and Crop	Ratio of Yield to Control, %	
	Radium 1×10^{-9} C/kg	Uranium 1×10^{-4} C/kg
1947 Potatoes (tubers)	124.2	137.3
1948 Cabbage (marketable)	113.3	138.8
1949 Oats (grain)	133.9	123.6
1950 Clover-timothy mixture (hay)	110.9	118.1
1950 Same aftermath	112.7	168.9
1951 Spring wheat (grain)	133.8	135.5
1952 Red beet (roots)	110.0	129.6
1953 Sunflower (silo)	111.3	122.8
1954 Potatoes	132.1	121.9
Average for 8 years	120.2	132.9

In field experiments with sugar beet carried out in different years with the application of radioactive fertilizers, there was an increase in sugar content varying from 0.2% to 1.5% and a 10–18% increase in the yield. It will be noted that the results of the field experiments varied with the soil and vegetation conditions.

Many experiments conducted under laboratory and field conditions demonstrated a considerable stimulating effect of small doses of natural radioactive elements (uranium, radium, thorium), as well as of radioactive fragments, on reproduction, development, and nitrogen-fixing capacity of radicicola bacilli and *Azotobacter*. In some experiments the stimulating effect amounted to 60–100%.

The formation of tubercles on the roots was not observed in vegetation experiments with peas, in which nutrient mixtures were used specially purified of admixtures of natural radioactive elements. The addition to the medium of radium or uranium salts in concentrations of 10^{-10} and 10^{-5} curie per kg respectively resulted in abundant formation of tubercles and a considerable increase in the average weight of the air-dry aerial mass, particularly in the weight of dry seeds (by 60–80%).

The stimulating effect of radioactive substances on the growth of tubercles was likewise observed upon application of artificial radioactive isotopes. A series of experiments demonstrating the influence of radioactive phosphorus on the development of tubercles on the roots of aquatic and arenaceous pea cultures belonging to the "Capital" variety, may serve as an illustration (Table 8.21).

TABLE 8.21. EFFECT OF P^{32} ON THE DEVELOPMENT OF TUBERCLES

Doses of P^{32} in C/l	Number of Tubercles		Volume of Raw Tubercles		Weight of Air-dry Mass	
	Pieces	Per Cent	Cm^3	Per Cent	Gm	Per Cent
Aquatic cultures						
Control	276	100	0.19	100	0.103	100
8.3×10^{-7}	329	119	0.22	116	0.155	150
8.3×10^{-6}	1122	406	0.44	233	0.358	347
8.3×10^{-5}	844	305	0.58	307	0.551	534
Arenaceous cultures						
Control	225	100	0.14	100	0.100	100
8.3×10^{-7}	275	122	0.13	94	0.125	125
8.3×10^{-6}	887	394	0.46	317	0.371	371
8.3×10^{-5}	476	212	0.61	417	0.582	582

The results of five-year large-scale production tests of the action of fertilizers consisting of natural radioactive rocks on various crops demonstrated a steady 15–25% rise in the yield. Moreover, the possibility of a positive effect being produced by other non-radioactive microelements contained in the specimens at issue is not excluded. However, a comparison of the results obtained with other data pertaining to the effect of radioactive elements corroborates the decisive role played by radioactivity in these experiments too.

Although the introduction of radioactive microfertilizers into the soil has justified itself in numerous experiments, it cannot yet be recommended for practical application, since the selective adsorption of the radioactive elements by the harvest and the effect produced upon the animal and human organism have not been sufficiently studied.

This method of application of radioactive fertilizers raises the extremely serious problem of protecting the working personnel. Until this problem is definitely solved, the method cannot be recommended for practical purposes.

Continuous Irradiation of Agricultural Crops with Cobalt Gamma Rays

Research work in this field was started in 1954 and the results thus far obtained are therefore only preliminary. The effect of continuous exposure to weak external Co^{60} gamma radiation (activity of 1 curie) of the growing culture of sugar beet and buckwheat was studied in field conditions on a leveled-out plot. Observations were made along radial zones at various distances from the source. During the two months of the experiment the dose received by the plants varied between 0.9 and 116 r, depending on the distance from the plant to the radiation source. This experiment conclusively showed an acceleration of buckwheat growth. In the zones of maximum radiation (4.3–116.7 r) the buckwheat blossomed forth five days in advance of control seeds.

The number of blooming plants per zone (in per cent) on the 65th day of vegetation is given in Table 8.22. The buckwheat was grown for rutin and

TABLE 8.22. EFFECT OF GAMMA RADIATION ON THE INITIATION OF FLUORESCENCE IN BUCKWHEAT

Zone	Dose, roentgens	Number of Blooming Plants
Control zone (outside of radiation)		6 ± 0.72
0 zone	116.7	51 ± 5.2
1st zone	21.5	41 ± 1.4
2nd zone	4.3	33 ± 1.78
3rd zone	1.7	10 ± 0.84
4th zone	0.9	7 ± 0.84

the harvest was gathered in the form of a green mass at the moment of florescence. The harvest estimates are summarized in Table 8.23. The experiment showed that plants of the first zone, treated with 21.5 r during the period

TABLE 8.23. EFFECT OF GAMMA RADIATION OF BUCKWHEAT ON THE HARVEST OF VEGETATION MASS

Zone	Dose, roentgens	Yield of 1 Plot, kg	Ratio to Control, %
0 zone	116.7	5003	112
1st zone	21.5	6323	145
2nd zone	4.3	5675	123
3rd zone	1.7	5047	110
4th zone	0.9	4515	99.3
Control	Natural conditions	4569	100

between sowing and the moment of florescence, increased the harvest yield by 45% as compared with the control.

Under similar conditions sugar beet plants were irradiated at a moment when the rosette had 4–5 leaves. The result was an increase in sugar content varying from 0.6% to 1.2% in the zones with the lowest intensity of irradia-

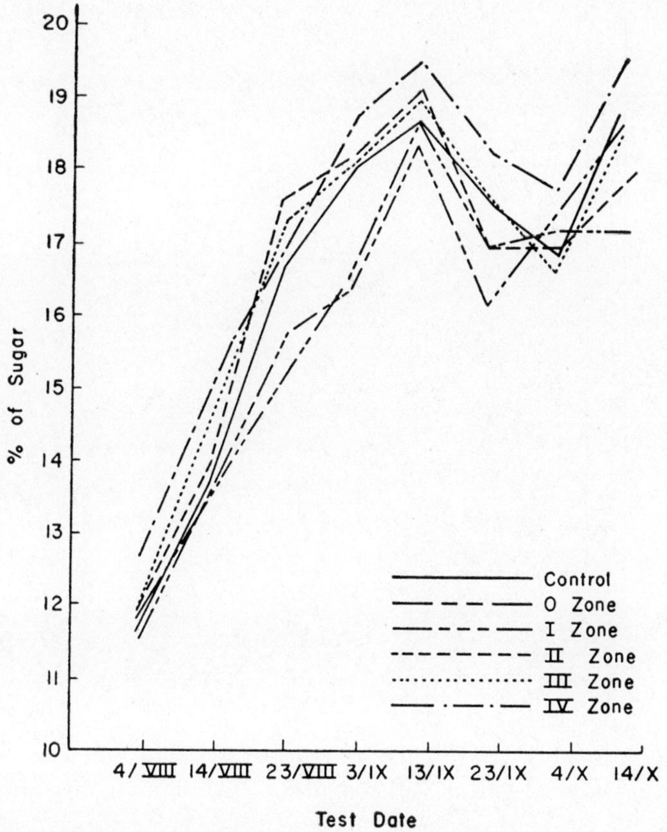

Fig. 8.2. Sugar content in the roots of irradiated and non-irradiated sugar-beet plants.

tion. The results of two repeated experiments with sugar beet plants are presented in Figs. 8.2 and 8.3.

It will be noted that numerous experiments have been carried out on sugar beet, in which the seeds were soaked in solutions of radioactive elements P^{32}, Zn^{69}, Ca^{45}, and S^{35}, and the plants received radioactive phosphorus as extra nutrition via the leaves. In these cases as well as in vegetation experiments with micro quantities of uranium and radium salts, there occurred a 0.3–1.2% increase in sugar content above 16.8–19.4% control. The results obtained are in conformity with the conclusions drawn by Granhall and associates (Ref.

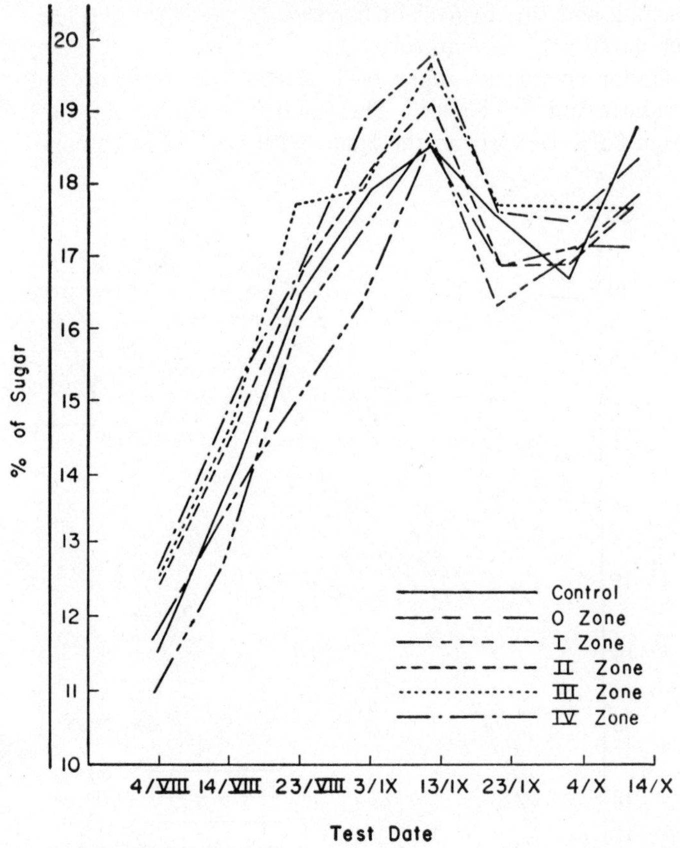

FIG. 8.3. Sugar content in the roots of irradiated and non-irradiated sugar-beet plants.

22), who likewise observed a stimulating effect of small radiation doses. The preliminary results obtained with small-dose irradiation of growing plants throughout the vegetation period open new prospects for further research along these lines, particularly with valuable crops grown in the areas where acceleration of the harvest may prove of decisive significance.

The above experiments demonstrated the possibility of raising harvest yields by using the stimulating effect of small doses of ionizing radiation. This makes it expedient to elaborate concrete methods of applying ionizing radiation for agricultural purposes. However, considerable fluctuations mark the degree of effect as manifested by experiments conducted under different soil and climatic conditions, and insufficient elaboration of the theory of stimulation point to the necessity of combined efforts of scientists of all countries for the solution of this problem.

References for Chapter 8

1. Körnicke, *Ber. Deutsch. botan. Ges.* (1904, 1905); *Jahrb. wiss. Botan.* (1915); *Fortschr. geb. Röntgenstr.* (1919).
2. Evler, *Jahrb. Wiss. Botan.* 5:65, 116 (1906).
3. Miège and Coupe, *C.R.Ac. Paris* 15 Ser. 4:338 (1914).
4. Rochlin and Gleichgewicht, *Fortsch. geb. Röntgenstr.* 33:971 (1925).
5. A. B. Doroschenko, *Trudy biogeokhim. prikl. bot., i selek.,* 118:2 (1939).
6. I. Stoklasau Penkava, *Biologie des Radiums u. radioaktiven Elemente,* Berlin 1932.
7. N. V. Chekhov, *Trudy Tomskovo Gos. Universite* 35:67 (1934).
8. L. P. Breslavets and A. I. Atabekova, *Semenovodstvo* 1 (1934).
9. L. P. Breslavets and A. S. Afanasyeva, *Tr.Vs. in-ta udobr. i agropichvoved. vyp.* 7:245 (1935).
10. L. P. Breslavets, A. S. Afanasyeva, and G. B. Medvedeva, *Vestn. rentgenologii i radiologii i radiologii* 10:288 (1935).
11. T. Longaa and H. Karsten, *Plant Physiology* (1936), 11, 615.
12. G. Frolov, *Trudy s. kh. akademii im. Timiryazeva,* 1,2:189 (1936).
13. A. I. Atabbekova, *Biolog. zhurnal* 5,1:99 (1936).
    A. I. Attabbekova, *Biolog. zhurnal* 6,2:38 (1937).
    A. I. Atabbekova, *Priroda* 7,8:56 (1938).
14. L. P. Breslavets, *Trudy Botan. sada MGU,* 111, 1940.
    L. P. Breslavets, *Rastenie i luchi Rentgena,* AN SSSR M.L., 1946.
15. A. A. Drobkov, *Trudy biogeokhim. lab.* 5 (1939).
    A. A. Drobkov, *Izv. A.N. SSSR,* Ser. Biol. 5 (1940).
16. W. Dietrich, *Strahlenther.* 80:17 (1949).
17. A. H. Sparrow and E. Christensen, *Am. J. Bot.* 37:667 (1950).
18. R. Bossi, *Science* 21:379 (1951).
19. M. Yamada, *J. Phys. Therapy* (1917), 6.
20. S. Makamura, *J. Agr. Univ. Tokyo* 8:2 (1918).
21. Komuro, *Bot. Mag. Tokyo* (1922).
22. Granhall, Ehrenberg, and Borenius, *Botan. Notizer* 2:155 (1953).
23. E. Johnson, *Plant Physiol.* 23,4:544 (1948).
24. Schwartz, Czepa, and Schindler, *Fortsch. geb. Röntgenst.* 31,6:665 (1923).
25. A. V. Koltsovy and L. P. Zapiski, *Len. s. khoz. in-ta* 2 (1925).
26. R. Patten and S. Wigoder, *Nature* 123:606 (1939).
27. A. A. Drobkov, *DAN* 17, No. 4 (1937).
28. A. A. Drobkov, *DAN* 32, No. 2 (1940).
29. A. A. Drobkov, *DAN* 69, No. 3 (1947).
30. R. Gillerne, *J. Landu.* 89,3:233 (1943).
31. R. S. Russel, *Nature* 164:993 (1949).
32. G. Bould, *Nature* 167:140 (1951).
33. P. A. Vlasyuk, *Trudy Inst. fiziologii rast. i agrokhimii AN Ukrainskoi SSR,* No. 5 (1942).
34. L. P. Alexander, *Agronomy J.* 42:252 (1950).
35. C. Dion, *Nature* 163:906 (1949).
36. A. Lepape, *Annales Agronom.* 4:319 (1943).
37. *Chem. Engng. News* 30,8:750 (1950).

Chapter 9

ROOT GRAFTING IN THE TRANSLOCATION OF NUTRIENTS AND PATHOGENIC MICROORGANISMS AMONG FOREST TREES *

Forest trees constitute an important and replaceable natural resource. Intensified efforts are being made not only to improve existing stands but also to provide reforestation. Thus, factors influencing tree developments have special interest. An understanding of the movement of materials within the sap streams of trees helps to explain their responses to nutrients, silvicides, and pathogenic microorganisms. The function and frequency of natural root grafting among forest trees influence the translocation of water, nutrients, and disease-inducing organisms from one tree to another. Experiments with radioactive isotopes have clarified these relationships.

The purpose of the present study was to determine the rate of movement and the distribution of chemicals in the sap streams of certain forest trees, the environmental conditions which directly affected such movement, and both the incidence and the role of root grafting among forest trees. This information has helped to interpret the movement of materials within and between oak trees infected by the oak wilt fungus, *Endoconidiophora fagacearum* Bretz.

Materials and Methods

Naturally occurring northern pin oaks (*Quercus ellipsoidalis* Hill) and bur oaks (*Q. macrocarpa* Michx.) were used in most studies. Many were scrub oaks of stump sprout origin, growing in mixed oak and pine stands on relatively infertile Plainfield sand in central Wisconsin, U.S.A. Other forest tree species both in natural stands and in plantations also were studied.

The radioactive isotopes employed were sodium iodide-131 and rubidium-86 carbonate. Both had satisfactory gamma radiation, were readily soluble in

* This chapter is taken from Geneva Conference Paper 105, "The Use of Radioactive Isotopes to Ascertain the Role of Root Grafting in the Translocation of Water, Nutrients, and Disease-Inducing Organisms Among Forest Trees" by J. E. Kuntz and A. J. Riker of the United States.

water, and could be detected easily at low concentrations, and had half-lives of 8 and 19.5 days, respectively. Thus, no disposal problem occurred. Standard precautions were taken in handling these materials. Radioactivity was detected with a portable, Geiger-type monitoring instrument with a usable sensitivity of approximately 0.05 milliroentgens per hr. The working background was 0.05 to 0.07 mr per hr. An activity twice this background was set as a minimum positive reading.

Treatments of trees were made in various ways. In branch treatments, a cut branch was inserted immediately into a bottle containing the isotope. Roots were treated after excavation in a similar manner. Trunks were treated by fastening waterproof paper cones with asphalt cement to the main bole. A dilute solution of iodine-131 was poured into the cone and a half-inch chisel cut was made under the liquid through the last three annual rings. However, with rubidium-86, a 1.5% KCl solution first was poured into the cone to reduce tissue adsorption or absorption of the rubidium. Uptake was rapid in the summer on sunny days. About one-half a millicurie of rubidium-86 in water then was added. Readings were taken immediately and at suitable intervals thereafter.

The movement of isotopes was checked by using 0.1% aqueous solutions, respectively, of eosin, safranin O, and fast green as well as a concentrated sodium arsenite solution containing an excess of sodium hydroxide.

Experimental Results

The normal rate of upward movement of the isotopes in the sap streams of oaks in full sunlight and with low relative humidity was between 1.5 and 3 ft per min. Downward movement also was rapid. However, the extent of downward movement into the roots was limited except in those roots which were grafted to other roots from adjacent trees.

In northern pin oaks, the isotopes were distributed throughout the crown. Radioactivity was detected within 20 min in most branches, twigs, and leaves of oaks 35 ft high. In bur oaks and white oaks the rate was comparable. However, radioactivity appeared only in narrow, vertical streaks which originated at the chisel cuts and in branches which had vascular connections with such trunk streaks. Thus, certain branches were radioactive; others were not.

The movement of isotopes in the sap stream varied considerably throughout a 24-hr period. Rapid movement occurred during the day from the time the first sun shone on dry leaves until the sun dropped below the crown canopy. At this time, light intensity diminished rapidly, temperature decreased, and relative humidity increased. A greatly reduced rate (0.05 to 0.1 ft per min) occurred throughout the night. Comparable results were obtained when trees were covered with light-proof canvas. During daylight hours, dense clouds over the sky reduced movement to 0.4 ft per min. Movement was only 0.2 ft per min when the leaves were wet from dew, rain, or artificial sprays with water. No movement occurred at 0° C or below.

Seasonal variations in the rate of upward movement were examined through three calendar years. The average rate of movement remained uniformly high during the growing season until fall coloration appeared. During the dormant period and until the buds swelled in the spring, movement continued at a much

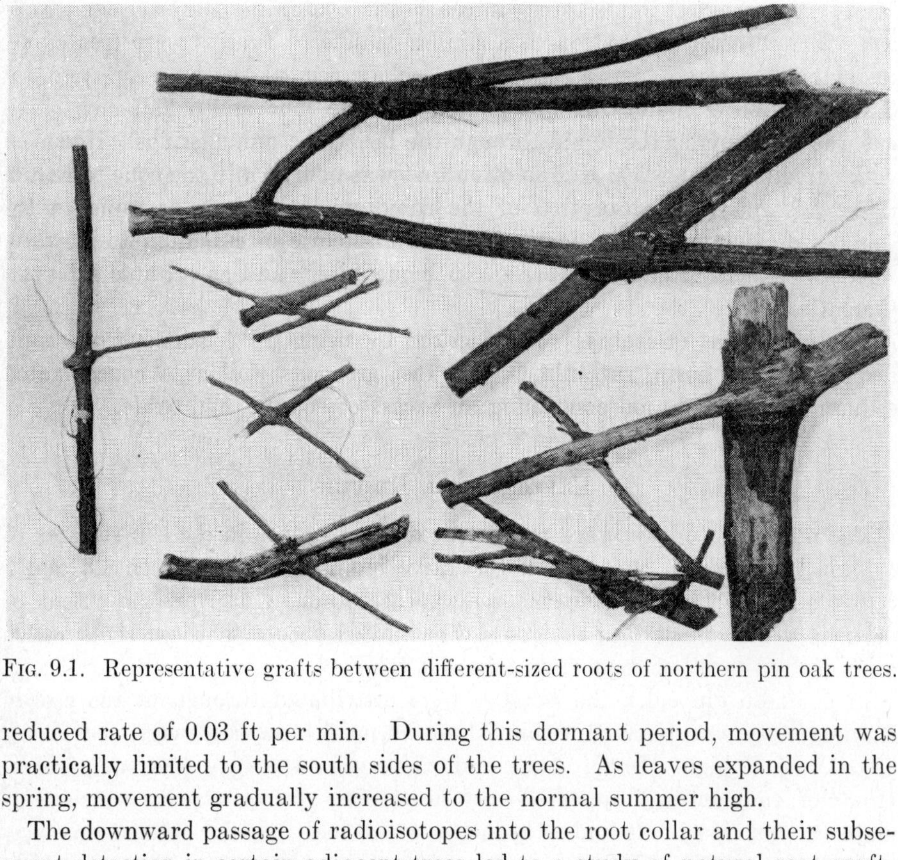

FIG. 9.1. Representative grafts between different-sized roots of northern pin oak trees.

reduced rate of 0.03 ft per min. During this dormant period, movement was practically limited to the south sides of the trees. As leaves expanded in the spring, movement gradually increased to the normal summer high.

The downward passage of radioisotopes into the root collar and their subsequent detection in certain adjacent trees led to a study of natural root grafting among forest trees (Fig. 9.1). Radioactive potassium bromide-82, because of its extremely short half-life and its strong gamma radiations, proved to be especially effective in tracing movement through roots and root grafts. To insure maximum downward movement, trees were cut above the cone just prior to treatment. Because of the expense of isotopes and because of their rapid dilution in passage through trees, comparable studies also were made with dyes and poisons (Fig. 9.2).

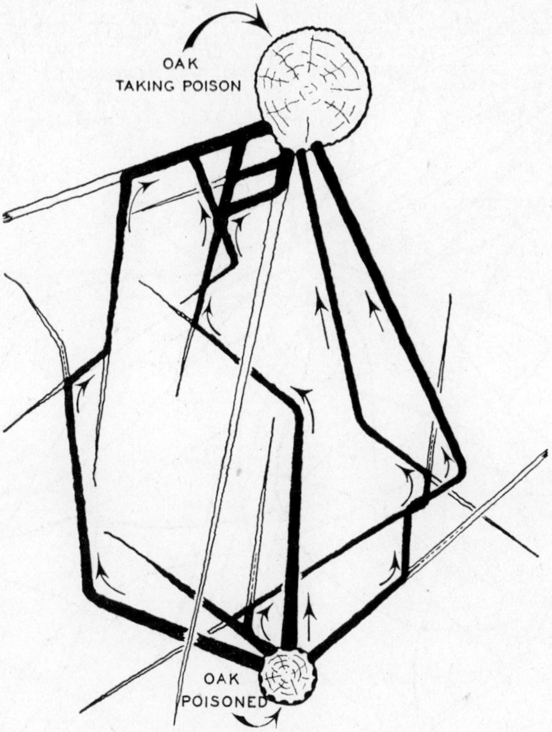

Fig. 9.2. Diagram of the root systems of 2 northern pin oaks, showing the movement of sodium arsenite from one treated tree to a nearby tree. Grafted roots through which the poison passed are solid black.

In a representative experiment where trees were about 10 ft apart, treatment of 1 tree disclosed that 5 nearby trees were grafted to it. When these were treated, 21 additional trees proved to be joined. When these were treated in turn, 10 more trees were involved. Thus, after only 3 successive treatments, 36 trees were found to be grafted directly or indirectly. Occasionally, some trees near a treated tree were skipped; however, such trees usually became involved following successive treatments. From a series of comparable experiments, each tree appeared grafted to 3 or 4 adjacent trees. In one case, at least 8 trees within 52 ft of the treated tree were connected.

Among hundreds of trees investigated, root grafting was common among northern pin oaks, but was infrequent among bur oaks or among white oaks. In only one instance was a graft discovered between different species of oaks; namely, a northern pin oak and a bur oak.

Roots of all sizes, ⅛ in. or larger, were found grafted together. Trees in dense stands were united much more frequently than were those in open stands. Washing and digging out of tree root systems confirmed these results. For

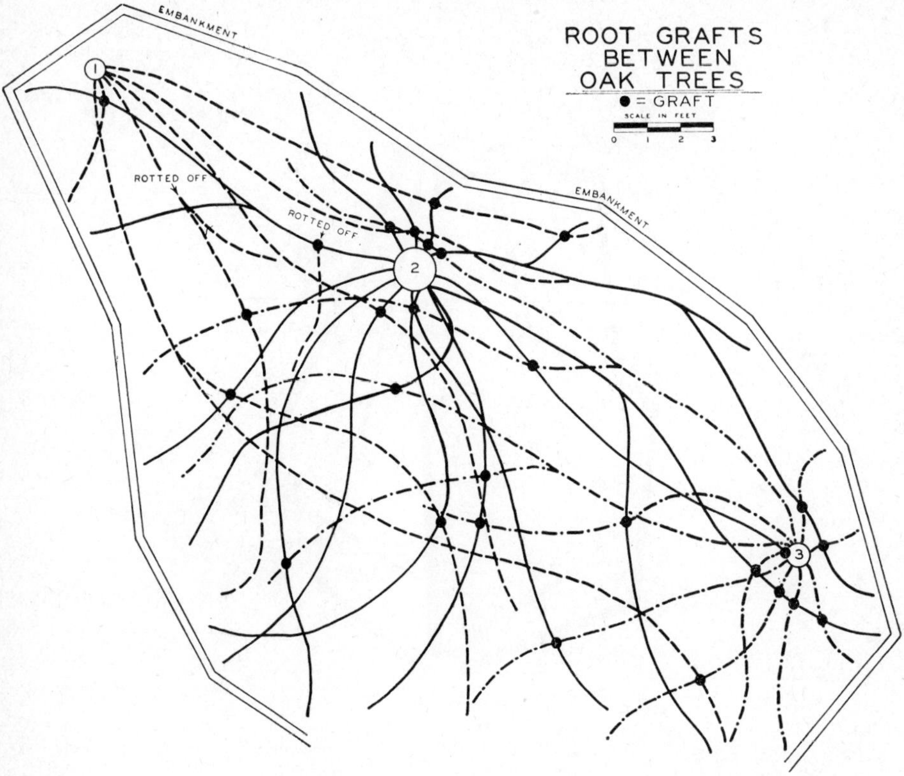

FIG. 9.3. Diagram of the partly exposed root systems of 3 northern pin oak trees. Small solid black circles indicate 28 graft unions between roots of different trees. The numerous grafts between roots of the same tree are not shown.

example, when the root systems of 3 nearby northern pin oaks were partly excavated, 28 root grafts among the 3 trees were revealed (Fig. 9.3). Grafts between roots of the same tree were common. By means of root grafts, trees often "adopted" parts of the root systems of other trees which had been destroyed (Fig. 9.4).

The direction of sap flow between dominant and suppressed trees was determined in limited studies. While isotopes moved in both directions, they more often passed from the dominant to the suppressed tree. This phenomenon perhaps assists the survival of suppressed trees.

Root grafting within other given species was observed as follows: large-toothed aspen and red pine trees were grafted to a considerable extent; red oak, sugar maple, and white pine trees were grafted frequently; quaking aspen trees were grafted occasionally; white birch and jack pine trees were grafted seldom. No grafts were found among white spruce, black spruce, or balsam fir trees. Root grafting was common among 4-yr-old red pine seedlings in nurs-

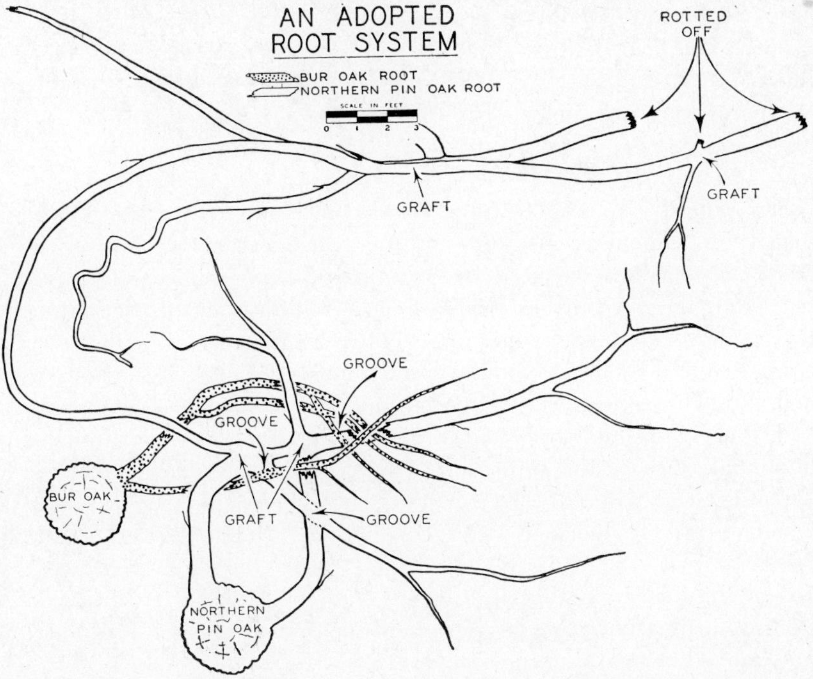

FIG. 9.4. Diagram of several roots from a tree previously destroyed which have been "adopted" through root grafts by an adjacent healthy tree. Bur oak roots (stippled), though tightly intertwined and deeply grooved by mutual pressure with roots of the same northern pin oak, did not graft.

ery transplant beds. These results confirm the occasional reports in the literature concerning root grafting between trees.

In a similar manner, radioactive isotopes have clarified the manner in which the oak wilt disease develops and spreads from tree to tree in local areas. Oak wilt is a vascular-wilt disease which results when the fungus, *Endoconidiophora fagacearum* Bretz, invades the water-conducting vessels of oak trees. Sudden wilt, premature defoliation, and death of infected northern pin oaks usually follow. Bur oaks and white oaks more often show only wilting of the leaves on scattered branches and usually live for several seasons after initial symptoms. Oak wilt is the most serious disease threatening the vast hardwood timber forests of eastern United States (Ref. 3). Hence, oak wilt studies involving radioisotopes are of special significance.

The movement not only of solutions but also of particulate material was traced with isotopes. The rate and distance radioactive spores of the oak wilt fungus were carried in sap streams were measured. A concentrated spore suspension first was treated for 1 day with 1 millicurie of $Ag^{110}NO_3$. Next, the spores were washed repeatedly until the supernatant liquid had little radioactivity. The spores were resuspended and treated with 1 millicurie of NaI^{131}. The spores, now containing insoluble $Ag^{110}I^{131}$, were doubly radioactive. They again were washed repeatedly. Although their weight was increased, the radioactive spores were soon detected in the terminal branches of an oak 5 ft high. Fungus spores thus passed readily through the water-conducting vessels of oaks.

Studies of host responses to fungus invasion demonstrated that the upward movement of radioactive isotopes in the transpiration streams of infected northern pin oaks was reduced by 85% 3 to 4 days before foliage wilt appeared (Ref. 1). A further decrease occurred during the development of moderate and severe wilt symptoms. Movement in trees with severe wilt often was reduced by 99%. Similar reductions were found in the flow of water through branch sections cut from wilting trees. Microscopic examinations showed that tyloses and gums had formed in the xylem vessels throughout the outer annual rings of the sapwood. They were consistently associated with and preceded leaf wilt (Ref. 6). These results indicated that vascular plugging had effectively blocked water conduction. Permanent wilt and death followed.

Critical differences in the reaction of other oak species to infection were detected immediately by means of radioisotopes. In infected bur oaks, radioisotopes moved normally until three days prior to incipient foliage wilt. After wilt development, movement continued only in certain unobstructed portions of the outer annual ring. Radioactivity was detected only in the symptomless branches. Microscopic examinations of the trunk and branches of infected bur oaks showed that vascular plugging had developed only in narrow, vertical sectors of the trunk and in branches having vascular connections with these sectors. Thus, as with infected pin oaks, radioisotopes demonstrated that fungus distribution, vascular plugging, and symptom development were closely associated (Ref. 4).

The continued downward movement of isotopes into and through grafted roots of diseased oaks indicated that vascular plugging was limited in the roots. Through roots, isotopes moved even after aerial portions of the tree had died.

This basic information expedited effective control measures (Refs. 2, 5). In lawn and park areas where individual trees were of great value, severing all root connections to a depth of 36 in. or more between wilting and adjacent healthy trees usually prevented further local spread. Since the oak wilt fungus frequently moved from infected trees soon after wilt development into nearby trees, severing root connections between two or three rows of trees appeared advisable. In woodland areas where individual trees were of relatively low value, poisoning healthy oaks within 25 ft of the wilted trees usually confined the infection. Root kill was essential. The width of the poisoned barrier depended on species, age, stand density, and accessibility.

Discussion

Radioactive isotopes proved to be effective tools for studying the movement of materials within and between forest trees. Even at low concentrations, their presence could be detected immediately and for some time. As used, they caused no visible disturbance to the treated trees. They facilitated root-graft studies by indicating which roots should be excavated. Their chief disadvantages were their expense and their rapid dilution in moving through the trees.

In relatively dense stands of northern pin oaks in central Wisconsin, practically all trees appeared to be united to their neighbors through root grafts. Thus, trees of a forest stand might be considered as a united "community" rather than as independent individuals. Normally, such unions could be beneficial. However, when a disease such as oak wilt appeared, the fungus had direct vascular pipelines from tree to tree. Even-aged stands of the same species are especially vulnerable to such attack.

Determinations of the extent and functions of natural root grafts clarified, in part, the survival and persistence of suppressed trees, the increased growth of the trees remaining after a partial thinning, the longevity of certain stumps, and the vigor of many stump sprouts.

In oak wilt investigations, radioactive isotopes opened the way for studying certain host reactions to fungus invasion, the basic reasons for symptom development, and the manner in which the causal fungus spread from an infected tree to neighboring trees. Doubtless many of these findings may be helpful with other serious vascular wilts of forest trees.

Summary

Radioactive isotopes were used to trace the movement of materials in the sap streams within and between both healthy and diseased trees. Iodine-131 and rubidium-86 proved satisfactory. Upward movement in oaks under conditions favoring transpiration averaged 1.5 to 3 ft per min. Very low light intensities, free moisture on the leaves, very cold temperatures, and the absence of functional leaves all greatly reduced movement. Little movement occurred in the dormant period. Downward movement into the roots was limited except in those roots which were grafted to roots of adjacent trees.

When dominant and suppressed trees were connected by root grafts, isotopes moved both ways, but most commonly from the dominant to the suppressed trees.

The diffuse movement of isotopes throughout the trunks and the crown of northern pin oaks was readily traced. Such movement was strikingly different from the limited linear flow in the trunk and certain branches of bur oak trees.

In studies of the oak wilt disease, radioactive fungus spores were carried rapidly for some distance in the sap stream. The failure of iodine-131 and rubidium-86 to move in trees with oak wilt led to the discovery that the xylem vessels in above-ground parts were plugged with tyloses and gums. Critical differences between oak species were found. Few root vessels were obstructed, however. Thus, the isotopes helped to demonstrate a close association of fungus distribution, vascular plugging, symptom development, and the spread of infection from tree to tree through root grafts. This basic information advanced the development of practical control measures for oak wilt.

References for Chapter 9

1. C. H. Beckman, J. E. Kuntz, A. J. Riker, and J. G. Berbee, "Host responses associated with the development of oak wilt," *Phytopathology* 43:448–454 (1953).
2. J. E. Kuntz, "The control of local spread of oak wilt and certain other tree diseases with herbicides," Northcent. Weed Control Conf. Proc., 11th Ann. Meeting, 1954.
3. J. E. Kuntz, "Recent progress in oak wilt research," *Soc. Amer. Foresters Proc.*, 1954, pp. 176–179 (1955).
4. J. R. Parmeter, Jr., J. E. Kuntz, and A. J. Riker, "Oak wilt development in bur oaks," *Phytopathology* 46:423–436 (1956).
5. A. J. Riker and J. E. Kuntz, "Oak wilt and its control in the United States" (Abstract) *Cong. Internatl. de Bot. 1954*, Sect. 20, p. 165 (1954).
6. B. Esther Struckmeyer, C. H. Beckman, J. E. Kuntz, and A. J. Riker, "Plugging of vessels by tyloses and gums in wilting oaks," *Phytopathology* 44:148–153 (1954).

Chapter 10

THE EFFECTIVENESS OF FOLIAR ABSORPTION OF PLANT NUTRIENTS *

Future progress in world agriculture depends to an increasing degree upon exactness and efficiency in cultural operations, accurate evaluation of new procedures, and reduction of chance hazards imposed by environment. To this end, familiar and long-established concepts in plant nutrition and fertilizer practices need to be re-examined in the light of new knowledge and new tools.

Roots are commonly accepted as the principal organs through which plants absorb nutrients, which are then transported to other portions of the plant—stems, leaves, flowers, and fruits. That the reverse of this process could and does occur, namely, the uptake of nutrients by stems, leaves, flowers, and fruits—has long been indicated by certain horticultural practices. Yet only recently has this truth been demonstrated experimentally.

Several factors have recently prompted interest in the possibilities of applying mineral nutrients to the above-ground parts of plants. Thus, current progress in the formulation of chemical fertilizers has been characterized by greater water solubilities and higher analyses of the major mineral elements utilized by plants. Coincident with the commercial availability of these materials has been the wide, general use of sprays and dusts as pesticides and herbicides, accompanied by remarkable engineering advances in spraying and dusting equipment. Finally, radioisotopes of the elements commonly applied as fertilizers have made it possible to follow and evaluate the absorption of nutrients by above-ground parts of plants in comparison to absorption by roots. A remarkable efficiency of uptake by leaves and many important facts relative to mineral nutrition and fertilizer use have been revealed.

ENTRY OF NUTRIENTS INTO DORMANT BRANCHES

Radioactive mineral nutrients applied to leaf, stem, and fruit surfaces are readily absorbed as measured by subsequent assay of non-treated parts. In

* This chapter is taken from Geneva Conference Paper 106, "Utilization of Radioactive Isotopes in Resolving the Effectiveness of Foliar Absorption of Plant Nutrients" by H. B. Tukey, S. H. Wittwer, F. G. Teubner, and W. G. Long of the United States.

fact, nutrients have been shown to enter even through the bark of dormant fruit trees. Cotton gauze was dipped into solutions of K^{42} potassium carbonate and P^{32} orthophosphoric acid, and the impregnated gauze wrapped around branches of apple trees (*Malus domestica*) and peach trees (*Prunus persica*) (Ref. 9). Within 24 hr of application, during February and March even with temperatures below freezing, radioactivity was detected within the branches 18 to 24 in. both above and below the points of application. Similar applications, made just as the buds were commencing to swell, showed that the material moved in through the bark and up through the branches and concentrated near the buds, available for the flush of new growth.

A proved method is now available for treatment of wounded and injured fruit trees through application of nutrients to the trunk and branches during the dormant and early spring seasons, and the practices of correcting zinc deficiency and applying nutrients by dormant applications is rationalized.

Further, the surface area of the trunk, branches, and twigs of fruit trees is substantial. For a 25-yr-old McIntosh apple tree, the figure is 800 sq ft, equivalent to a wall 8 ft high and 100 ft long. It has been shown that as much as 3 lb of high-analysis fertilizer may adhere to such a surface area when applied as slurries or pastes.

Uptake and Movement of Mineral Nutrients Applied to Foliage

Although nutrients may enter woody plants through the bark, even during the dormant season, it is of more importance that they may enter through the leaves. It is now established that the surface of a leaf is not the impervious structure described in many textbooks. Instead, it is structurally well equipped to absorb materials through both the upper and lower surfaces (Refs. 6, 7). In addition, the leaf area of a plant is considerable. Thus, it has been determined that a 12-yr-old apple tree provides in its foliage, both upper and lower surfaces, an area of $\frac{1}{10}$ of an acre or approximately ten times the spread of the tree.

By isotope techniques, the highly mobile elements nitrogen (Ref. 10), phosphorus (Ref. 8), potassium, and rubidium applied to leaves are shown to be freely transported, both acropetally and basipetally, at a rate comparable to that which follows root absorption. Calcium (Ref. 1), strontium, and barium do not move from the absorbing plant part, and basipetal transport is negligible. The patterns of phosphorus, potassium, and calcium absorption and transport in the bean plant (*Phaseolus vulgaris*), as illustrated by autoradiograms, are duplicated without exception in many other crop species and under many unusual conditions. Further, acropetal and basipetal movement of phosphorus is not restricted by reciprocal grafting of roots and tops of tomato (*Lycopersicum esculentum*) and tobacco (*Nicotiana tabacum*). In contrast, no basipetal movement of calcium has been detected in either intact or grafted plant segments.

The absence of basipetal transport of radiocalcium is exemplified by the strawberry (*Fragaria* spp.). Following application to the roots of the mother plant, active upward transport occurs. Lateral movement then follows into the stolons and aerial parts of daughter plants. The almost complete absence of basipetal transport into the roots of daughter plants is striking (Fig. 10.1).

FIG. 10.1. Autoradiograms showing the distribution of Ca^{45} in a series of strawberry runner plants following initial root absorption by center plant. Top, autoradiogram of treated plants. Note lack of appreciable movement of Ca^{45} into roots of runner plants. Bottom, photograph of plants prior to treatment (leaves which appear dark are turned with upper surface towards camera).

Obviously as revealed by the radioisotope technique, the roots of the runner plants must depend upon their own absorption for a source of calcium. The apparent absence of basipetal transport of calcium in plants does not, however, exclude the usefulness of calcium as a foliage spray for the correction of certain nutritional disorders (Ref. 3). Radiocalcium is readily absorbed by leaves and fruit from external sprays, but there is no transport or redistribution from these organs to other plant parts.

The transport and distribution of radiophosphorus is similar within the plant following either foliar or root absorption. Accumulation occurs in rapidly growing meristematic regions (Ref. 11) such as root tips, vegetative growing points, flowers, fruits and seeds, and even embryos within seeds. The autoradiograms further suggest that foliar-absorbed phosphate is metabolized and functionally serves the same needs as that absorbed by roots.

By using isotope techniques it has been demonstrated conclusively that with many herbaceous crops, foliar-applied nutrients can make a significant contribution to the total nutrient needs of the plant. Numerous carefully controlled experiments have also shown that the percentage of phosphorus utilized from applied fertilizer is highest in foliage sprays—up to 95% of the fertilizers applied, in some instances. In fact, the application of phosphorus to the leaves of plants represents the most efficient method of fertilizer placement yet devised.

For example, beans and tomatoes were grown in pot culture in the greenhouse on three soil types and at variable levels of phosphate in the soil. Measured quantities of P^{32}-labeled phosphoric acid (0.3% solutions) were applied to the soil in a band in the root area and as a foliage treatment during early flowering. Accumulation of the applied phosphorus in the developing fruit during a 3-week period from the two methods of placement was ascertained. In Table 10.1 are found the essential data for the two crops. The foliage

TABLE 10.1. COMPARATIVE EFFECTIVENESS OF FOLIAGE AND ROOT APPLICATIONS AS SOURCES OF PHOSPHORUS FOR DEVELOPING FRUIT OF THE BEAN AND TOMATO AS INDICATED BY APPLICATIONS OF P^{32} ORTHOPHOSPHORIC ACID

Crop	Part Treated	Phosphorus Accumulated in Developing Fruit		
		Micrograms	Per Cent of Total	Per Cent of Applied
Bean	Foliage	91	1.10	6.98
	Roots	36	0.38	0.27
Tomato	Foliage	33	0.40	1.23
	Roots	24	0.27	0.18

applications were far more effective for the micrograms of applied phosphorus accumulated in the developing fruit, for the percentage of the total phosphorus in the fruit derived from that applied, and for the percentage of the applied phosphorus found in the developing fruit. With the bean, the foliage was more than 25 times as efficient as the roots in phosphate uptake based on the percentage applied which accumulated in the fruit.

Typically, many crops, especially fruits and vegetable crops, during normal development pass through critical periods associated with flowering and fruiting when nutrient demands are high and when availability from soil sources is low or at a minimum. It has been demonstrated by the isotope technique that during these critical periods leaves can supplement the function of the roots as nutrient-absorbing organs. Up to 25% of the phosphate needed in fruit growth can be supplied by properly timed foliage spray applications.

Factors Which Affect Foliar Absorption

It may be anticipated that foliar absorption of nutrients will be affected by external and internal factors. Temperature, light, the pH and the carrier of the treating solutions, and various additive chemicals may be important, as well as the species of plant involved, the morphological nature of the absorbing organ, and the nutritional status of the plant (Ref. 5).

Of the many phosphates (organic and inorganic) which have been tested, orthophosphoric acid at a concentration of 0.30% (2½ lb per 100 gal of water) was found most desirable for stimulating growth and early fruiting. Three cations (carriers) of phosphate have been compared with orthophosphoric acid with respect to the rapidity of uptake through the leaves of the bean plant. The three carriers included ammonium, potassium, and sodium, the first two being commonly found in the high-analysis fertilizers recommended for nutritional sprays, and sodium often accompany natural phosphate deposits. The hydrogens of orthophosphoric acid were replaced by adjusting the pH of the treating solutions with the hydroxides of ammonium, sodium, and potassium.

Several important facts concerning phosphate absorption by leaves are evident from the data presented in Table 10.2. All phosphate salts were

TABLE 10.2. EFFECT OF pH AND ACCOMPANYING CATION ON FOLIAR ABSORPTION OF PHOSPHORUS BY BEAN PLANTS AS INDICATED BY MICROGRAMS DERIVED FROM TREATING SOLUTION CONTAINING P^{32} (6 HR AFTER APPLICATION)

Cation	pH					
	2	3	4	5	6	7
K	1.47	.96	.16	.11	.41	.80
Na	2.03	2.97	.31	1.59	1.21	.25
NH_4	3.70	3.94	1.59	2.44	.33	.26

absorbed less effectively than orthophosphoric acid to which only a slight amount of ammonium, sodium, or potassium was added (pH 2–3). The slow-est penetration through the leaf was with monopotassium phosphate (pH 4–5). Of the three di-substituted phosphates at pH 7, the potassium salt was absorbed most rapidly. This has been confirmed in later experiments at a pH of 8; however, at pH 7 and 8 the rate of absorption was only half that of orthophosphoric acid (pH 2). The lack of stability of di-ammonium phosphate may account for its low rate of uptake at pH 7 and 8. A mixture of the mono-substituted phosphates and orthophosphoric acid (pH 2–3) resulted in 2- to 10-fold increases in uptake through bean leaves as compared with other inorganic phosphates.

The effects of pH and ionic carrier on absorption of radiopotassium and radiorubidium by bean and tomato leaves were in marked contrast to those for radiophosphorus. Phosphate, citrate, and chloride were the accompanying anions, with pH levels of 2, 4, and 8. The absorption of rubidium was accelerated 10 to 20 times at a pH of 8, compared with a pH of 4 when phosphate was the carrier. In general for rubidium, chlorides or citrates were less effective than phosphates. However, at similar pH levels and with the same carriers, potassium uptake when applied as the citrate at pH 8 was twice that of any other form.

Leaf absorption of radiophosphorus and other nutrients is facilitated and possible leaf burning is avoided at recommended dosages if a wetting agent is added to the spray formulation. This results in a more uniform film of liquid on the leaves. Certain detergents, however, will reduce the speed with which phosphorus is absorbed. Young and rapidly expanding leaves are more efficient in absorption than are leaves fully matured, although all green leaves seem to absorb some radiophosphorus.

Both sides of the leaf blade, as well as the petiole, will absorb nutrients. When the primary leaf blade of the bean plant is treated with a radioactive phosphoric acid solution, and subsequent accumulation of radioactivity in the roots, is ascertained, absorption is slightly greater through the upper leaf surface. This is most evident when applications are made at the base of the leaf blade, as indicated in Table 10.3. Since the frequency of stomata is seven times as great in the lower epidermis of the bean leaf (Ref. 2), it would appear that phosphate entry is not facilitated by these structures. Placement of the treating solution at the tip, along the outer margin, and in the center of the leaf along the midrib gave only a slight increase in absorption when compared to similar positions on the lower surface. On the other hand, application to the petiole of the leaf resulted in very rapid uptake. It was found, however, that the region of placement on the leaf surface was of no importance in determining the level of phosphorus which had accumulated in the roots by 12 hr after treatment.

Furthermore, the opening and closing of stomata is of little significance, since foliar absorption is not confined to daylight hours, but is equally apparent

TABLE 10.3. RELATIVE ABSORPTION OF PHOSPHORUS BY DIFFERENT REGIONS OF BEAN LEAVES AS SHOWN BY MICROGRAMS IN ROOTS DERIVED FROM SOLUTIONS CONTAINING P^{32} (2 HR AFTER APPLICATION)

Position on Leaf	Surface of Leaf	
	Upper	Lower
Tip	.104	.062
Margin	.130	.102
Midrib	.148	.104
Base	.206	.090
Petiole	.516	

during the night. Slight diurnal fluctuations are revealed, however, where continuous recordings of nutrient uptake have been made. Initially, light has a slight but significant effect in increasing the foliar absorption of phosphorus, as is evident from studies on bean stem tissue presented in Table 10.4. This

TABLE 10.4. THE EFFECT OF LIGHT AND SUCROSE ON ABSORPTION OF PHOSPHORUS BY BEAN LEAVES AS INDICATED BY MICROGRAMS IN THE STEMS DERIVED FROM SOLUTIONS CONTAINING P^{32}

Solution	Environment	Hours After Treatment		
		3	6	12
.2% H_3PO_4	Light	.684	1.046	1.172
.2% H_3PO_4	Dark	.382	.670	1.390
.2% H_3PO_4 plus 5% sucrose	Light	.346	.614	1.142
.2% H_3PO_4 plus 5% sucrose	Dark	.334	.566	1.070

initial stimulation disappears under prolonged periods of light or dark; and the actual accumulation of foliar-applied phosphorus by roots and buds is favored by prolonged darkness. The depression of absorption (or possibly the movement from epidermal cells into the vascular system of the leaf) by dark may be duplicated by the addition of sucrose to the treating solution. However, this does not greatly impair the ability of plants in the dark to accumulate foliar-applied phosphorus at some distal point. This occurs regardless of the inclusion of sucrose in the treating solution.

An effect similar to that obtained with sucrose occurs when boron is added to the phosphate solution (Table 10.5). Not only did borate (0.005%) depress

TABLE 10.5. EFFECT OF BORON AND TEMPERATURE ON ABSORPTION AND TRANSLOCATION OF PHOSPHORUS APPLIED TO BEAN LEAVES AS INDICATED BY MICROGRAMS IN THE ROOTS DERIVED FROM SOLUTIONS CONTAINING P^{32}

Treating Solution	Hours After Treatment	Temperature		
		14° C	21° C	25° C
Phosphate [a]	3	.015	.307	.243
Phosphate plus boron [b]	3	.013	.202	.183
Phosphate	6	.433	1.040	.560
Phosphate plus boron	6	.230	.975	.625
Phosphate	12	1.230	1.675	.738
Phosphate plus boron	12	1.225	1.162	.868

[a] 0.2% H_3PO_4.
[b] 0.005% boron.

the initial absorption of phosphate, but this initial depression appeared to be independent of temperature. Maximum uptake of radiophosphorus occurred at 21° C with and without boron, and was less at both 14° C and 25° C. The similarity in levels of radiophosphorus in the roots after 12 hr at the two lower temperatures (14 and 21° C) with and without boron, suggests that absorption rather than translocation or accumulation is the limiting factor.

The many factors which have been shown to influence the absorption of foliar-applied nutrients provide evidence for the existence of more than a mere passive entry through leaf surfaces into the plant. The effects of pH and cation carrier when considered in the light of the ionic species of phosphate and the degree of molecular dissociation which occur, suggest an exchange mechanism in the entry of foliar-applied phosphorus. This is confirmed by the apparently competitive action of the borate ion. That the exchange is active rather than passive would seem to follow also from the temperature relationships already discussed. Ionic penetration of phosphate would preclude diffusion through the cuticle, and the now demonstrated widespread occurrence of epidermal plasmodesmata by Lambertz (Ref. 6) provides most probable sites for protoplasmic exchange at the surface of the leaf.

Foliar Applications of C^{14} Urea

Urea is one of the most useful and well-known nitrogenous fertilizers for leaf application. With some horicultural crops the entire requirement for nitrogen can be satisfied by a few appropriately timed sprays. Radioisotopes provide a specific and sensitive tool to ascertain the rate and extent of utilization and to predict crop tolerance of foliar applications of urea nitrogen. The first step in the utilization of the nitrogen in urea by the leaves of plants presumably is hydrolysis by the enzyme urease, splitting the urea molecule and giving ammonia and carbon dioxide. As a measurement of urease activity, and thereby the rate of hydrolysis and possible utilization of urea applied to the leaves of horticultural plants, radioactive C^{14} urea may be employed and the rate of evolution of radioactive carbon dioxide determined (Ref. 11).

The apparent relative rates of urea hydrolysis (utilization) of foliar-applied C^{14} urea by a number of vegetable and fruit crops have been determined and plotted against field and greenhouse tests of the tolerance of the foliage of the same crops to urea sprayed in pounds per 100 gal of water (Ref. 11). From such studies it is possible to predict the tolerance of the leaves of various plant species to concentrations of urea applied as nutrient sprays. It is of interest that the leaves of plants which hydrolyze (utilize?) urea most rapidly—cucumber (*Cucumis sativus*) and bean—are those which are the most responsive to treatment and which show the lowest spray concentration tolerance. Conversely, crops showing the greatest tolerances—cherry (*Prunus cerasus*), peach (*Prunus persica*), celery (*Apium graveolens*), and potato (*Solanum tuberosum*) —are those which have the lowest rates of urea hydrolysis or utilization. Intermediate crops are apple (*Malus domestica*), strawberry (*Fragaria* spp.), grape (*Vitis labrusca*), tomato, raspberry (*Rubus* spp.), corn (*Zea mays*), and plum (*Prunus domestica*).

Leaching of Nutrients from Leaves

Since, as has been shown, nutrients may enter the leaf, it would seem plausible that under appropriate conditions they may be lost from the leaf. The leaching of nutrients from plants during times of heavy rainfall may be, indeed, as fully responsible for lack of productivity as a deficiency in sunlight. This hypothesis has been subjected to experimental evaluation by allowing plants to absorb isotopically labeled nutrients through their roots or through cut stems and then exposing them to simulated rainfall ("foliage leaching") from a mist atomizer in a propagation chamber.

Measured quantities of radiophosphorus were applied to the roots of bean, sweet potato (*Ipomaea batatas*) or poinsettia (*Euphorbia pulcherrima*). After 48-hr absorption, the foliage was leached for 48 hr, and no loss of P^{32} was observed. However, if stem cuttings of these plants were supplied P^{32} through the bases of the cut stems and then subjected to foliage leaching, 1.5% to 12.8% of the absorbed P^{32} was lost.

Results with radiopotassium and radiorubidium were different (Table 10.6).

TABLE 10.6. LOSS OF K^{42} AND Rb^{86} FROM BEAN LEAVES BY LEACHING FOR 4 HR UNDER SIMULATED RAINFALL AS INFLUENCED BY LIGHT AND NUTRIENT INTENSITY

Experimental Conditions	Percentage Absorbed Ions Lost	
	Potassium	Rubidium
Daylight: full nutrient solution	5.1	4.9
Daylight: 1/10 nutrient solution	12.2	4.9
Dark: full nutrient solution	71.0	14.4
Dark: 1/10 nutrient solution	42.5	6.6

Following a 12-hr absorption of these ions, subsequent leaching for 4 hr removed up to 71% of the K^{42} and 14% of the Rb^{86}. The plants were grown either in nutrient solutions at full strength (Ref. 4), or at 1/10 dilution. This loss was greatest from plants grown in the dark (covered with black cloth) at the full nutrient level. With those grown in daylight, the reverse was true. More potassium than rubidium was lost by leaching.

These data show that loss of nutrients by leaching from leaves does occur. The magnitude of loss with crops grown under field conditions is not known. The results suggest, however, that the leaching of nutrients from plants in humid areas or during periods of frequent heavy rainfall may be comparable to the losses of some ions from soil.

References for Chapter 10

1. R. W. Bledsoe, C. L. Comar, and H. C. Harris, "Absorption of radioactive calcium by the peanut fruit," *Science* 109:329–330 (1949).
2. S. F. Eckerson, "The number and size of stomata," *Bot. Gaz.* 46:221–224 (1908).
3. C. M. Geraldson, "The control of blackheart in celery," *Proc. Amer. Soc. Hort. Sci.* 63:353–358 (1954).
4. D. R. Hoagland and D. I. Arnon, *The Water-Culture Method for Growing Plants Without Soil,* California Agr. Expt. Sta. Circular 347, 1950.
5. K. Von Kaindl, "Untersuchung über die Aufnahme von P^{32}-markiertem primärem Kaliumphosphat durch die Blattoberfläche," *Bodenkultur* 4:324–353 (1953).
6. P. Lambertz, "Untersuchungen über das Vorkommen von Plasmodesmen in den Epidermisaussenwänden," *Planta* 44:147–190 (1954).

7. E. A. Roberts, M. D. Southwick, and D. H. Palmiter, "A microchemical examination of McIntosh apple leaves showing relationship of cell wall constituents to penetration of spray solutions," *Plant Physiol.* 23:557–559 (1948).
8. C. A. Swanson and J. B. Whitney, Jr., "Studies on the translocation of foliar-applied P^{32} and other radioisotopes in bean plants," *Amer. J. Botany* 40:816–823 (1953).
9. H. B. Tukey, R. L. Ticknor, O. N. Hinsvark, and S. H. Wittwer, "Absorption of nutrients by stems and branches of woody plants," *Science* 116:167–168 (1952).
19. Richard Volk and Clayton McAuliffe, "Factors affecting the foliar absorption of N^{15} labeled urea by tobacco," *Proc. Soil Sci. Soc. Amer.* 18:308–312 (1954).
11. S. H. Wittwer and H. B. Tukey, "Isotopic Tracers in Plant Nutrition" in *Mineral Nutrition of Fruit Crops,* Horticultural Publications, New Brunswick, N. J., 1954, Chap. 15, pp. 758–774.

Chapter 11

UPTAKE AND TRANSPORT OF MINERAL NUTRIENTS IN PLANT ROOTS *

The passage of solutes between cells and the ambient chemical pool forms one of the principal topics of physiology. The permeability of cells to many types of substances has been studied—to large and small molecules, dyes, and organic and inorganic electrolytes. Autotrophic plants do not have to obtain organic materials from the environment; hence, interest in the flux of solutes between plants and their medium centers around the inorganic electrolytes present in the solution bathing the roots.

Experimental biology advances in proportion to the degree in which phenomena are quantitatively described in terms of physicochemical reactions and processes. It is natural, therefore, that investigators of permeability in cells and tissues have focused attention on recognized physicochemical reactions of molecules and ions that might result in transfer of these substances into and out of cells—diffusion, ion exchange, Donnan equilibria, membrane potentials, and others. A measure of success has certainly attended these endeavors, for every one of the phenomena mentioned has been observed in various types of cells and tissues, ranging from unicellular microorganisms through the roots of higher plants to the kidney tubules, to mention but a few examples.

However, it is being increasingly recognized that none of these mechanisms adequately describe ion absorption and secretion in many systems, including plant roots. These mechanisms are all of such a nature as to lead to equilibria, and can fairly readily be duplicated in the laboratory by model systems (Ref. 17). But one of the chief characteristics of ion transport in biological systems is its dependence on active metabolism, and another, the fact that the living cell is part of an open system not in equilibrium with the environment (Ref. 18). Yet another feature difficult to account for on the basis of simple model systems is the high degree of selectivity characteristic of biological ion transport.

The dissatisfaction with earlier proposed mechanisms has found expression

* This chapter is taken from Geneva Conference Paper 112, "Uptake and Transport of Mineral Nutrients in Plant Roots" by E. Epstein and S. B. Hendricks of the United States.

in a subtle change of terminology, as noted by Steinbach (Reg. 18). Whereas papers dealing with this general topic used to bear titles referring to the "permeability" of some cell or tissue to the substance in question, now reference is made more often to the "uptake," "absorption," or "transport" of the substance. This does not, of course, signify a return to vitalistic or other nonmechanistic ideas. It merely reflects an increased awareness of the complexity of the phenomena under investigation.

Tracer Isotopes

Before giving an account of current findings and concepts regarding ion transport by the roots of higher plants, it may be appropriate, in the present context, to review briefly the usefulness of radioactive isotopes for research in this field. In addition to carbon, hydrogen, and oxygen, thirteen mineral elements have been identified as being essential to the growth of higher plants: potassium, calcium, magnesium, nitrogen, phosphorus, sulfur, iron, manganese, zinc, copper, molybdenum, boron, and chlorine. Of these thirteen, only nitrogen and boron have no radioactive isotopes suitable as tracers. Several other elements, though not known to be essential, are of great physiological interest, and suitable radioactive isotopes exist. In this group belong sodium and cobalt. It may be mentioned that the first biological application of radioactive tracers was in the very field discussed here. We are referring to Hevesy's use in 1923 of thorium B (Pb^{212}) in a study of the uptake of lead by plants.

The usefulness of radioactive isotopes for research in ion transport by plants is two-fold. They serve as superior analytical tools, extending the sensitivity and specificity and increasing the ease of analysis beyond what can be achieved by conventional analytical methods. And, secondly, they enable the investigator to measure ion fluxes in a given direction even when there is no net flux of that ion, or when the net flux is in the opposite direction.

Of these two, the first, or "analytical" use of radioisotopes, represents an extension and improvement of existing techniques. But the second, or "exchange" type of application, is a new departure entirely, undreamed of before the advent of tracer isotopes, and has made possible new insights into the dynamics of living things that could be achieved in no other way. We shall have occasion to demonstrate both types of application of radioactive isotopes in the study of ion transport in plant roots.

Experimental Methods

The method used in our laboratory is essentially a modification of the excised-root technique described by Hoagland and Broyer (Ref. 9). The modifications consist largely of a reduction in the scale of the experimental setups, and in the amount of root tissue used. These modifications permit quicker and more flexible experimental procedures and enable us to reduce experimental periods to very short intervals when desired—of the order of a minute. Briefly, barley seeds are germinated and grown for five days under strictly controlled conditions. The nutrient solution is incomplete, consisting of a single calcium salt, usually the sulfate; but when sulfate absorption is to be studied, calcium phosphate is used during this preliminary stage.

Just before the experiment, the roots are excised and suspended in water, and 1.00-gm portions are weighed out and transferred to aeration tubes containing 50 ml of water. (The roots are blotted before weighing, to remove water adhering to the surface.) The water is then replaced by an equal volume of experimental solution containing a radioactively labeled ion. During the absorption period, the solution is constantly aerated and the temperature maintained at 30° C. The experimental period may vary from 1 min to 4 hr. After the absorption period, the roots may be assayed directly for the radioactively labeled ion. Alternatively, in exchange studies, the roots may subsequently be exposed to water or non-radioactive solutions containing the same or other ions. In this way, information is obtained concerning diffusible, exchangeable, and non-exchangeable fractions.

Concentration Effects

In many physicochemical systems, increasing concentrations of a variable result in progressive saturation of the system, until eventually complete saturation may be achieved. Familiar examples are the adsorption of gases on charcoal and other solids and the exchange reactions between solutes and solid exchangers, such as clays or synthetic resins. The usual interpretation in such cases is that the system possesses a finite capacity for holding the substance taken up. A familiar instance is the base exchange capacity of soils and other cation exchangers (Ref. 11). When the rate of absorption of an ion by excised roots is measured as a function of the concentration of that ion in the solution, the same type of curve we have been discussing is frequently obtained. Upon plotting the rate as the ordinate and the concentration as the abscissa, the curve rises steeply and almost linearly at first; but at the higher concentrations, the curve flattens out and asymptotically approaches a limiting value. The conclusion suggests itself that in this case, also, some finite system is being progressively occupied by the ions in question. However, there is this important difference between the latter case and the others we have mentioned, such as cation exchange on clays. The clay, on being immersed in a solution, quite rapidly comes to equilibrium with the solution, and thereafter no net movement of ions between the clay and the solution occurs. We measure *amounts* taken up at equilibrium. In the case of ion absorption by roots, on the other hand, we measure *amounts per unit time,* or *rates.* Once the rate of absorption of a given ion, at a certain concentration of that ion, is established, it may be maintained for many hours.

If we nevertheless wish to maintain the idea that at increasing concentrations, the ions progressively occupy some finite system within the root, additional assumptions are necessary. It is necessary to assume, first, that the ions, having occupied the system, are subsequently released again into a compartment other than the ambient solution (for in that case, no net transport would occur). And, secondly, in order to account for the typical relation between concentration and absorption rate, it must be assumed that the postulated release of the ions from the system is the rate-limiting step in the process, so that the rate is proportional to the extent to which the system is occupied by the ions at the given concentration. The maximal rate approached at high concentrations indicates complete saturation of the system by the ions under those conditions.

The kinetics we have just outlined are precisely those that have been found to apply in many cases of enzymatic catalysis. It is only necessary to substitute "substrate" for "ions," and "enzyme" for what we have vaguely called a "system," and which henceforth we shall speak of as "carriers." The Michaelis-Menten equation applies (Ref. 14):

$$v = \frac{V(S)}{K_s + (S)} \qquad (1)$$

where v = the observed rate of absorption of an ion present at concentration (S),
V = the maximal velocity attainable at complete saturation of the carriers, and
K_s = the Michaelis constant corresponding to the substrate (ion) concentration at which half the maximal velocity is attained.

Lineweaver and Burk (Ref. 13) showed that when the reciprocal of the above equation is written, straight lines are obtained in a plot of $1/v$ against $1/(S)$.

Effect of Interfering Ions

In this treatment, interfering ions assume the role of inhibitors or alternate substrates. Interference may be competitive or otherwise, depending on whether or not the interfering ions attach themselves to the same binding sites on the carriers that are, or may be, occupied by the substrate ions. In the double reciprocal, or Lineweaver and Burk plot, the presence of a competitive ion results in an increase in the slope of the line obtained by plotting $1/v$ against $1/(S)$, with no change in the intercept. The increase in the slope is by the factor $1 + (I)/K_i$, where (I) is the concentration of the interfering ion and K_i its Michaelis constant. If the interference is not competitive, the intercept is increased.

Fig. 11.1 shows the results of an experiment on sulfate absorption by excised barley roots (Ref. 12). The concentration of potassium sulfate (labeled with S^{35}) was varied over the range 0.005 to 0.05 m.e./1. Selenate interfered competitively, i.e., selenate and sulfate are bound by the same binding sites on the carriers effecting the transport. Neither nitrate nor phosphate (as the potassium salts) compete with sulfate. In fact, in the presence of these salts, the rate of sulfate absorption was increased somewhat.

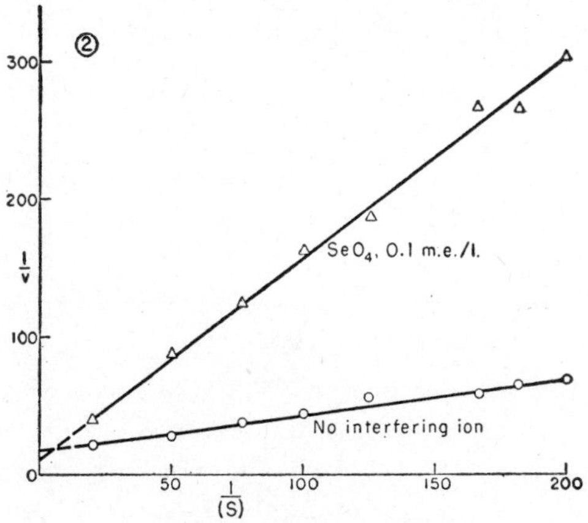

FIG. 11.1. Double-reciprocal plot of sulfate absorption by excised barley roots. Concentration of $K_2S^{35}O_4$, (S), in m.e./1. Rate of absorption, v, in m.μ.e./gram fresh weight/3 hours.

Generality of the Carrier Mechanism

To recapitulate, the evidence on concentration effects and the effects of interfering ions indicates that the absorption of ions by plant roots is mediated by carriers to which the ions are temporarily attached. The carriers are equipped with different binding sites characterized by differential affinities for different ions. The following table indicates that this general concept applies

TABLE 11.1. COMPETING AND NON-COMPETING IONS

Substrate Ion	Competing Ions	Non-competing Ions *	Reference
Rb^+ (Rb^{86})	K^+, Cs^+	Na^+, Li^+	6
Br^- (Br^{82})	Cl^-, I^-	NO_3^-	3, 4
Sr^{++} (Sr^{89})	Ca^{++}, Ba^{++}	Mg^{++}	7
SO_4^{--} (S^{35})	SeO_4^{--}	$H_2PO_4^-$, NO_3^-	12

* At high concentrations of the ions, the indicated specificities break down in some instances, and ions listed as non-competing may become competitive under these conditions.

to a wide variety of ions. The radioisotope used is indicated for each substrate ion.

In addition, Hagen and Hopkins (Ref. 8), in a study of phosphate absorption by excised barley roots, concluded that the two ionic species, $H_2PO_4^-$ and HPO_4^{--}, are bound by different sites, and that OH^- ions compete with both phosphate species.

It would appear on the basis of these findings that the carrier concept of active ion transport in plant roots is quite generally useful. However, as indicated earlier, it should be recognized that the active mechanism we have outlined is not the only one whereby ions may enter plant roots.

Passive Entry of Ions into Plant Roots

Cation exchange. Fig. 11.2 shows the time course of the uptake of Sr^{89} from a solution of $Sr^{89}Cl_2$ at 1 m.e./l. The curve may be resolved into two components. The first is rapid and approaches completion in 30 min. The second is linear with time, and no equilibrium is attained in the experiment. The ions taken up by the former mechanism are readily exchangeable with ambient non-radioactive Sr (see the dotted line). Other cations also exchange with this fraction, but the amount lost to water is only a fraction of the amount displaced by salts. The exchange proceeds even under anaerobic conditions. In contrast, the ions taken up by the second mechanism are non-exchangeable, and this uptake is virtually abolished under anaerobic conditions. This second mechanism is active transport as discussed above. The rapid process is straightforward cation exchange, in which the root acts as a solid cation exchanger, like clays or synthetic exchange resins. This mechanism does not result in a net uptake of ions.

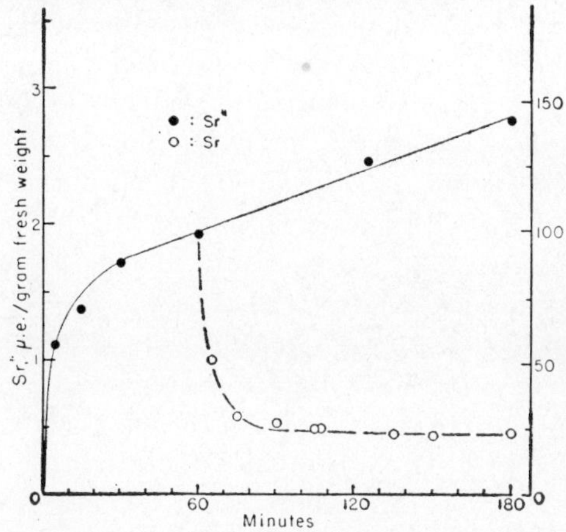

Fig. 11.2. Uptake of Sr^{89} by excised barley roots and its loss to non-radioactive Sr. Concentration of $Sr^{89}Cl_2$:1 m.e./1. Concentration of non-radioactive $SrCl_2$:1 m.e./1.

The observation that it takes approximately 30 min for the cation exchange equilibrium to be reached indicates that the negative exchange sites are not merely superficial, but lie within the tissue, so that uptake by this mechanism does indeed represent penetration of the tissue by the ions in question. Fig. 11.2 shows that there may be a loss of ions from this exchangeable fraction while simultaneously there is a net influx of the same ion into the root—a finding that would not be possible without tracer isotopes.

Diffusion. When roots are removed from a solution containing $K_2S^{35}O_4$, blotted to remove the solution adhering to the surface, and transferred to water or to a solution containing non-radioactive sulfate, they lose a considerable amount of their radiosulfate (Fig. 11.3). The amount lost to water is identical

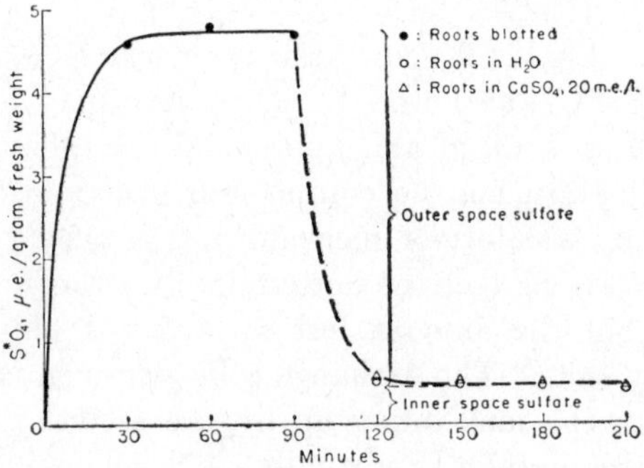

FIG. 11.3. Uptake of $S^{35}O_4$ by excised barley roots, and its loss to non-radioactive sulfate and to water.

with the amount lost to non-radioactive sulfate, indicating that there is no need for replacing or exchanging ions to effect this removal. Rather, there seems to be a fraction of the volume of the tissue (the "outer" space) to which the ions have free and reversible access by diffusion.

Equilibration with the ambient solution or water is essentially complete in 60 min, and no further loss to water or non-radioactive sulfate takes place thereafter. A fraction of the radiosulfate taken up during the initial period in the radioactive solution is retained by the roots after their transfer to water or non-radioactive sulfate. This represents the amount absorbed by the active transport mechanism, and we shall refer to it as being in the "inner" space, to contrast it with the diffusible "outer" space sulfate (Ref. 5).

As pointed out, the loss of radiosulfate to water was identical with the loss to non-radioactive sulfate (Fig. 11.3). Nevertheless, these two processes differ in a way which could not be demonstrated without the use of tracer isotopes. The exit of radiosulfate into water denotes a net loss of sulfate from the roots. The movement of radiosulfate into the solution of non-radioactive sulfate, on the other hand, represents no net movement of sulfate at all. The concentration of the initial radiosulfate solution was 20 m.e./l., and that also was the concentration of the solution of non-radioactive sulfate to which the roots were transferred at the 90-min point.

We may make the reasonable assumption that at equilibrium, the concentration of sulfate in the outer space of the root is the same as in the ambient solution. In the experiment shown in Fig. 11.3, the concentration of $K_2S^{35}O_4$ was 20 m.e./l. Sulfate in the outer space (i.e., the amount lost by diffusion) was 4.45 μe./gm fresh weight. On the basis of the above assumption, the volume of the outer space of the root tissue is $4.45/20 = 0.22$ ml/gm fresh weight. The water content of this tissue is 94%, so that over 25% of the total tissue water is accessible to the ions by diffusion. This magnitude of the space has been verified over a wide range of sulfate concentrations. It is independent of pH over the range 4.0 to 7.7.

It is of some interest to compare the outer space of barley roots, i.e., the fraction of the volume of the tissue accessible to ions by diffusion, with the corresponding values for other types of cells and tissues. Conway and Downey (Ref. 2), for a variety of solutes, found a value of approximately 0.1 for the "outer region" of yeast. For *Escherichia coli,* Roberts, Abelson, Cowie, Bolton, and Britten (Ref. 15), using $S^{35}O_4$, found a "water space" essentially identical with the total cell water. In frog nerve, Shanes (Ref. 16) measured spaces of the order of 0.62 to 0.65, using chloride tagged with Cl^{36}. In Hope's experiments with bean roots (Ref. 10) the "apparent free space" varied with the concentration of KCl used in measuring it, indicating a Donnan effect. The true space was of the order of 0.14. Butler (Ref. 1) found values for wheat roots that varied from about 0.25 to 0.34. He used chloride, phosphate (P^{32}), and mannitol.

Conclusion

Radioactive isotopes have vastly accelerated research into the mineral nutrition of plants, and have made possible an understanding of the dynamics of biological ion transport that would be inconceivable without them. Specifically, studies on ion transport in the roots of higher plants have shown that ions freely move into and out of an "outer" space of the roots, by diffusion and exchange, independent of the simultaneous active transport of the same ionic species which results in their transfer into an "inner" space where they are no longer exchangeable with the same or other ionic species. The exchange and diffusion processes are reversible, non-selective, non-metabolic, and come to equilibrium within an hour, approximately, after immersion of the roots in a new solution.

Sustained, selective absorption of ions due to metabolic activity of the cells, and rendering the ions non-exchangeable with ambient ions of the same or other species, results from the activity of carrier molecules which operate in a turnover fashion, like enzymes. The carriers possess different binding sites more or less specific in their affinity for various ions. This concept of active ion transport applies to a wide variety of both cations and anions.

References for Chapter 11

1. G. W. Butler, "Ion uptake by young wheat plants: II. The 'apparent free space' of wheat roots," *Physiol. Plant.* 6:617–35 (1953).
2. E. J. Conway and M. Downey, "An outer metabolic region of the yeast cell," *Biochem. J.* 47:347–55 (1950).
3. E. Epstein, "Ion absorption by plant roots," *Proc. Fourth Ann. Oak Ridge Summer Symp.*, U.S. Atomic Energy Commission, pp. 418–34 (1953).
4. E. Epstein, "Mechanism of ion absorption by roots," *Nature* (London) 171:83–4 (1953).
5. E. Epstein, unpublished data.
6. E. Epstein and C. E. Hagen, "A kinetic study of the absorption of alkali cations by barley roots." *Plant Physiol.* 27:457–74 (1952).
7. E. Epstein and J. E. Leggett, "The absorption of alkaline earth cations by barley roots: kinetics and mechanism," *Am. J. Bot.* 41:785–91 (1954).
8. C. E. Hagen and H. T. Hopkins, "Ionic species in orthophosphate absorption by barley roots," *Plant Physiol.* 30:193–9 (1955).
9. D. R. Hoagland and T. C. Broyer, "General nature of the process of salt accumulation by roots with description of experimental methods," *Plant Physiol.* 11:471–507 (1936).
10. A. B. Hope, "Salt uptake by root tissue cytoplasm: The relation between uptake and external concentration," *Austral. J. Biol. Sci.* 6:396–409 (1953).
11. W. P. Kelley, *Cation Exchange in Soils,* Reinhold Publishing Corp., New York, 1948.
12. J. E. Leggett and E. Epstein, unpublished data.
13. H. Lineweaver and D. Burk, "The determination of enzyme dissociation constants," *J. Am. Chem. Soc.* 56:658–66 (1934).
14. L. Michaelis and M. L. Menten, "Die Kinetik der Invertinwirkung," *Biochem. Z.* 49:333–69 (1913).
15. R. B. Roberts, P. H. Abelson, D. B. Cowie, E. T. Bolton, and R. J. Britten, *Studies of Biosynthesis in Escherichia coli,* Carnegie Institution Publication 607, Washington, D. C., 1955.
16. A. M. Shanes and M. D. Berman, "Penetration of intact frog nerve trunk by potassium, sodium, chloride and sucrose," *J. Cell. and Comp. Physiol.* 41:419–50 (1953).
17. K. Sollner, "A physiochemical cell model which simultaneously accumulates anions and cations against concentration gradients," *Arch. Biochem. Biophys.* 54:129–34 (1955).
18. H. B. Steinbach, "Permeability," *Ann. Rev. Plant Physiol.* 2:323–42 (1951).

Chapter 12

TRACING FUNGICIDAL ACTION IN PLANTS *

Plant diseases, mostly caused by fungi, result in annual crop losses in the United States alone of almost three billion (10^9) dollars. Control measures which keep such losses from being much higher are aimed almost exclusively at the prevention of the germination and subsequent growth of fungus spores. These spores are very small; from 100,000 to 200,000,000, depending upon the species, are required to weigh 1 mg. Before the ready availability of radioisotopes quantitative studies on interactions between fungus spores and toxicants could be carried out only with great difficulty. As a result almost no information was at hand on how much toxicant fungus spores took up when exposed to solutions or suspensions of fungicidal agents or how much was required on a weight basis to prevent germination. Radioisotopes have also been useful in determining possible penetration of applied fungicides into host plants. Here, too, very small quantities are involved, often of the same elements already present in plants.

In the work reported in this paper the toxicants used included C^{14}-labeled 2-heptadecyl-2-imidazoline and 2,3-dichloro-1,4-naphthoquinone, S^{35}-labeled ferric dimethyldithiocarbamate, elemental S^{35} and ions of Ag^{110}, Hg^{203}, and Ce^{144}. Rates of uptake by suspensions of spores from dilute solution (1–10 ppm) were determined for most of the toxicants and spores of a number of species of fungi as well as the quantities in micrograms per gm of spore weight required to prevent germination of 50% of the spores (ED50 values). In general the spores took up the toxicants very rapidly in large amounts and ED50 values ranged from 85 to over 10,000 micrograms per gm of spore weight. Sulfur was not accumulated by the spores, but was reduced to hydrogen sulfide which was then released (Ref. 1).

In order to obtain information on possible common receptor sites, spores were exposed to more than one toxicant, both simultaneously and consecutively, and uptake determined. Results previously published (Ref. 2) have shown that the three toxicants, 2-heptadecyl-2-imidazoline, silver, and cerium do not

* This chapter is taken from Geneva Conference Paper 100, "Use of Radioisotopes in Tracing Fungicidal Action" by L. P. Miller and S. E. A. McCallan of the United States.

seem to have any receptor sites in common. Saturation of spores with one or two of these toxicants does not lessen uptake of the others. Extension of these studies to include many other ions has shown some unexpected competition between ions not closely related chemically and also brought out some instances in which previous exposure to or the simultaneous presence of one ion accelerates the rate of uptake of a second one. Thus silver increases the rate of uptake of mercury but does not influence the total quantity taken up. Pretreatment with fairly large amounts of mercury reduces subsequent silver uptake. Experiments with spores labeled with radioactive phosphorus have shown that silver has a marked effect on membrane permeability as measured by the release of phosphorus compounds into the ambient solution. Studies have also been made on the ease with which some of the toxicants taken up will exchange with non-radioactive toxicants added to the spore suspension. The effect of foreign ions on the release of ions previously taken up was also studied.

Materials and Methods

The isotopes and labeled compounds, mentioned in the introduction, were obtained either from the Oak Ridge National Laboratory, Oak Ridge, Tennessee, U.S.A., or from commercial sources. All the materials were available at sufficiently high specific activity in relation to the quantities needed to inactivate the spores so that radioactivity determinations of the treating solutions could be made directly without any correction for self-absorption even with the C^{14}-labeled compounds. Samples of spores, because of their small size, could also be taken in small enough quantities for direct counting without loss from self-absorption.

Conidia of the following species of fungi have been included in these studies: *Neurospora sitophila* (Mont.) Shear and Dodge, *Monilinia fructicola* (Wint.) Honey, *Alternaria oleracea* Milbraith, *Aspergillus niger* van Tiegh, *Rhizopus nigricans* Ehr., *Cephalosporium acremonium* Corda, *Glomerella cingulata* (St.) Sp. and von S., *Myrothecium verrucaria* (Alb. and Schw.) Ditm, ex Fr., *Stemphylium sarcinaeforme* (Cav.) Wilts., *Venturia inaequalis* (Cke.) Wint., and *V. pyrina* Aderh. Cells of *Saccharomyces cerevisiae* Hansen were also used in some of the tests. Methods employed for culturing the fungi, harvesting the spores, determining the spore weights, and the nutrients added for germination tests are given in previous papers (Refs. 1, 3).

In experiments on the interaction between the spores and various toxicants, known weights of spores were suspended in solutions of the toxicant or toxicants in 15-ml centrifuge tubes with conical bottoms. The toxicants were in solution either in water or in 2% acetone with organic compounds not readily soluble in water. A series of 12 tubes could be run at the same time. At appropriate intervals the suspensions of spores were centrifuged and aliquots of the supernatant and of the supernatant plus spores, when required, were taken for determination of the radioactivity. When placed in the centrifuge, the spores descended to the small area at the bottom of the tubes in a few seconds. In time studies the spores were considered out of contact with the solutions as soon as they aggregated at the bottom of the tubes.

Whenever desired, the remaining supernatant was removed from the tube and the spores resuspended in water or various solutions for studies on exchange or release of absorbed toxicant by various treatments. In many experiments the effect of a series of ions on the uptake or release of a radioactive ion was determined. The interfering ions which were not labeled as a rule were either used in pretreatments followed by exposure to the radioactive ion, or the spores were subjected to the ions simultaneously, or the ions were added after uptake of the radioactive ion and the subsequent release determined of the toxicant previously taken up.

Samples to be counted were transferred to 1-in. cupped planchets, of either nickel-plated steel, stainless steel, or glass, and dried under a heat lamp. With Hg^{203} and 2,3-dichloro-1,4-naphthoquinone special precautions had to be taken to prevent loss by volatilization. To prevent loss of mercury, 0.5 to 1.0 mg of Na_2S was added to each planchet before drying and loss of the naphthoquinone was prevented by adding similar quantities of KOH. Radioactivity was determined by the use of conventional scaling equipment and end-window type Geiger-Müller tubes with thin windows.

To determine effects of the toxicants on the outward movement of cell constituents, spores of *Aspergillus niger* were grown on potato dextrose agar in 100-ml tubes containing 25 ml of media each plus about 9 microcuries of P^{32} as phosphate. After 7 to 10 days the spores were harvested as usual and centrifuged with several changes of water. Spores grown in this way usually gave about 10,000 counts per min per mg of spores and released only about 1.5% of the count to the ambient solution. The effect of added toxicant on the rate of release of phosphorus compounds to the suspending media compared to distilled water could then be determined.

The presence of small quantities of C^{14} in green plants from absorbed labeled fungicides was determined by the use of the methods of Van Slyke *et al.* (Ref. 4) for wet combustion to CO_2 and the $C^{14}O_2$ was determined by proportional counting in Bernstein-Ballentine tubes (Ref. 5).

Rate of Uptake of Toxicants

The rate of uptake of toxicants as determined by the method described above was found to be exceedingly rapid. For practically all toxicants studied, depending somewhat on the quantities supplied in relation to amount required to inhibit germination and to saturate the spores, over 50% of the quantity available was taken up by the time the first samples were taken, in ½ to 5 min. Typical curves showing cumulative uptake with time for 2-heptadecyl-

2-imidazoline, cerium, mercury, 2,3-dichloro-1,4-naphthoquinone, and silver are shown in Fig. 12.1. The quantities available at the start of the test in terms of micrograms of toxicant per gm of spores were selected on the basis of information concurrently obtained on the quantities required to inhibit germination. Spores usually will take up two to three times the amount required to reduce germination by 50%. This will become evident by comparing the amounts used in the experiments shown in Fig. 12.1 with the corresponding

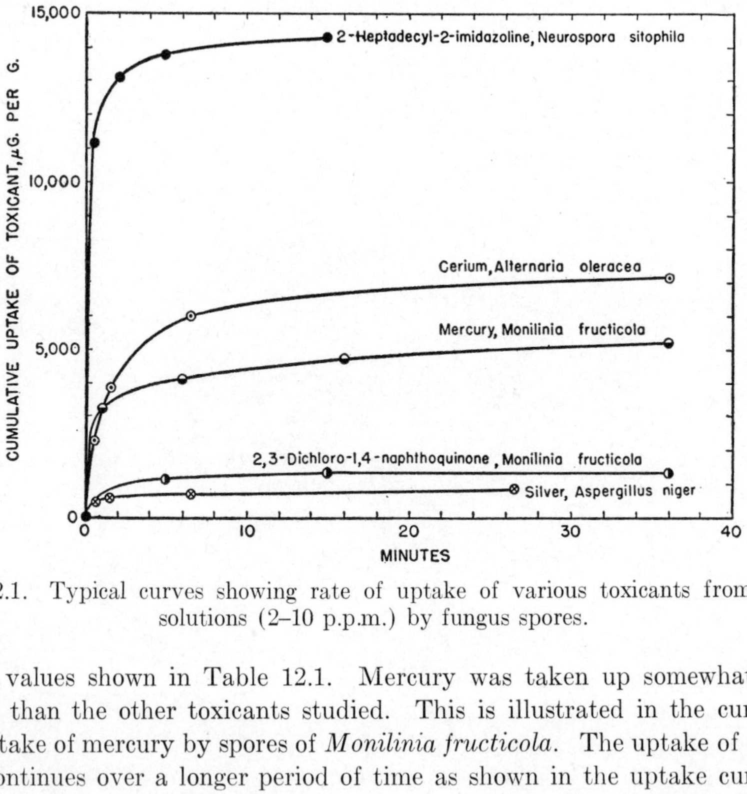

FIG. 12.1. Typical curves showing rate of uptake of various toxicants from dilute solutions (2–10 p.p.m.) by fungus spores.

ED50 values shown in Table 12.1. Mercury was taken up somewhat more slowly than the other toxicants studied. This is illustrated in the curve for the uptake of mercury by spores of *Monilinia fructicola*. The uptake of cerium also continues over a longer period of time as shown in the uptake curve for spores of *Alternaria oleracea*. As can be seen in Table 12.1 cerium is not toxic to most of the species studied.

Quantities Required to Inhibit Germination

By determining the quantity taken up by the spores and the effect on germination at several dose levels, a series of values are obtained which can be plotted to give a straight line on logarithmic-probability paper. For comparative purposes doses giving a 50% inhibition of germination are usually used. ED50 values in micrograms per gm of spore weight for 2-heptadecyl-2-imidazoline, cerium, mercury, 2,3-dichloro-1,4-naphthoquinone, and silver and spores of a number of species of fungi are shown in Table 12.1.

TABLE 12.1. ED50 VALUES IN MICROGRAMS PER GRAM OF SPORE WEIGHT FOR A NUMBER OF TOXICANTS FOR SPORES OF REPRESENTATIVE FUNGI

Toxicant	Species	ED50 Value
2-Heptadecyl-2-imidazoline	*Neurospora sitophila*	5800
	Venturia pyrina	9300
Cerium	*Neurospora sitophila*	>970 [a]
	Monilinia fructicola	4600
	Alternaria oleracea	>7100 [a]
	Aspergillus niger	>8440 [a]
Mercury	*Neurospora sitophila*	5030
	Monilinia fructicola	2830
2,3-Dichloro-1,4-naphthoquinone	*Neurospora sitophila*	560
	Monilinia fructicola	385
	Alternaria oleracea	400
	Rhizopus nigricans	680
	Myrothecium verrucaria	>1400 [a]
Silver	*Neurospora sitophila*	165
	Monilinia fructicola	216
	Alternaria oleracea	360
	Aspergillus niger	540
	Venturia pyrina	85

[a] No effect on germination at these doses.

It is seen that the ED50 values range from 85 to 9300 micrograms per gm of spores. Values for sulfur are not included in the table since sulfur is not accumulated by the spores but is given off as hydrogen sulfide. If the quantity of hydrogen sulfide produced is considered as the dose then the ED50 value for *Neurospora sitophila* is about 11,500 micrograms per gm (Ref. 1). *Cephalosporium acremonium* was especially resistant to sulfur; 28% of the spores still germinated after 53,600 micrograms of hydrogen sulfide had been given off per gm of spores.

Silver is the most toxic of the fungicides tested. 2,3-Dichloro-1,4-naphthoquinone has an effect approaching that of silver and is the most toxic of the organic fungicides included in the studies. An examination of the table reveals some interesting differences between species. Although the naphthoquinone is quite toxic to spores of four species of fungi it has no effect on the germination of spores of *Myrothecium verrucaria* at the highest dose employed. Cerium is taken up in large amounts by spores of three species; it is toxic only to one of these three species. Although spores of *Neurospora sitophila* are active in taking up most of the toxicants studied, they took up comparatively small amounts of cerium. The ED50 values for 2-heptadecyl-2-imidazoline are the highest in the table at 5800 and 9300 for the two species listed. These values are reached rapidly, however, as shown in Fig. 12.1, from solutions containing only 2 ppm. This fungicide is used commercially for the control of apple scab (*Venturia inaequalis*) and pear scab (*V. pyrina*). The effectiveness of a given toxicant in preventing the germination of fungus spores depends upon the actual dose required to inactivate the spores and the ease with which this dose is obtained. It can thus follow that a toxicant inherently less effective on a weight basis may be more useful if it is taken up sufficiently more readily by the spores than a second innately more toxic chemical.

The range of ED50 values of 85 to over 10,000 (if sulfur is included) shows that the innate toxicity of the fungicides studied is quite low. This is brought out clearly by examining the data given in Table 12.2 showing the LD50 values for a number of biocidal agents.

TABLE 12.2. APPROXIMATE LD50 VALUES FOR A NUMBER OF BIOCIDAL AGENTS

Toxicant	Subject	Approx. LD50, Micrograms per Gm
Atropine	Man	1.4
Botulinum toxin	Mouse	0.23×10^{-6}
o,o-Diethyl *o-p*-nitrophenyl phosphorothionate	Housefly	0.9
Diethyl *p*-nitrophenyl phosphate	Housefly	0.5
2,4-Dichlorophenoxyacetic acid	Tomato	10
Penicillin	Staphylococci	2
Various fungicides	Spores	85–10,000

Some of the more effective toxicants listed, such as the phosphorus-containing insecticides, 2,4-dichlorophenoxyacetic acid, and penicillin are of comparatively recent origin. There is no reason to expect that fungi are more difficult to inactivate than other organisms. The search for new fungicides would therefore be expected to disclose toxicants much more effective on a weight basis than those now known.

Exposure of Spores to More Than One Toxicant

When spores are exposed to various toxicants either singly, simultaneously, or successively, the effects on uptake are determined by the nature of the toxicants and by the quantities used. If quantities insufficient to saturate receptor sites are presented, competition between toxicants for receptor sites may not be evident. In studies on the exchange of absorbed radioactive toxicant after the addition of more toxicant, the results may be confusing if much of the additional toxicant is taken up by the spores to saturate receptor sites and is therefore not available for exchange. Results previously published (Ref. 2) have indicated that the toxicants 2-heptadecyl-2-imidazoline, silver, and cerium do not have receptor sites in common. When the experiments were extended to include mercury and silver, which would be expected to have some receptor sites in common, it was found, somewhat unexpectedly, that pretreatments or simultaneous treatment with silver increased the rate of uptake of mercury. Experiments with spores of *Neurospora sitophila* have shown an increase in the uptake of mercury as large as 75% when silver was supplied simultaneously or previously (Ref. 6). The rate of uptake of silver was retarded when mercury was also present, but after 20 min the amount of silver taken up was the same as if there had been no simultaneous exposure to mercury at the levels used in these tests.

These results have been verified in further experiments with spores of *Neurospora sitophila* and also other species. In Table 12.3 are shown results of pretreatments with silver or cerium on the subsequent uptake of mercury by spores of *Monilinia fructicola*. Spores treated with silver took up 105 micro-

TABLE 12.3. EFFECT OF PRETREATMENT WITH SILVER OR CERIUM ON UPTAKE OF MERCURY AND GERMINATION OF SPORES OF *Monilinia fructicola*

Pretreatment			Cumulative Uptake of Mercury, μg per Gm Spore Weight, After 10 and 85 Min			Germination After 10 and 85 Min, %	
Ion	Amount Taken Up, μg per Gm	Germination, %	Quantity Available	10	85	10	85
Ag	105	19	10,000	7,200	7,375	3	3
			5,000	4,200	4,400	9	3
			2,500	2,175	2,340	11	5
Ce	970	100	10,000	4,270	5,730	29	7
			5,000	3,440	4,185	41	49
			2,500	2,030	2,345	92	92
None	—	100	10,000	4,170	6,030	47	9
			5,000	3,490	4,305	65	46
			2,500	2,245	2,410	74	72

grams per gm, which reduced germination to 19%. Pretreatment with cerium did not reduce germination, although 970 micrograms per gm of spores were taken up. On subsequent exposure to mercury the spores pretreated with silver took up about 70% more mercury at the first sampling period at the highest dosage level than control spores or those pretreated with cerium. At the lowest dosage level any effect of silver was obscured by the fact that about 90% of the available mercury was taken up by all three lots of spores at the time the first sample was taken.

Subsequent exposure to mercury following pretreatments with silver or cerium does not result in any appreciable release of toxicant previously taken up.

In the experiment presented in Table 12.3 pretreatment with cerium had no effect on subsequent uptake of mercury. In other tests with *Neurospora sitophila,* spores which had taken up about 500 micrograms of cerium per gm took up as much mercury as the controls but spores which had received 980 micrograms per gm of cerium took up only about one-half as much. Further experiments have shown that there is competition for receptor sites between cerium and mercury.

The effect of silver in increasing the uptake of mercury is perhaps the result of an effect of some kind on cell permeability and does not necessarily indicate that there is no competition for the same receptor sites between these two ions. As a matter of fact in simultaneous treatments with silver and mercury in which the presence of the silver increases the rate of uptake of mercury, there is a lesser uptake of silver than when no mercury is present.

When copper and mercury are presented to spores at the same time there is some indication that copper, like silver, increases the uptake of mercury. This is more pronounced when spores have been pretreated with copper than when simultaneous treatments are used. If the amounts of copper are large, uptake of mercury is reduced when copper is simultaneously present.

Mercury and cerium were found to show marked antagonism in uptake studies when fairly large quantities of interfering ion were used. This is true for spores of all species tested. Some common receptor sites for these two ions are apparently indicated.

Exchange of Toxicant Taken Up

Studies on the rate and degree of exchange of fungicide previously taken up by the spores when an excess of non-radioactive toxicant is added to the ambient solution, have shown that of the materials under study silver is exchanged most rapidly and completely. The data indicate that the exchange is close to 100% of the theoretical. Typical data for spores of *Neurospora sitophila* are shown in Table 12.4. Exchange was equally complete regardless of length of contact between silver and the spores. Some silver is also released

TABLE 12.4. EFFECT OF LENGTH OF EXPOSURE OF SPORES OF *Neurospora sitophila* TO SILVER ON SUBSEQUENT EXCHANGE WITH NON-RADIOACTIVE SILVER

Treatment with Ag^{110}		Percentage Exchanged on Adding 1000 μg Non-radioactive Silver	
Exposure Time, hr	Amount Taken Up, μg	In 5 Min	In 65 Min
28.3	16.8	96	95
24.8	20.8	77	75
4.8	19.9	95	90
1.8	20.3	93	105
0.17	19.3	92	90
0.0014	6.0	90	90

when mercurous ion is added, no doubt because of replacement of the silver at some of the receptor sites.

Exchange with most of the other toxicants was usually less than 50%. More radioactivity was released from spores previously treated with mercuric ion when an excess of mercurous rather than mercuric ions was added. This would indicate that the mercuric toxicant on contact with the spores has been reduced to the mercurous state.

2,3-Dichloro-1,4-naphthoquinone was the only toxicant in which practically no exchange took place when non-radioactive toxicant was added subsequently. Some data are given in Table 12.5. These indicate strongly that the toxicant

TABLE 12.5. EXCHANGE ON ADDITION OF NON-RADIOACTIVE 2,3-DICHLORO-1,4-NAPHTHOQUINONE TO TREATED SPORES

Species	Spore Weight, mg	Radioactivity, c/m		Percentage Released
		Treated Spores	Released on Addition of Cold Toxicant	
Monilinia fructicola	10	1099	63	5.7
	30	1843	63	3.4
	90	2019	27	1.3
Myrothecium verrucaria	10	1080	36	3.3
	30	1720	36	2.1
	60	1940	54	2.8

has undergone chemical change and does not exist in the spores as the naphthoquinone. This is also shown by the fact that on the addition of spores to solutions of the naphthoquinone, uptake occurs for only a short time, usually not more than 5 min, after which the remaining toxicant will not be taken up even by a fresh lot of spores. If fresh toxicant is added uptake is resumed.

Effects on Outward Movement of Cell Constituents

Effects of various toxicants on cell permeability were studied by using spores labeled with P^{32} and determining the degree of outward movement of phosphorus compounds subsequent to exposure. An experiment was carried out with silver, copper, mercury, zinc, cadmium, and cobalt as indicated in Table 12.6. Spores of *Aspergillus niger* which gave 10,000 counts per min per mg were used. All the metal ions were tested at a dose of 10^{-5} equivalents per 9.8 mg of spores and in addition, silver, copper, and mercury were also tested at lower doses. The particular dose chosen was based on the relative toxicity

TABLE 12.6. EFFECT OF VARIOUS METAL IONS ON THE PER CENT GERMINATION AND THE OUTWARD MOVEMENT OF PHOSPHORUS COMPOUNDS FROM SPORES OF *Aspergillus niger*

Treatment		Germination and Cumulative Percentage of P Content Released by Spores After Various Time Intervals, hr						
		0.33		17.0		66.5	170	
Metal Ion	Quantity in Equivalents per 9.8 mg Spores	Germination	P	Germination	P	P	Germination	P
Ag^{+}	2×10^{-7}	0	1.5	0	26.9	39.3	0	35.3
	1×10^{-5}	0	1.7	0	34.3	40.1	0	36.1
Cu^{++}	3×10^{-6}	100	1.7	15	1.3	5.6	0	9.0
	1×10^{-5}	6	2.2	0	2.8	14.6	0	45.3
Hg^{++}	1×10^{-6}	3	0.8	0	1.0	1.8	0	1.7
	1×10^{-5}	0	0.4	0	1.9	4.5	0	8.3
Zn^{++}	1×10^{-5}	6	0.3	0	1.5	3.4	0	2.3
Cd^{++}	1×10^{-5}	100	0.9	12	1.7	5.6	0	4.9
Co^{++}	1×10^{-5}	100	0.4	100	0.4	4.7	42	2.3
Distilled water		100	−0.2	100	0.2	1.3	39	1.2
Spores killed by heat		0	26.7	0	27.6	26.8	0	21.9

of the ions. Spores removed after 0.33 hr of exposure were unable to germinate at both doses of silver and at the highest mercury dose. The 10^{-5} equivalent dose of copper and zinc also seriously affected germination after 0.33 hr. Effects on release of phosphorus compounds from the cells were minor at this time but were noticeable with the spores treated with silver and copper.

Further samples were taken after 17.0, 66.5, and 170 hr. After 17 hr, effects on germination were greater and the release of cell contents as a result of treatment with silver became very pronounced. It is clear that copper, mercury, and zinc brought about inability to germinate without any marked effect on release of cell contents. Examination at later time intervals up to 170 hr showed that a release of cell contents equivalent to 35–45% of the phosphorus compounds was brought about by silver and the highest dose of copper. Killing the spores by heat released less than 30% of the cell contents. Although after 170 hr the spores were unable to germinate in all the other lots except the control and that treated with cobalt, there was little outward movement of phosphorus compounds. Spores suspended in water lost only 2.2% of their cell contents and those exposed to 1×10^{-5} equivalent of mercury, zinc, cadmium, and cobalt lost only up to 8.3%.

The effect of silver in releasing cell contents and therefore altering permeability of the membrane is of interest in view of the finding that pretreatment with silver increases subsequent uptake of mercury. It would appear that the effect on permeability is not merely the result of killing of the spores, since the spores were also killed by mercury, zinc, and cadmium without a comparable effect on cell permeability.

At both doses of silver used in the tests represented in Table 12.6, germination was completely inhibited at the first sampling period. To study the effect of silver over a wider dose range, an experiment was set up in which the quantity of silver was varied from 200 to 25,600 micrograms per gm of spores (9×10^{-7} to 2.4×10^{-4} equivalents per gm) and the effect on the germination and release of cell contents studied over a 116-hr period. The results on release of phosphorus compounds up to 67 hr are plotted in Fig. 12.2. The numbers on the curves represent doses of silver expressed as micrograms per gm $\times 10^{-3}$. At the highest doses of silver (3200 to 25,600 micrograms per gm) the final effects on release of cell contents were the same, but there were differences in the speed with which the release set in. At all these doses germination was completely inhibited at the first sampling period. At 1600, 800, and 400 micrograms per gm (the values for 200 micrograms were not plotted, since they were essentially the same as the control for the first 67 hr) some germination response persisted even up to 67 hr even though cell permeability was affected to the extent that from 4.4% to 37.0% of the phosphorus compounds had leached out. The results, compared with those in Table 12.6 suggest that silver has more effect on cell permeability than some other toxicants, quite apart from its greater toxicity. This follows, since cell contents are released

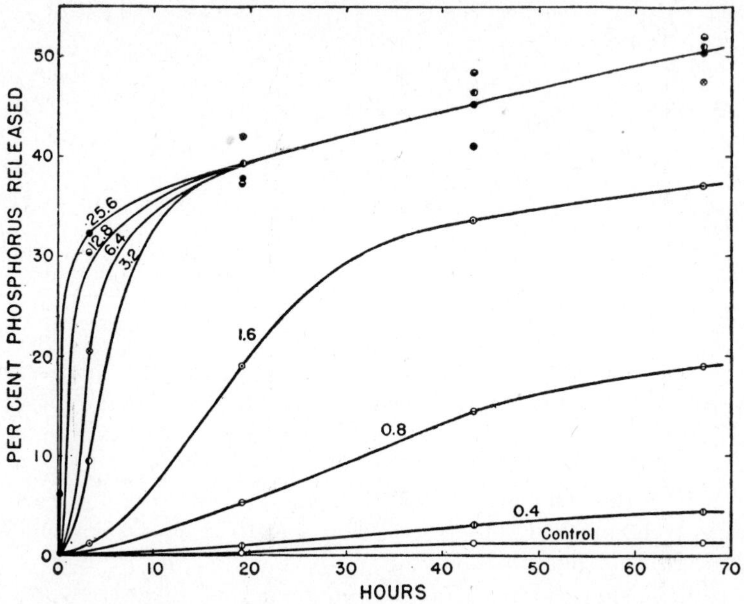

Fig. 12.2. Effect of various doses of silver on the permeability of the spore membrane determined by the outward movement of phosphorus compounds. Numbers on curves represent μg. per g. of spores $\times 10^{-3}$.

at doses giving less than 100% kill, while with cadmium, zinc, mercury, and low doses of copper there is very little release of cell contents even though all spores are unable to germinate.

Uptake of Toxicants by Host Plants

The absorption of fungicides by host plants has been investigated in a preliminary way. Experiments with S^{35}-labeled ferric dimethyldithiocarbamate have shown that a very small amount of S^{35} can be found in leaves of plants near treated leaves (Ref. 7). This has been found to be the result of absorption of CS^{35}_2, a decomposition product of the dithiocarbamate. Whether

TABLE 12.7 UPTAKE OF 2,3-DICHLORO-1,4-NAPHTHOQUINONE THROUGH ROOTS OF BEAN PLANTS

Time After Addition of Quinone, hr	Content of Tops, μg per gm Fresh Weight
2.5	0.4
23.0	2.2
43.0	8.2
74.0	15.4

any other portion of the molecule is absorbed has not been determined since C^{14}-labeled toxicants have not been available. Some uptake through the root system of small tomato and bean plants occurs when 2-heptadecyl-2-imidazoline or 2,3-dichloro-1,4-naphthoquinone is added to the medium. The amounts are small, however, as shown in Table 12.7 for bean plants and the naphthoquinone. In view of the known instability of this toxicant it is very unlikely that the absorbed material has remained unaltered chemically. Fungicides that are active systematically will no doubt be found in the future and the use of isotopes should aid in discovering them and in studying their action.

Summary

Rates of uptake and the quantities required to reduce germination by 50% have been determined for spores of representative species of fungi with the use of S^{35}, S^{35}-labeled ferric dimethyldithiocarbamate, C^{14}-labeled 2-heptadecyl-2-imidazoline and 2,3-dichloro-1,4-naphthoquinone, and Ag^{110}, Ce^{144}, and Hg^{203}. Germination was inhibited 50% after 85 to 10,000 micrograms per gm of spore weight had been taken up. Sulfur is an exception in that it is not accumulated by spores but is reduced to hydrogen sulfide and released. Lethal doses were removed from dilute solutions (1–10 ppm) within ½ to 5 min. The fungicides studied are much less toxic on a weight basis than many animal poisons, insecticides, herbicides, and bactericides. The results suggest that further search for better fungicides should be rewarding, since there is no reason to believe that fungi are more difficult to kill than other organisms.

Studies on interferences among various toxicants on uptake and subsequent release by fungus spores have yielded information on competition or lack of competition for receptor sites. Certain toxicants, such as 2-heptadecyl-2-imidazoline, cerium, and silver, seem to have no receptor sites in common, and saturation of spores with one of these toxicants does not interfere with subsequent uptake of the others. Toxicants competing for the same receptor sites interfere with each other and rates of uptake are reduced when they are used simultaneously. The presence of a second toxicant may increase the rate of uptake. When silver and mercury are used together the rate of uptake of mercury is increased and that of silver decreased. The results indicate that silver has an effect in increasing permeability of the cell membrane. Studies on effects of various substances in decreasing or increasing uptake should lead to more efficient practical use of fungicides and thereby lessen present losses caused by fungus diseases.

Effects on the permeability of cell membranes were studied by determining the release of phosphorus compounds into the ambient solutions when spores labeled with P^{32} were exposed to various toxicants. Silver was the most active in releasing cell contents, even at doses which did not completely inhibit germination, while other ions such as mercuric, cadmium, zinc, and copper could produce 100% inhibition of germination with comparatively little effect on cell permeability.

Silver which had been taken up by spores exchanged rapidly and practically completely with silver subsequently added to the medium. The other toxicants did not exchange so readily; usually the degree of exchange was less than 50%. With spores treated with mercuric ion, more exchange occurred on adding mercurous ion than when more mercuric ion was supplied. Practically no exchange occurred with spores treated with 2,3-dichloro-1,4-naphthoquinone. This indicates that the naphthoquinone has undergone chemical change on contact with the spores.

Preliminary experiments have indicated that there is little uptake of toxicants by host plants in tests with ferric dimethyldithiocarbamate, 2-heptadecyl-2-imidazoline, and 2,3-dichloro-1,4-naphthoquinone. Radioactive tracers are especially useful in studying uptake by host plants and possible activity of toxicants as systemic fungicides.

References for Chapter 12

1. L. P. Miller, S. E. A. McCallan, and R. M. Weed, "Quantitative studies on the role of hydrogen sulfide formation in the toxic action of sulfur to fungus spores," *Contribs. Boyce Thompson Inst.* 17:151–71 (1953).
2. L. P. Miller, S. E. A. McCallan, and R. M. Weed, "Accumulation of 2-heptadecyl-2-imidazoline, silver, and cerium by fungus spores in mixed and consecutive treatments," *Contribs. Boyce Thompson Inst.* 17:283–98 (1953).
3. L. P. Miller, S. E. A. McCallan, and R. M. Weed, "Rate of uptake and toxic dose on a spore weight basis of various fungicides," *Contribs. Boyce Thompson Inst.* 17:173–95 (1953).
4. D. D. Van Slyke, R. Steel, and J. Plazin, "Determination of total carbon and its radioactivity," *J. Biol. Chem.* 192:769–805 (1951).
5. W. Bernstein and R. Ballentine, "Gas-phase counting of low energy β-emitters," *Rev. Sci. Instruments* 21:158–62 (1950).
6. L. P. Miller, S. E. A. McCallan, and R. M. Weed, "The use of radioisotopes in studying the affinity of various toxicants for fungus spores," *Proc. 2nd Radioisotope Conf., Oxford, I, Med. and Physiol. Applications,* 381–90 (1954).
7. R. M. Weed, S. E. A. McCallan, and L. P. Miller, "Factors associated with the fungitoxicity of Ferbam and Nabam," *Contribs. Boyce Thompson Inst.* 17:299–315 (1953).

Chapter 13

EFFECTS ON PLANTS OF CHRONIC EXPOSURE TO GAMMA RADIATION *

Although the use of kilocurie radiation sources in experimental botany is a relatively new development, results already obtained show that such a tool can be useful in the study of both normal and abnormal plant growth. During the past six years the authors have investigated the effects of chronic exposure to gamma radiation in over 100 species of plants representing 35 different families. Several papers dealing with various aspects of this work have already been published, as well as a number of preliminary reports. However, since the technique of chronic exposure of plants to ionizing radiation has previously been investigated by only a few workers using small radiation sources, a great deal of new information is still available concerning the cytogenetic and morphogenetic responses of plants to long continued exposures to ionizing radiation. Since a general review of the morphogenetic abnormalities induced by ionizing radiation has already been published (Ref. 11), this chapter will deal with a few new specific examples.

Methods

Radioactive cobalt in the form of metallic cylinders or hollow tubes has been used consistently as the source of gamma radiation. For comparative purposes, plants have been exposed frequently to acute radiation with X rays, and in a few experiments beta radiation from P^{32} and fast electrons (800 kvp) from a resonant transformer electron beam generator were also utilized. The cobalt-60 sources were used in both outdoor and indoor installations. A special greenhouse utilized 1½ to 14 curie sources, while two different outdoor gamma radiation fields varying in size from about 2.6 to 9 acres used 16–1800 curie sources. In all cases the radioactive cobalt could be remotely controlled in such a fashion that when personnel desired to enter the field or greenhouse

* This chapter is taken from Geneva Conference Paper **266**, "The Effects on Plants of Chronic Exposure to Gamma Radiation from Radiocobalt" by A. H. Sparrow and J. E. Gunckel of the United States.

the source could be lowered into a shield which reduced the level of radiation well below the permissible dose of 50 mr per 8-hr day. Except where special experiments required some deviation, the standard procedure was to expose plants for 20 hr out of every 24. This allows 4 hr each day for planting, cultivating, examination and collection of the material exposed. Although somewhat inconvenient at times, this schedule seems to be a practical one

The daily dose rate received by any given plant depends on the size of the radiation source (i.e., number of curies) and the distance from the source to the plant. Except for a minor correction due to air absorption, the dose rate falls off according to the inverse square law. Thus, all plants growing equidistant from the source receive the same daily dose rate. However, plants at any given position receive slightly less radiation per day at the end of the season than at the beginning, owing to the decay of the radiocobalt. Since the decay is only about 1% a month, the change produced during the summer is of minor consequence for most of our work. In this report the daily dose rates given are those determined by measurement and calibration of the source at the beginning of the growing season each year, usually in April or May. More detailed techniques for operation of a gamma radiation field have been published elsewhere (Ref. 34).

Tolerance in Certain Plant Species of Chronic Gamma Radiation

Prior to the establishment of the Brookhaven gamma radiation field, there was essentially no data on the tolerance of plants to chronic gamma irradiation. During the six years since the first gamma field was put into operation the tolerance of over a hundred species of plants has been determined. Some of this information has been reported elsewhere (Refs. 30, 34), but since the available information on tolerance may be of value to other institutions who wish to carry on similar work, it was decided to present a brief summary of tolerance data so far available (Table 13.1).

The dose rates specified in Table 13.1 indicate the range in which severe radiation damage (usually severe growth inhibition or dwarfing) would occur. The dose rate which causes severe growth inhibition varies greatly in the different species studied. Some species, i.e., *Tradescantia paludosa* and *Lilium longiflorum*, show severe morphogenetic effects with dose rates as low as 30 to 40 r per day, while others, such as *Gladiolus*, show little or no effect at dose rates of 5000 to 6000 r per day. Very few plants so far studied have shown gross morphological changes at dose rates below 15 r per day. However, chromosome aberrations and mutations have been noted in significant numbers at dose rates below 1 r per day (Refs. 27, 34).

There are undoubtedly a considerable number of factors which determine the radiosensitivity of a given species of plant. The results of our investigations seem to indicate clearly (1) that plants with very large chromosomes

TABLE 13.1. DAILY DOSE RATES REQUIRED TO PRODUCE SEVERE RADIATION EFFECTS ON 79 SPECIES OF PLANTS *

Dose Rate, r/day		
30–50	*Lilium longiflorum*	*Tradescantia paludosa*
	Taxus media	*Tradescantia ohiensis*
51–100	*Cornus florida*	*Setcreasia* sp. (4n)
	Impatiens sultanii	*Vicia faba*
101–200	*Acer rubrum* (6n or 8n)	*Ilex* (4n)
	A. spicatum (?)	*Magnolia* sp. (?)
	Commelina coelestis (?)	*Pyrus malus*
	Cosmos	*Rhododendron* (hybrid)
201–400	*Antirrhinum majus*	*Mirabilis jalapa*
	Canna generalis	*Nicotiana bigelovii* (4n)
	Capsicum frutescens (2n, 4n)	*N. glauca*
	Chrysanthemum nipponicum	*N. glauca* × *langsdorffii* (6n)
	Coleus blumei (4n)	*N. langsdorffii*
	Dahlia (hybrid) (8n)	*N. rustica* (4n)
	Datura stramonium	*Phytolacca decandra* (4n)
	Gossypium hirsutum (4n)	*Pisum sativum*
	Kalmia latifolia	*Prunus persica*
	Liriodendron tulipifera	*Vicia angustifolia*
	Luzula purpurea	*Vicia tenuifolia* (4n)
	Melilotus officinalis	*Zinnia elegans* (?)
401–800	*Allium cepa*	*Lycopersicon esculentum*
	Althea rosea (6n or 8n)	*Petunia hybrida*
	Celosia cristata (4n)	*Pieris japonica*
	Chenopodium album (4n)	*Ricinus communis*
	Chrysanthemum ircutianum (4n)	*Rosa* (Hybrid Tea Rose)
	Helianthus annuus	*Saintpaulia*
	Ipomoea noctiflora	*Sedum aizoon* (?)
	Kalanchoë daigremontiana	*Stachyurus* sp. (?)
	Lactuca sativa	*Xanthium* sp. (4n)
801–1600	*Chrysanthemum arcticum* (8n)	*Lenophyllum pusillum* (?)
	C. lacustre (22n)	*Linum usitatissimum*
	C. yezoense (10n)	*Mollugo verticillata* (8n)
	Cucurbita (pumpkin) (4n)	*Phaseolus vulgaris* (?)
	Iris (hybrid) (4n)	*Sedum acre* (12n)
	Kalanchoë blossfeldiana	*S. album* (16n)
1601–6000	*Digitaria sanguinalis* (4n)	*Kalanchoë tubifolia* (4n)
	Gladiolus (hybrid) (6n)	*Lenophyllum texanum* (?)
	Graptopetalum bartramii (2n)	*Luzula acuminata* (8n)
	Graptopetalum MacDougallii (22n)	*L. multiflora* (4n)
		L. pallescens (4n)

* Chromosome number is diploid unless otherwise stated. Question mark (?) indicates degree of polyploidy uncertain.

have a high radiosensitivity, (2) that plants with smaller chromosomes tend to be less sensitive than those with large chromosomes, and (3) that polyploid species within a genus tend to be less sensitive than diploid species.

Growth Inhibition and Stimulation

Previous work, with both acute and chronic exposures of ionizing radiation, has shown clearly that growth inhibition or stunting is a very common effect. Growth stimulation, however, is much more controversial, and the question has been much discussed in the literature for many years (Ref. 25).* The interest in this phenomenon is not surprising, since any treatment that can stimulate plant growth is not only of considerable scientific interest but is also of potential economic value. We report, therefore, the following observations even though the cause for the enhanced growth is not clear.

Plants of *Antirrhinum majus* have been grown in the gamma field over a wide dose range for four years. The first two years it was noted that toward the end of the summer certain plants growing at the higher dose rates grew much taller than average. The following year, measurements of plant heights were taken at dose rates varying from 14 to 600 r per day. The data indicate that at dose rates above about 125 r per day there is a gradual increase in plant height, reaching a maximum at 230 r per day. At 285 r per day the plants are still much taller than average, but at 330 r the average plant height has decreased appreciably to less than that of normal plants. Associated with this increase in plant height is an increase in average stem diameter. Increased leaf thickness has also been observed (see below).

A number of pure species of *Nicotiana,* as well as certain interspecific hybrids, have been grown in the gamma field and the gamma greenhouse for several years. On several occasions it has been found that plants growing at a certain dose rate grew taller than plants at higher or lower dose rates and also that they bloom earlier. The exact dose rate at which this apparent stimulation occurs varies from experiment to experiment and is apparently influenced by the external or internal environment of the plant, both of which could be expected to vary considerably from experiment to experiment and from year to year. A typical experiment with the hybrid *Nicotiana glauca* × *langsdorffii* is illustrated in Fig. 13.1 which shows a plant of average size from each dose rate row as well as a control plant. The photograph clearly indicates that the plant which grew at 15 r per day was much smaller and bloomed well in advance of either the control or any of the plants grown at the higher dose rates.

The above two examples of apparent stimulation are of considerable interest, but they are not typical responses for plants in general. More commonly, the

* See also Chapter 8 of this volume.

FIG. 13.1. Tobacco plants of the cross *N. glauca* × *Langsdorffii* after 47 days of exposure in the gamma greenhouse. Irradiation was begun in the first true leaf stage. Stem and root growth are inversely proportional to the dosage. Note enhanced flowering in the plant which received 15 r per day.

growth of plants exposed to chronic gamma irradiation is inversely proportional to the dose rate (Fig. 13.2). This relationship between dosage and degree of growth inhibition is the basis for the inhibition of sprouting reported

FIG. 13.2. Plants of *Vicia Faba* after 39 days exposure to a wide range (15–255 r per day) of chronic gamma rays. Both top and root growth are inversely proportional to the dose rate.

for potatoes following treatment with X and gamma rays (Refs. 24, 29, 31) and with fast electron beams (Sparrow, Schairer and Lawton, in press) and for onions following X radiation (Ref. 7).

Morphological Effects

As indicated above, one of the most common effects of irradiation upon plant growth is growth inhibition and, more rarely, growth stimulation. In addition, the roots, stems, leaves, buds, and flowers of most plants may have their normal expression altered by exposure to radiation (Refs. 12, 23). Most of the early work, particularly X-ray effects, was reviewed by Johnson (Ref. 15) and later work, particularly with gamma radiation, was reviewed by Gunckel and Sparrow (Ref. 11). The present chapter will be confined to a few unpublished examples of leaf, bud, stem, and flower development following chronic exposure to gamma radiation. In all cases, the response will vary, depending upon the level and duration of the dosage and upon the species, age, and physiological condition of the plants irradiated.

Leaves. In most plants, the leaves already present on the plant at the beginning of the exposure are changed relatively little. Rarely, these older leaves may have a restricted blade or show some growth of the blade between the veins; more commonly the leaf texture may be changed, becoming dry, stiff, and coarse or thickened and leathery.

Young leaves formed during irradiation show a wide range of responses. In general, the leaf abnormalities include dwarfing, asymmetrical development of the leaf blade, distorted venation, a change in texture and thickening of the blade. This latter is most interesting and has been studied in some detail in *Antirrhinum majus* at 17 dosages ranging from 0.5 to 600 r per day. At the lowest dosages, the leaf was not significantly thicker than that of the controls. However, with increasing dosage there was a progressive thickening of the leaf up to at least three times that of the controls at 600 r per day. The leaves at dose rates above 240 r per day had a thick, leathery aspect, quite different from the controls. Sections were made from representative leaves and from these measurements showed that two-thirds of the increase in thickness was due to repeated cell divisions in the palisade layer. Enlargement of spongy mesophyll cells contributed about one-third to the leaf thickening.

An additional leaf response may be reported from the 1954 gamma field studies. Plants of *Graptopetalum MacDougallii* were placed at various dosages and observed for morphological changes. Leaves which received 590 r per day during their development not only became slightly thickened but underwent considerable enlargement in length and width by comparison with controls.

Root development. We have previously reported that chronic gamma irradiation of *Kalanchoë* at 1280 r per day seems to have a stimulating effect on the production of roots from stems (Ref. 1). It has been shown that acute X radiation of localized stem areas will cause the formation of roots above the irradiated zone (Ref. 5; Sparrow, unpublished). Observations on plants grown at fairly high dose rates also indicate that the root system frequently suffers severe damage (Fig. 13.2). Inhibition of the formation of new roots has also been observed in irradiated sectors of the stem cuttings of *Impatiens*. In one experiment with this plant, cuttings placed in vermiculite showed root development from the basal node of irradiated stem sectors at 2000 r, whereas sectors irradiated at 4000, 8000, 16,000, and 32,000 r showed no root development at all in the irradiated zone. Roots did develop, however, above the irradiated zone at all dosages and were at least as numerous, if not more numerous, at 16,000 and 32,000 than in the lower dosages or in the controls.

Flowers. With chronic gamma irradiation, flowering is generally retarded at the higher dosages and approaches that of the controls with decreasing dosages. However, in some plants, as in snapdragons and tobaccos already cited, flowering may be stimulated in a critical dosage range which varies with the species. Since most of this has been reviewed previously (Refs. 11, 12, 13, 15) it will suffice here to mention just one further observation on *Tradescantia paludosa*. This species is of considerable interest because it has been used so frequently for experimentation and is one of the most sensitive plants so far studied. In a critical range between 13 r and 37 r per day the plants showed many abnormal axillary buds and inflorescences due principally to the induction of multiple growth centers. Inflorescences receiving 20–24 r per day show,

FIG. 13.3. Irradiated flower stalk of *Tradescantia paludosa*. Proliferated inflorescence of leaves and modified flowers after 24 r per day for 32 weeks. Removed from the gamma field and held in the gamma greenhouse, the inflorescence was producing a large number of vegetative shoots.

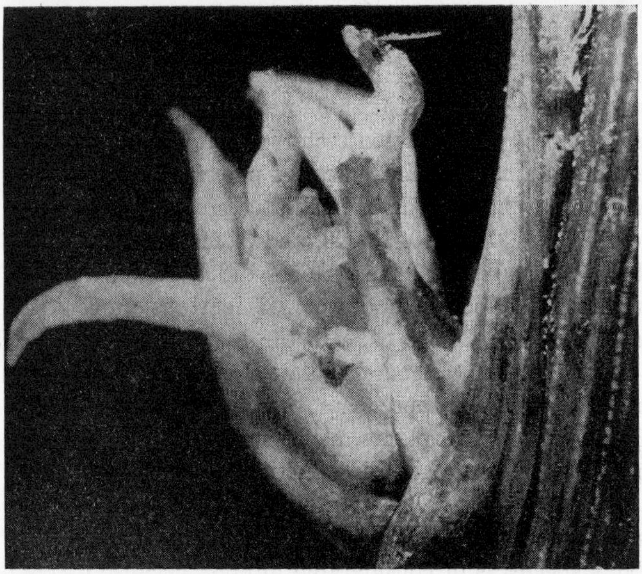

Fig. 13.4. Irradiated flower stalk of *Tradescantia paludosa*. Plant which received 34 r per day for about 3 months. Multiple leafy shoots were developing from one locus in the internodal region. Such shoots are found usually in axillary positions and are frequently chlorotic.

Fig. 13.5. Irradiated flower stalk of *Tradescantia paludosa*. Terminal inflorescence on plant in gamma greenhouse which received 34 r per day for 32 weeks. Note the formation of modified leaves in a floral position.

after 8 weeks, proliferation by leaf-like structures and modified flowers so that by 3 weeks a globose flower head 3–5 cm in diameter results. Removed from the gamma field and brought into the greenhouse, these globose flower heads commonly generate a large number of vegetative shoots (Fig. 13.3). At the higher dosages in this range (e.g., 34 r per day) radiation favors vegetative rather than floral production. Accordingly, multiple, leafy shoots are borne in each leaf axil and in internodal positions (Fig. 13.4) and modified leaves occur in floral positions (Fig. 13.5). These observations are of interest, particularly in suggesting that one might induce proliferations where the normal stem, leaf, or flower character might be lost. Such is indeed the case, and this topic is of sufficient importance to discuss independently of the morphological effects already described.

Tumor Induction

Plants of the amphidiploid hybrid *Nicotiana glauca* × *langsdorffii* have long been known to develop spontaneous tumors on leaves, stems, and roots (Ref. 16). Plants of this hybrid were exposed to 20 hr per day of chronic gamma radiation at dose rates of 385, 320, 210, 105, 50, and 26 r per day. The percentage of plants showing one or more tumors of significant size was recorded after 37, 44, and 64 days. The results show that dose rates above 210 r promote more tumor development than lower rates, and that longer exposures at any

Fig. 13.6. Plant of the tobacco amphidiploid *N. glauca* × *Langsdorffii* which received 9 weeks of gamma radiation at 325 r per day. This is a close-up view of terminal and axillary tumors (leaves removed). The tumor mass consists of a large number of induced growth centers consisting of numerous meristematic areas and a series of modified leaves. The axillary bud tumors undergo some elongation while the buds of the terminal tumors are very restricted in growth.

given rate also promote tumor development (Ref. 32). For example, after 37 days' exposure 47% of the plants at 385 r showed tumors, whereas none showed in these at 210 r or less. In the 210 r per day plants, the percentage of plants with tumors increased from none at 37 days to 33% at 44 days and up to 80% at 64 days. This increase in the number of plants with tumors was paralleled by an increase in the average amount of tumor per plant. The tumor weight is greater at the higher dose rates, reaching a high of 47% and 23% of the wet weight respectively for tops and roots at 385 r per day, compared with 0.1% and 1.0% wet weight for controls. Similar but less extensive work with acute X irradiation and internal beta irradiation from P^{32} indicates a similar response.

Morphologically, at least six types of induced tumor growth patterns may be described. Histologically, at least two growth patterns are evident. First, a tumor may be a composite of a large number of induced growth centers, each consisting of an apical meristem, derived from the surface and subsurface layers, and a number of associated leaves (Fig. 13.6). Such leafy tumors are characteristically found in axillary positions. Secondly, tumors may increase by subepidermal cell divisions and by cell enlargement, the surface layers remaining fairly discrete and nearly glabrous (Fig. 13.7). After growing for some months in sterile culture, each of the tumor types may still be recognized, indicating that the induced morphological changes are not due merely

FIG. 13.7. Plant of the tobacco amphidiploid *N. glauca* × *Langsdorffii:* axillary plant tumor induced by 2100 r of x-rays. In addition to some leafy outgrowths there was a yellow, glabrous, central mass. This tumor differed from the leafy in that the surface remained fairly discrete and the mass increased by subepidermal divisions and overall cell enlargement.

to temporary physiological disturbances induced by the radiation. The specific cause of tumor induction is not known, but work is under way to determine the chemical constituents and the nutrient requirements of each of these morphological types of tumors.

Peaceful Applications of Atomic Energy in Plant Science and Agriculture

The preceding sections of this paper have been devoted to brief descriptions of some of the results that we have obtained from irradiating plants with chronic gamma rays from a cobalt-60 source, with fast electron beams, or with acute X rays. It might be well to mention some of the problems raised by these investigations and to note briefly some of the prospective practical applications of ionizing radiation in plant science and agriculture.

Applications in agriculture. By using the techniques described for growing plants in the gamma field one may obtain a significant increase in mutation rates, in some cases (e.g. *Antirrhinum*) at dose rates below 1 r per day. The potentialities of the application of ionizing radiation in agriculture and plant-breeding programs have recently been evaluated (Refs. 25, 26, 36) and are considered in Chapter 26, so will not be considered here.

Another potential use of ionizing radiation which may prove of practical value is the inhibition of sprouting of certain vegetables. Sparrow and Christensen (Ref. 31) and Sawyer and Dallyn (Ref. 24) have reported that exposures of potato tubers to gamma-ray dosages in the 10,000- to 20,000-r range is effective in the inhibition of sprouting. Certain other undesirable changes which normally accompany sprouting are also reduced in the irradiated material. It has also been shown (Ref. 7) that sprouting in sweet Spanish onions can be effectively controlled by 4000 r of gamma rays. There is every reason to believe that the same technique will prove useful in the inhibition of sprouting of other vegetables. With the advent of megacurie radiation sources it seems not improbable that certain vegetable crops may be irradiated on a commercial scale within the next few years (Refs. 3, 17). If the process is commercially adopted, the economic value of the method would be considerable. In addition to the improved storage quality, recent work from the Brookhaven National Laboratory indicates that an X-ray dose of approximately 20,000 r may be sufficient to prevent reproduction of the golden nematode of potato (Fassuliotis and Sparrow, Ref. 37). If this work is confirmed, and perhaps extended to other plant diseases, the application of ionizing radiation in plant disease control might be considerable.

The difficulty in rooting cuttings of certain plants, even after chemical pretreatment, makes it desirable to investigate further the commercial possibilities of using radiation for the induction of adventitious roots. The method might prove to be a useful and practical one.

Applications to basic problems in the plant sciences. On theoretical and practical grounds the work reported on *Tradescantia* has intriguing possibilities. The results suggest that in critical dosage ranges one may selectively discriminate between a dosage level which favors a vegetative response and another which favors a floral response (Ref. 12). The mechanisms of vegetative and floral development are little understood and might be effectively studied by such methods.

During the life of the plant various internal systems or cells are more easily modified by radiation than others (Ref. 14). Further, it has been shown that radiosensitive systems may be modified by changes in the internal or external plant environment (Refs. 4, 6, 19). A study of these possible mechanisms of effect may help to solve some of the physiological problems of growth and differentiation.

The development of irradiated leaves is interesting in that it shows that the rather characteristic thickening may be due to cell division or to cell enlargement. It is suggested that irradiated leaves might be good material for studying the relative roles and the mechanics of cell division and cell enlargement in the growth process, and for trying to elucidate the manner in which radiation initiates abnormal proliferation.

The work on tumors deserves special comment (Ref. 32). We feel that our studies may throw some light on the regenerative capacity of cells. More information on the growth of normal cells is necessary before too much can be said about abnormal cell growth. However, the studies under way on the chemical constituents of these tumor cells may enable us not only to grow these tumor cells more effectively in sterile culture but also to alter experimentally the growth pattern of these tumors. By additives to the nutrient medium we hope to determine the effectiveness of various compounds against radiation damage, or conversely, to induce, by additives, radiation effects in non-irradiated plants.

A recent review of the literature on plant growth following irradiation reports many tentative conclusions which may contribute toward a solution of some of the problems raised here (Ref. 32). For example, the relationship between chromosome breakage, mitotic inhibition, and growth inhibition has been considered by several workers (e.g., Refs. 9, 33). A number of physiological disturbances are induced in plants by ionizing radiation. Some of these disturbances, e.g., auxin metabolism, have specific morphogenetic effects. Any combination of the following factors may contribute to the morphogenetic effects described in this paper: (1) growth substance production and distribution (Refs. 8, 28), (2) the nutritional level of the plant (e.g., Ref. 10), (3) an effect upon the mechanisms of assimilation (Ref. 22), (4) local mobilization of nutrient materials (e.g., Ref. 20), (5) organic phosphates (Ref. 18), (6) accumulation of free amino acids (Ref. 35), (7) changes in enzyme activity (Refs. 1, 2), and (8) inhibition of synthesis (Ref. 21). With the exception of mutations, most of the irradiation effects described are probably secondary effects of physiological disturbances only quantitatively different from those which occur in unirradiated plants. Radiation may thus be a useful tool for the analysis of many problems in plant growth.

References for Chapter 13

1. J. G. Aldous and K. D. Stewart, "Action of x-rays upon some enzymes of the living yeast cell," *Rev. Canad. Biol.* 11:49 (1952).
2. L. R. Blinks, "Effects of radiation in marine algae," *J. Cellular Comp. Physiol.* 39, Suppl. 2:11–18 (1952).
3. L. E. Brownell, J. V. Nehemias, and J. J. Bulmer, "Designs for potato irradiation facilities," *Univ. of Michigan, Engineering Research Institute Report,* No. 1943: 7-23-P, 1954.
4. R. S. Caldecott and L. Smith, "The influence of heat treatments on the injury and cytogenetic effects of x-rays on barley," *Genetics* 37:136–57 (1952).
5. E. Christensen, "Root production in plants following localized stem irradiation, *Science* 119:127–28 (1954).
6. A. R. Cooke, "Effect of gamma irradiation on the ascorbic acid content of green plants," *Science* 117:588–89 (1953).
7. S. L. Dallyn, R. L. Sawyer, and A. H. Sparrow, "Extending onion storage life by gamma irradiation," *Nucleonics* 13(4):48–9 (1955).
8. S. A. Gordon, "Occurrence, formation, and inactivation of auxins," *Ann. Rev. Plant Physiol.* 5:341–378 (1954).
9. L. H. Gray and M. E. Scholes, "The effect of ionizing radiations on the broad bean root: Part VIII. Growth rate studies and histological analyses," *Brit. J. Radiol.* 24:82–92; 176–80; 228–36; 285–91; 348–52 (1951).
10. J. E. Gunckel, K. V. Thimann, and R. H. Wetmore, "Studies of development in long shoots and short shoots of *Ginkgo biloba* L.: IV. Growth habit, shoot expression and the mechanism of its control," *Am. J. Bot.* 36:309–16 (1949).
11. J. E. Gunckel and A. H. Sparrow, "Aberrant growth in plants induced by ionizing radiation." In *Abnormal and Pathological Plant Growth,* Brookhaven Symp. in Biol. No. 6, 1954, pp. 252–279.
12. J. E. Gunckel, A. H. Sparrow, I. B. Morrow, and E. Christensen, "Vegetative and floral development of irradiated and non-irradiated plants of *Tradescantia paludosa, Am. J. Bot.* 40:317–332 (1953).
13. J. E. Gunckel, I. B. Morrow, A. H. Sparrow, and E. Christensen, "Variations in the floral morphology of normal and irradiated plants of *Tradescantia paludosa, Bull. Torrey Botan. Club* 80:445–456 (1953).
14. P. S. Henshaw and D. S. Francis, "A consideration of the biological factors influencing the radiosensitivity of cells," *J. Cellular Comp. Physiol.* 7:173–195 (1935).
15. E. Johnson, "Effects of x-rays upon green plants." In B. M. Duggar, ed., *Biological Effects of Radiation,* McGraw-Hill Book Company, Inc., New York, 1936, pp. 961–85.
16. A. E. Kehr and H. H. Smith, "Genetic tumors in *Nicotiana* hybrids." In *Abnormal and Pathological Plant Growth,* Brookhaven Symp. in Biol. No. 6, 1954, pp. 55–78.
17. O. A. Kuhl, A. H. Sparrow, and B. Manowitz, "Portable pilot plant for irradiating potatoes," *Nucleonics* 13 (No. 11):128–129 (1955).

18. W. J. McIlrath, "Cotton-leaf malformations induced by organic phosphate insecticides," *Bot. Gaz.* 112:221–25 (1950).
19. K. Mikaelsen, "Protective properties of cysteine, sodium hyposulfite, and sodium cyanide against radiation-induced chromosome aberrations," *Proc. Nat. Acad. Sci.* 40(3):171–78 (1954).
20. J. W. Mitchell, "Effect of naphthalene acetic acid and naphthalene acetamide on nitrogenous and carbohydrate constituents of bean plants," *Bot. Gaz.* 101:688-99 (1940).
21. S. R. Pelc and A. Howard, "Chromosome metabolism as shown by autoradiographs," *Exptl. Cell Research,* Suppl. 2:269–278 (1952).
22. H. Quastler and M. Baer, "Inhibition of plant growth by radiation: V. Radiation effects on initiation and completion of growth," *Cancer Research* 10:604–612 (1950).
23. E. Sankewitsch, "Untersuchung von Röntgenamorphosen bei *Nicotiana rustica* L.," *Beitr. Biol. Pflanz.* 29:1–74 (1952).
24. R. L. Sawyer and S. L. Dallyn, "The effect of gamma irradiation on storage life of potatoes," *Am. Potato J.* 32(4):141–43 (1955).
25. K. Sax, "The effect of ionizing radiation on plant growth," *Am. J. Bot.* 42:360–364 (1955).
26. W. R. Singleton, "The contribution of radiation genetics to agriculture," *Agron. J.* 47(3):113–17 (1955).
27. W. R. Singleton, C. F. Konzak, S. Shapiro, and A. H. Sparrow, "The contribution of radiation genetics to crop improvement." In *International Conference for the Peaceful Uses of Atomic Energy, 1955,* Columbia University Press, New York, 1956, Vol. 12, pp. 25–30.
28. F. Skoog, "The effect of x-irradiation on auxin and plant growth," *J. Cellular Comp. Physiol.* 7:227–70 (1935).
29. A. H. Sparrow and E. Christensen, "Effects of x-ray, neutron, and gamma irradiation on growth and yield of potatoes," *Am. J. Bot.* 37:667 (1950).
30. A. H. Sparrow and E. Christensen, "Tolerance of certain higher plants to chronic exposure to gamma radiation from cobalt-60," *Science* 118:697–98 (1953).
31. A. H. Sparrow and E. Christensen, "Improved storage quality of potato tubers after exposure to Co^{60} gammas," *Nucleonics,* 12(8):16–17 (August 1954).
32. A. H. Sparrow and J. E. Gunckel, "Tumor formation in hybrid *Nicotiana glauca* × *langsdorffii* exposed to chronic gamma irradiation from cobalt-60," *Reports and Communications, 8th Intern. Congr. Bot.,* Section 8, 1954.
33. A. H. Sparrow, M. J. Moses, and R. Dubow, "Relationships between ionizing radiation, chromosome breakage and certain other nuclear disturbances," *Exptl. Cell Research,* Suppl. 2:245–67 (1952).
34. A. H. Sparrow and W. R. Singleton, "The use of radiocobalt as a source of gamma rays and some effects of chronic irradiation on growing plants," *Am. Naturalist* 87:29–48 (1953).
35. R. A. Steinberg, J. D. Bowling, and J. E. McMurtrey, Jr., "A possible explanation of symptom formation in tobacco with frenching and mineral deficiencies," *Science* 110:714–15 (1949).
36. N. E. Tolbert and P. B. Pearson, "Atomic energy and the plant sciences," *Advances in Agron.* 4:279–303 (1952).
37. Fassuliotis and Sparrow, in *Plant Disease Reporter* 39 (7):572 (1955).

Chapter 14

CYTOLOGICAL AND CYTOCHEMICAL EFFECTS OF RADIATION *

In studying and interpreting radiation effects and the action of chemical agents, one is often confronted with the great diversity and variability which are inherent properties of living systems at different levels of integration. Despite this, a convenient basis by which the sensitivity of the cell to radiation (and the so-called radiomimetic substances) can be determined is to investigate the damage to the architecture of the chromosomes, coupled with a study of the changes in cytochemical properties wherever possible.

Material and Methods

The biological materials on which the present chapter has been based are: Ehrlich and Krebs ascites tumors, sarcoma 37 ascites, and Walker rat ascites tumor, the normally proliferating root-meristems of *Vicia faba* and *Luzula purpurea,* and the actively dividing meiotic tissues of the grasshopper, *Chorthippus.*

The methods and techniques of treatment with radiation and radiomimetic agents varied in dose rates and concentrations with different materials. Thus, in the case of Ehrlich ascites tumor, whole-body radiation was given in a single dose with 1400 r (165 kv, filter 1 Al, intensity 80 r/m); with Krebs ascites, Sarcoma 37, and Walker ascites tumors, the animals received intraperitoneally different concentrations of di-2-chloroethyl aminophenyl butyric acid; *Vicia* and *Luzula* meristems were similarly treated with different doses and intensities of the same chemical agent. Finally, the frequencies of breaks and interchanges in particular chromosomes of *Chorthippus* in meiosis were studied, employing varying doses of X rays. Before presenting the results of these experiments, it is necessary to review briefly certain important recent ideas on chemical and radiation "breakage" of chromosomes.

* This chapter is taken from Geneva Conference Paper 904, "Cytological and Cytochemical Effects of Radiation (and Radiomimetic Substances) in Actively Proliferating Biological Systems" by A. R. Gopal-Ayengar of India.

General Considerations

The action of X rays and other types of ionizing radiations on normally proliferating cellular systems like the root meristems of *Vicia faba* has been studied from time to time by many investigators. Although there has accumulated an extensive body of literature on the subject, the frontiers of radiobiology are widening out. The exciting developments in chemical mutagenesis during the last few years have opened out newer vistas and have led to a reorientation of our thought and outlook on many of the problems relating to chromosome breakage and cell dynamics.

In an attempt to clarify the position regarding the production of structural changes by ionizing radiations and certain chemical agents, Revell (Refs. 1, 2) has postulated a chromatid-exchange hypothesis based on a "contact first" mechanism, in preference to the orthodox breakage-and-reunion concept. The fundamental difference between these alternative hypotheses, as Gray (Ref. 3) has pointed out in his discussion on the subject, consists in the fact that whereas the orthodox view assumes that a chromosome may be broken along its length, Revell's postulate demands that breakage is secondary to exchange initiation and can supervene only at loci where the chromatids are already in contact. Accordingly "the 'breaks' would not represent the residue of primary aberration, the rest of which had either reunited in pairs as exchanges or else restituted. Interpreted in this way the visible changes thus provide no evidence of 'breakage first,' or indeed of breakage at all"—(Ref. 2).

In comparing the action of ionizing radiations with that of chemical agents like diepoxide, mustards, and other alkylating substances, it should perhaps be recognized that though the final expressions of these aberrations are, qualitatively speaking, indistinguishable from each other, the initial steps must be different, with different types of causal agents. It is true that chemical agents like diepoxide, di-2-chloroethyl aminophenyl butyric acid and related substances produce preferentially localised aberrations in particular regions of chromosomes in *Vicia faba*, as compared with the more nearly random distribution of radiation-induced abnormalities. Moreover, in *Vicia* meristems, as reported by Revell, the first half of the resting nucleus is most vulnerable to the action of chemicals as against ionizing radiations, which are maximally effective in the latter half of the interphase nuclear cycle. Though this quantitative difference is statistically distinguishable, it does not seem to warrant the assumption that "the two agencies are inducing exactly the same type of aberrations by two entirely different mechanisms."

It has often been tacitly assumed that since breaks were found to rise about linearly and interchanges as the square of the X-ray dose (under certain conditions) this should be considered as demonstrable proof that each interchange involves two independent breaks. Moreover "data from experiments where

X-ray dose was separated into two halves by varying times or where the dose intensity was varied, have been used to estimate the mean time which is supposed to elapse between breakage and reunion. But this kind of evidence, does not itself indicate that two breaks are the result of two events induced, because it cannot by itself give any information about the nature of the changes in the two chromosomes which lead them to exchange." Further, the triple processes of ionization, breakage of chromosomes and the reunion of broken chromosome ends are all causally connected though they are distinctly separated events in time (Ref. 4).

Effects of Whole-body Radiation on Ehrlich Ascites Tumor

It is now well recognized that the effects of acute whole-body exposure to penetrating radiations are manifold. These express themselves in a complex of syndromes involving nearly every system of the body. Although interpretations in terms of causality of events are beset with numerous difficulties it was thought worthwhile to examine the cytological and cytochemical action of acute whole-body X radiation on an ascites type of tumor cells. Ehrlich ascites was chosen for the purpose and the experimental animals (mice) received a single dose of 1400 r. At the time of radiation the tumor cells were at the peak of their growth phase.

The immediately recognizable effect consisted of a sharp drop in the mitotic activity to zero level, which persisted for a period of nearly 30 hr. The fall in the mitotic index was more or less exponential. During this amitotic phase after irradiation the cell population remained constant. This is in agreement with the findings of Klein and Forssberg (Ref. 5). Contemporaneously with this amitotic phase, the cells began to show permeability changes which were quickly reflected in the progressive increases in cell volumes which continued as a linear function of time. The nucleo-cytoplasmic ratio however remained unchanged throughout this phase of cell enlargement. Recovery followed after 30 hr and there was a gradual rise in the mitotic index reaching peak proportions 96 hr after radiation exposure. Even here, the rise was not as high as in the controls.

While the usual types of damage such as clumping and stickiness of the chromosomes were in evidence in earlier samples, the cells examined 72 and 96 hr after irradiation exhibited structural changes, particularly chromatid interchanges and chromosome breaks. Among others, the most commonly observed aberrations were diplochromosomes, precocious separation of centromeres at metaphase, disruption of the anaphase configurations giving rise to single or multiple bridges, interlocks, loops, fragments and laggards, multipolar mitoses with associated bridges and fragments, superfragmentation, and pycnotic degeneration.

The most pronounced cytochemical effect was that relating to nucleic acid metabolism as manifested by the marked accumulation and deposition of RNA in the cells as determined by Pappenheim and Unna stain. This was particularly noticeable in the abnormal nucleolar elaboration at all stages of the mitotic cycle. Increased concentrations of RNA during the post-radiation period have also been reported (Refs. 6, 7, 8). Mitchell (Ref. 6) holds that the accumulation of ribose nucleotides is causally related to the disturbances of the normal metabolic process of the cell, which may be due to increased rate of formation or decreased rate of removal.

Cytological Action of Di-2-chloroethyl Aminophenyl Butyric Acid on Ascites Tumors of Mice and Rats

Animals (mice) bearing Krebs ascites and Sarcoma 37 ascites tumors respectively at the peak period of their growth were injected intraperitoneally with aqueous solutions of di-2-chloroethyl aminophenyl butyric acid in two concentrations (0.5 mg per cc and 0.5 mg per .5 cc). Similarly, Walker rats with ascites tumors were administered (1 mg per cc) a single dose of the chemical. Samples of the ascitic fluid were withdrawn at different time intervals (3, 6, 12, 24, 48, 72, 96, 120, 144, 168, and 288 hr after injection) from the treated animals to study the cytological effect of the drug on the tumor cells. The following results, based on an examination of representative samples of animals, give an indication of the characteristic cytological responses of the tumors to the drug.

Krebs ascites. Comparatively few cells were in mitosis 12 hr after injection of the drug. The mitotic index was still low at the end of 48 hr. Several abnormally large cells were present from the samples taken after 24 hr. The 72-hr collection revealed a fair number of abnormal mitoses especially in the giant cells, which were all polyploid—the range of ploidy being from 4n to 8n. The 96-hr samples showed, among other things, many cells with bizarre-shaped nuclei, aberrant tri- and multipolar mitoses, multinucleated giant cells, irregular scattering of chromosomes, misdivision of the centromeres, attenuated chromosomes with extensive fragmentations, errors in anaphase separation, and failure of synchronization in the mitotic apparatus. In addition there were cells in mitosis with persistent micronuclei. Not infrequently the micronuclei were also in division. The general picture was one of a chaotic upset of the whole mitotic mechanism. Another interesting feature was the persistence of nucleoli at all stages of the mitotic cycle and the presence of distinct moieties in the cytoplasm giving a characteristic stain for RNA. However, the staining reaction was no longer evident when the cells were pretreated with RNAse, suggesting that the cytoplasmic inclusions were in the nature of ribonucleotides. The cytological architecture continued to reveal its abnormal configurations in a more or less accentuated fashion until the conclusion of the experiments, 288 hr after the original injection of the drug. Towards the terminal stages the number of degenerating cells as well as polymorphs and leucocytes appeared to have increased.

Sarcoma 37 ascites. There was the usual initial mitotic lag over the 12-hr range, but by the end of 24 hr a definite resumption of the normal mitotic process was evident. Like the Krebs ascites tumor, giant cells were also seen. The 48–72-hr samples exhibited many aberrant mitoses associated with fragmentation of chromosomes not unlike those encountered in the Krebs, but this was much less exaggerated. Spindle abnormalities and superfragmentation appeared after 96 hr, followed by a gradual disintegration of the tumor cells in the succeeding hours. The disruption and superfragmentation appeared after 96 hr, followed by a gradual disintegration of the tumor cells in the succeeding hours. The disruption of the abnormal cells was complete after 192 hr. While the over-all response pattern of Sarcoma 37 to the action of di-2-chloroethyl aminophenyl butyric acid in the doses administered was basically similar to that observed in the Krebs tumor there were nevertheless noticeable differences between the two tumors. Whereas there was a sort of protracted mitotic hysteresis in the case of the Krebs tumor, recovery was quicker with Sarcoma 37. Again the end result of the reactivity of the two tumors was also different.

Walker ascites tumor. The course of the tumor cells under the action of the drug was comparable to that of the Krebs in one respect—the absence of any mitosis up to a period of 72 hr. Giant cells were also present with associated micronuclei. After 96 hr the giant cells had increased in number but the mitotic index was still small. The mitoses were characterized by many types of abnormalities such as excessive fragmentation, anaphase bridges, and pycnotic degeneration.

Cytological Action of Di-2-chloroethyl-*p*-aminophenyl Butyric Acid on *Vicia faba* and *Luzula purpurea*

With a view to eliciting as much information as possible on the mode of action of chemical mutagens and radiomimetic substances on different biological systems it was decided to investigate the nature of chromosome breakage and sensitivity during the nuclear cycle of somatic mitosis, using certain plant systems.

Normally proliferating meristematic cells of *Vicia* root tips and actively dividing zones of *Luzula* seedlings were treated with the carboxylic nitrogen mustard derivative di-2-chloroethyl-*p*-aminophenyl butyric acid, under conditions of growth, treatment, and fixation procedures similar to those adopted by Revell (Refs. 9, 10). The cytological effects in so far as they related to chromosomal aberrations were in the main scored at metaphase following treatment with different dose intensities and at different time intervals.

The types of aberrations commonly encountered in *Vicia* consisted of both chromosome (B″) and chromatid (B′) breaks with their associated centric (C''_1) and acentric (C''_0 and C'_0) fragments. In addition, translocation, i.e., chromatid interchanges, as well as anomalous recombinations like triradials (≡ chromatid isochromatid interchange), were also present. Although the interchanges were generally between homologous chromosomes, translocations between heterologous chromosomes were also seen in a proportion of cases. Moreover, the later samples of cells showed an increasing proportion of T2 cells. Detailed quantitative analysis of the breakage frequency and translocation is in progress. The balance of evidence seems to support the view that di-2-chloroethyl-*p*-aminophenyl butyric acid like the diepoxide investigated by Revell has an indirect action on the chromosomes, and presumably the mechanism of action involves an interference with some precursor or precursors concerned with the synthesis of chromosomal material; for it is most unlikely that radiomimetic agents could be supposed to "break" chromosomes in the literal sense. They might produce an effect like breakage either by causing a local inhibition of chromosome synthesis (i.e., a sort of biochemical lesion) or by interfering with the proper functioning of certain postulated labile sites which normally break down at meiotic crossing over. Since a normal chemical bond is associated with 20,000 to 30,000 K per mol and an equally large energy of activation, it is difficult to see how the chemical mutagen can actually *break* such primary valences. In this connection there is also the collateral evidence of the experiments on intact salivary glands of *Chironomus* about which details have been given in an earlier paper by Ambrose and Gopal-Ayengar (Ref. 11).

The high incidence of interchanges and triradials between obviously homologous chromosomes as observed by Revell in his diepoxide-treated material and again in the present experiments would seem to imply an earlier association between homologues and a subsequent exchange amounting in essence to somatic crossing over induced by the carboxylic nitrogen mustard derivative. This concept gains credence especially in the light of the fact that the number of interchanges is independent of dose intensity. For a given concentration-time product, the effects of di-2-chloroethyl-*p*-aminophenyl butyric acid, like the diepoxide investigated by Revell, were the same, both qualitatively and quantitatively, over a wide range of intensities.

Only a brief mention will be made here of the other plant species investigated. *Luzula purpurea* is unique in its cytological behavior. Besides their low number ($n = 3$), the chromosomes are characterised by the absence of any distinct or visible centromeres. Castro *et al.* (Ref. 12) from X-ray studies came to the conclusion that the centromere was of a "diffuse or manifold" nature. However, on the basis of his investigations of the effect of X rays on the meiotic chromosomes of the species, La Cour (Ref. 13) has favored a "polycentric" rather than the "diffuse" centromere system. In addition to the unique nature of the centromere, *Luzula* has an unusual meiotic mechanism in that the I division is equational instead of being reductional.

The cytological aberrations that were observed after treatment with the chemical mutagen were mostly in the nature of chromosomal (B″) and chromatid (B′) breaks. Chromatid interchanges were very infrequent and anomalous recombinations rare. No micronuclei were found through several mitotic cycles. Presumably the fragment chromosomes by virtue of their complex centromeres are able to restitute themselves and are not lost as would be the case with acentric fragments. On this basis as well as the fact that occasionally configurations are seen wherein the chromosomes have distinctly organized centromeres it would appear that *Luzula* has a "polycentric" or "multicentric" type of structure. Further investigations are in progress.

Radiation Effects on the Meiotic Chromosomes of *Chorthippus*

Preliminary observations made on the effect of variation of doses of X rays on the frequencies of breaks in particular chromosomes (the L bivalents) of *Chorthippus* has given interesting results in terms of similar aberrations occurring spontaneously in nature.

References for Chapter 14

1. S. H. Revell, "A new interpretation of the chemical induction of chromosomal aberrations," *Br. Emp. Cancer Camp. Ann. Rep.* 30:42–54 (1952).
2. S. H. Revell, "A new hypothesis for 'chromatid' changes," in *Radiobiology Symposium (Liége)*, Butterworth's Scientific Publication, London, 1955, pp. 243–253.
3. L. H. Gray, "The 'contact first' hypothesis." Discussion, *ibid.*, p. 277.
4. H. J. Muller, "An analysis of the process of structural change in chromosome of *Drosophila*," *J. Genetics* 40:1–66 (1940).
5. G. Klein and A. Forssberg, "Studies on the effect of X-rays on biochemistry and cellular composition of ascites tumours," *Exptl. Cell. Research* 6:211–220 (1954).
6. J. S. Mitchell, "Distribution of nucleic acid metabolism produced by therapeutic doses of X- and γ-radiations: Part III. Accumulation of pentose nucleotides in cytoplasm after irradiation," *Brit. J. Exp. Path.* 23:296–309 (1942).
7. B. P. Kaufman, *et al.*, "Patterns of organization of cellular materials," *Carnegie Inst. Washington. Yr. Bk.* 51:224 (1952).
8. K. Paigen, B. N. Kaufman, and R. Wood, "Cytoplasmic organization at the Biochemical level," *ibid.*, pp. 227–229 (1952).
9. S. H. Revell, "Chromosome breakage by X-rays and radiomimetic substances in *Vicia*" (Symposium on Chromosome Breakage), *Heredity* Supp. 6:107–124 (1953).
10. S. H. Revell, *A Cytological Investigation of the Action of X-rays and Radiomimetic Chemicals,* Thesis, London University, 1952.

11. E. J. Ambrose and A. R. Gopal-Ayengar, "Molecular orientation and chromosome breakage" (Symposium on Chromosome Breakage), *Heredity* Supp. 6:277–292 (1953).
12. De Castro, A. Camara, and N. Malheiros, "X-rays in the centromere of *Luzula purpurea* Link." *Genetica Iberica* 1:48–54 (1949).
13. L. F. La Cour, "The *Luzula* system analysed by X-rays" (Symposium on Chromosome Breakage), *Heredity* Supp. 6:77–81 (1953).

Part IV

SOILS AND FERTILIZERS

Chapter 15

USE OF RADIOISOTOPES IN SOIL AND FERTILIZER STUDIES *

The theoretical and applied aspects of soil-plant relationships, soil fertility, and fertilizer use have been under consideration for many years. The application of radioisotopes to these fields of study, although relatively new, has done much to advance our knowledge in these areas, and this trend will undoubtedly continue. During the past fifteen years about 300 titles have appeared in the scientific literature which relate to the application of radioisotopes in the study of soils and fertilizers. This chapter will review most of these studies in an attempt to evaluate progress resulting from the application of radioisotopes. A number of reviews and general discussions dealing with isotope usage in studies of soils and fertilizers have preceded this consideration on the subject. The reader is referred to Refs. 122, 24, 96, 23, 77, 53, 11, 38, 29, 104, 55, 9.

The majority of the investigations which have utilized radioisotopes have dealt with some phase of the phosphorus problem. In addition to the relative importance of this problem the isotope P^{32} has a number of desirable properties from the standpoint of an experimenter. Among these are its relatively short half-life and strong beta particle of 1.71 mev, which tend to minimize disposal and measurement problems. However, as experience and facilities have developed, research projects involving a number of isotopes have become relatively common; these include Ca^{45}, S^{35}, Zn^{65}, K^{42}, C^{14} and others.

The employment of radioisotopes allows the undertaking of experiments and measurements where ordinary chemical and physical measurements are not practicable. Two underlying principles are involved. (1) Populations of ions or molecules can be labeled and readily detected, traced, or a quantitative determination made of dilution by change in specific activity. (2) Ready identification and measurement at extremely low concentrations are possible. Radioisotopes have been used to advantage in a variety of investigations, in-

* This chapter is taken from Geneva Conference Paper 104, "Applications of Radioisotopes to the Study of Soils and Fertilizers: A Review" by L. A. Dean of the United States.

cluding ion mobility in soils, distribution and growth of roots, uptake and exchange phenomena, plant nutrition, and certain problems of analytical chemistry involving the soil system.

Ion Mobility in Soils

Radioisotopes are convenient tools for studying the movement of elements in the soil. Their use for this purpose seems to have been largely neglected. Substances move through soils by diffusion and as the result of mass flow of the soil solution. On the other hand, various processes may restrict mobility. Of primary interest is the movement of the constituents of soil additives such as liming and fertilizer materials.

The earliest report on the use of radioisotopes to study movement in soils was in 1941 (Ref. 52). Monocalcium phosphate labeled with P^{32} was placed on the surface of soils and water applied to simulate 2.5 in. of rain. Movement of phosphorus varied from 1.25 in. in a Cecil clay to 4 in. in a Crosby silt loam. Not more than 5% of a potassium application labeled with K^{42} moved over 1.6 in.

The movement of phosphate ions away from the site where they are introduced into a soil is limited by the phosphate fixation processes. Ordinary chemical methods do not permit a study of this movement in adequate detail because of the relatively large quantity of phosphorus already present in most soils. Through the use of radiophosphorus it has been possible to trace the movement of phosphatic fertilizers under actual field conditions. Researchers (Ref. 125) added 2300 lb P_2O_5 per acre of P^{32}-labeled phosphoric acid to a soil in conjunction with 4 in. of irrigation water. The phosphorus penetrated about 12 in. and the water about 20 in. After 11 and 43 days, 99% and 86% respectively of this phosphorus were in the first 6 in. Others (Ref. 89) compared the distribution of band-placed superphosphates with that of phosphoric acid applied in irrigation water. The movement of phosphorus from the superphosphate was much less than for the liquid phosphoric acid. Very little superphosphate moved beyond 3 in. The phosphoric acid moved to a depth of 12 in., but 85% of the total application was found in the top 4 in. Increasing increments of irrigation water were shown (Ref. 66) to increase the downward movement of broadcast applications of superphosphate. Another field study (Ref. 35) showed that only 2.5% to 5.2% of the superphosphate top-dressed on permanent pasture moved more than 1 in.

The dispersion of salt solutions labeled with P^{32} injected in the soil was studied (Ref. 75), and graphs were prepared for the distribution in different soils. Rainfall did not appear to affect the distribution of the P^{32}. Other studies have been made (Refs. 33, 54) of the leaching of phosphate fertilizers. Procedures have been suggested (Ref. 65) for applying radioautographic techniques to soil sections in the study of fertilizer phosphorus movement.

The foregoing discussion has dealt with the movement of fertilizer phosphorus under the influence of the penetration of rain and irrigation waters. The dissolution and migration in the absence of mass flow have also been studied; and (Ref. 56) the one-dimensional diffusion of phosphorus from solid phosphatic fertilizer into the adjacent soil was studied by experiments employing fertilizers tagged with P^{32}. The extent of diffusion of different fertilizer materials increased with the fraction of the fertilizer phosphorus that was in water-soluble form. The distance of diffusion increased with time and rate of phosphorus application. In 4 weeks the fertilizer phosphorus diffused only 3 to 4 cm. Under otherwise uniform conditions, diffusion of fertilizer phosphorus differed between soils and was considerably less in calcareous soils than in acid soils. The rapid dissolution of phosphorus from granular superphosphate in contact with moist soil was experimentally vertified (Ref. 72). Even in soils as low as 2–4% moisture, 20–50% of the phosphorus moved from the granules to the soil in one day. Tests were made (Ref. 18) of the validity of activity measurements as estimates of phosphorus diffusion from tagged phosphate sources. Although significant changes in the apparent specific activity of diffusing phosphorus were found, the over-all picture of phosphorus diffusion was not substantially different from that found by total phosphorus analysis. Irregularities in phosphorus distribution by diffusion were found, and the possibility of periodic precipitation was postulated.

The leaching of calcium in soils and its relation to problems of soil acidity have long been investigated. Two sets of laboratory experiments (Refs. 99, 12) dealing with the movement of calcium from Ca^{45}-labeled compounds under severe leaching conditions gave essentially the same findings. Apparently calcium moves downward in the soil by a continuous series of displacements, the calcium appearing in the leachate being that originally present in the lower layers of the soil. In lysimeter studies using Ca^{45}-labeled calcium carbonate and a calcium silicate slag (Refs. 50, 51, 25), varying rates were used and the additive calcium compounds were mixed throughout the soil mass. The leachates were examined for Ca^{40} and Ca^{45}. The interpretation of the findings was complicated by the rates of dissolution of the added calcium and by incomplete isotopic exchange with the native soil calcium.

Thoroughness of mixing soil amendments such as lime greatly affects their efficiency. Little information is available as to the soil-mixing characteristics of various tillage implements, although their tilth-creating characteristics may be well known. Radioactive phosphorus was effectively used as a tracer in an experiment (Ref. 58) where a series of tillage operations were compared.

Root Distribution and Activity

A thorough understanding of the development and activity of plant roots in their soil environment is needed in the development of scientific crop production. Most of the root studies conducted in the past have entailed some system whereby a direct observation of root systems is made. Some workers

(Refs. 48, 49) developed a tracer technique to measure growth and activity of plant root systems. The method consisted of strategically locating a small quantity of P^{32} at given distances from and below plants growing in the field and measuring its uptake by plants at various time intervals. Corn, cotton, peanuts, and tobacco were studied. Cotton was found to be the most shallow-rooted of these crops, while tobacco showed the greatest root activity at 16 in. below the surface. A similar technique was employed (Ref. 20) to investigate root penetration of grasses in relation to drought tolerance. Radiophosphorus was placed in the soil at depths up to 8 ft, and various species of grasses were planted. Root penetration measured in this way was found to be related to drought tolerance. Tracer techniques were employed (Ref. 76) to study the cross feeding in a muscatine grape vineyard.

A technique involving the use of tracers in studying the penetration into and absorption of nutrient from dry soils was explored (Refs. 59, 60).

Exchange Phenomena

There are a number of exchange reactions which are of importance to soil chemistry and plant nutrition. The study of these is frequently facilitated through the use of radioisotopes. The general expression for an isotopic exchange reaction is given by the equation

$$A^*Y + AX = AY + A^*X$$

where A^* represents the radioactive isotope and A the stable isotope. Starting with A^*Y, measurement of the rate of exchange can be accomplished by noting the appearance of A^*X or disappearance of A^*Y. The first applications to soils were a series of studies dealing with the "contact" and "soil solution" theories of ion absorption (Refs. 63, 64). Through the use of Na^{24}, K^{42}, and Rb^{86} the rates of absorption of cations by barley roots was shown to be greater with soil and clay suspensions containing absorbed sodium, potassium, and rubidium than from the corresponding filtrates. Thus it was shown that there is no clear relationship between the uptake of cations in the presence of clay surfaces and ion absorption from corresponding salt solutions.

The equilibrium of cations between solutions and exchange materials such as soils, clay minerals and resins has been followed with radiocations. The exchangeability of trace concentrations of cations as influenced by degree of saturation and nature of complementary ion were studied (Ref. 129). Using K^{42} and Sr^{90}, the researchers found that a certain limit value is approached which depends on the exchange ion, complementary ion, and exchanger used. Wiklander (Ref. 130) has observed that the rate of exchange between the exchangeable and fixed potassium of soils is quite slow. Co^{60} was employed (Refs. 2, 112) to investigate the chemistry of cobalt absorption and release in cation-exchange systems. Easily exchangeable and strongly adsorbed cobalt were distinguished and cobalt was found to be harder to replace than calcium. K^{42}, Ca^{45}, Rb^{86}, Cs^{137} and La^{140} were also used (Ref. 70) to test the validity of several expressions for the equilibrium of cation exchange reactions.

Kinetic exchange studies on clays (Ref. 17) demonstrated that Ca^{45}, added to clay suspensions, equilibrated rapidly and completely with the exchangeable calcium adsorbed on the clay minerals kaolinite, halloysite, hydrous mica, beidellite, and montmorillonite. Advantage was taken of this principle and a number of workers (Refs. 5, 108, 14) have proposed isotopic dilution methods for determining exchangeable calcium, base exchange capacity, and the rate of reaction of liming materials in soils. Such methods have certain advantages, especially in the study of soils containing free calcium carbonate.

Only a fraction of the amount of phosphorus necessary for crop production exists in the soil solution at any given time. If, for the purposes of this discussion, it is assumed that the soil solution is the sole source of phosphate for plants, then it would be necessary for this solution to be renewed many times during the cropping season. P^{32} was applied (Ref. 78) to a study of the exchange between solution and solid-phase phosphate. A part of the soil phosphate was found to rapidly equilibrate by exchange with the ambient solution. The amount of this rapidly equilibrating surface phosphate was found to parallel other estimates of phosphorus availability. A slower, less reversible exchange of phosphates was also recognized.

The mechanism of phosphate sorption and the kinetics of phosphate exchange in soils were investigated with the aid of P^{32} (Ref. 131). It was concluded that the soil phosphorus occurs in essentially two different ways: namely, a fairly reactive fraction and a fraction of low solubility which is difficult to mobilize but tends to be in equilibrium with the first fraction. A series of investigations (Refs. 3, 4, 7, 8) dealt with the exchange of phosphate ions within the solid phase of soils. They have also demonstrated an exchange between several of the forms of soil phosphorus. Related studies have been reported (Refs. 106, 34).

The amount of phosphate on the surface of soil particles and apatites has been measured in several studies (Refs. 90, 91, 92, 22). The phosphate associated with the solid phase which readily equilibrated with P^{32} in the ionic form was termed surface phosphate. Good correlations were found between the amount of this surface phosphate in soils and the plant-available phosphorus. From 19% to 31% of the phosphorus added in long-time rotation experiments was found to accumulate as surface phosphate. Monolayer absorption of phosphate on calcium carbonate and calcareous soils was found from dilute solutions, but with higher concentrations the percentage of the phosphorus absorbed appearing as surface phosphate decreased. Similar findings, but using an iron-phosphate system, were reported by Fried and Dean (Ref. 40).

Absorption of Nutrients from Soil and Fertilizer

Starting in 1947 with experiments (Refs. 113, 26, 85) there have been a large number of greenhouse and field experiments conducted utilizing fertilizers labeled with radioisotopes. Fertilizer of known specific activity in respect to phosphorus was incorporated in soil and a crop grown. The ratio of the specific activity of the phosphorus in the crop to that of the original fertilizer has been taken as being indicative of the fraction of the total phosphorus absorbed by the crop that was derived from the fertilizer. This fraction has been shown to increase with increasing rate of fertilizer application and to be lowest for soils high in available phosphorus.

Four groups of workers have proposed that the phosphorus status of soils may be measured by comparing the specific activity of the phosphorus absorbed by plants with that of the fertilizer added to soil (Refs. 71, 37, 45, 6). The methods employed by these investigators were all essentially the same, however the mathematical treatment and interpretation of results varied. The expression derived for estimating the available phosphorus supply of the soil has several forms, the simplest being

$$A = B\left(\frac{S_f}{S_p} - 1\right)$$

where A and B are the amount of soil and fertilizer phosphorus and S_f and S_p the specific activities of the fertilizer and plant phosphorus, respectively. This method has been applied to a number of investigations dealing with the phosphorus fertility status of soils and has been one of the tools used in calibrating soil tests (Refs. 27, 93, 95, 124). This method for measuring the phosphate status of the soil which is based on a comparison of the specific activity of the fertilizer phosphorus with that of the phosphorus absorbed by crops has a number of experimental limitations as was pointed out (Ref. 37). Russell *et al.* (Ref. 103) from a consideration of isotopic equilibria between phosphates in soils consider the method to have serious limitations and suggest that an examination of the total phosphorus content of plants may be a better means. On the other hand, Dean (Ref. 28) reports that similar information regarding the soil's phosphorus supply is obtained with three methods—namely, the extrapolation of growth curves, the extrapolation of yield-of-phosphorus curves, and from radiochemical data. All three methods offer both advantages and limitations.

Phosphorus is applied to soils as a variety of materials ranging from commercial fertilizers to manures. Since the availability to plants of the phosphorus in these materials is known to vary, an accurate evaluation of each material is of considerable economic importance. Radiophosphorus has been utilized in a study of this problem; the principles involved are discussed in

Ref. 39. In general, the procedures are a counterpart of the techniques described previously for assessing the phosphorus status of soils. Instead of applying the same fertilizer to a variety of soils, a number of fertilizers are applied to the same soil. Special techniques for preparing a number of different tagged fertilizer materials for this purpose are described in Ref. 57. Fertilizer evaluation by this method has both advantages and limitations. The possibility that isotopic exchange may interfere has been suggested (Ref. 80). The method is quite sensitive, and differences between fertilizers as measured by the relative amount of soil and fertilizer phosphorus absorbed may not, in turn, always be reflected in crop yields.

A number of factors are known to affect the availability of the different phosphate fertilizer materials. Soil properties, crop growth, time and method of application have all been shown to influence the availability. In all instances, superphosphate has been adopted as a basis of comparison. With but few exceptions, this material has been found to supply as much if not more phosphorus to plants than other materials, as indicated by tests utilizing P^{32}.

The chemistry of phosphorus in neutral and alkaline soils, where calcium predominates in the system, differs from that of acid soil systems. Thus, the behavior and effectiveness of the various fertilizer materials added to these two systems may not be the same. P^{32}-labeled fertilizers were used (Refs. 88, 89) in the comparison of a number of materials on alkaline soil; wheat, barley, sugar beets, and alfalfa were among the test crops. The phosphorus available for plant growth was generally in the order: superphosphate = monoammonium phosphate > calcium metaphosphate > dicalcium phosphate > tricalcium phosphate. Dicalcium was always inferior to superphosphate, but the results with calcium metaphosphate varied with time, soil, and crop. Superphosphate, monoammonium phosphate and liquid phosphoric acid were compared (Refs. 44, 89, 1) and little difference was found in the effectiveness of these sources. On the other hand, studies (Refs. 31, 32) with wheat suggest that ammonium and sodium phosphate are more effective than monocalcium phosphate. Investigations (Ref. 105) of the effectiveness of the Rhenania type phosphate on calcareous soils have shown it equal to superphosphate except when applied broadcast. A pot experiment comparing the relative efficiency of various phosphatic fertilizers on both an acid and an alkaline soil (Ref. 110) has shown forms of low water solubility to be less effective on the alkaline soil.

A number of fertilizer source comparisons, similar to those discussed above but dealing with acid soils, have been reported. These include studies with cotton and corn (Ref. 47), oats and alfalfa (Ref. 118), pasture species (Ref. 10), and soybeans (Ref. 19). The findings were in some respect different from those reported for the alkaline soil region. In general, calcium metaphosphate was a more effective material when applied to acid soils. Dicalcium phosphate, when mixed with acid soils, was also relatively more effective. Studies with neutron-irradiated phosphate rock (Ref. 36) have demonstrated the marked influence of soil acidity on the effectiveness of this material as a phosphate fertilizer. Reports dealing with the availability of granulated phosphatic fertilizers (Refs. 116, 16, 68) show an interaction between particle size, water solubility of the fertilizer, and soil acidity. Generally speaking, all of the conclusions drawn from the experiments with P^{32}-labeled fertilizer materials parallel those arrived at by a study of crop yield. However, the P^{32} experiments do show a relatively greater difference between the materials compared.

Radiophosphorus has also been used in studies designed to compare the availability of the phosphorus contained in organic farm residues with that of superphosphate. Because of the limited amount of suitably labeled material that it is practical to produce, all comparisons were made in greenhouse pot culture. Fuller and co-workers (Refs. 41, 42, 43), White *et al.* (Ref. 128) and Nielsen *et al.* (Ref. 87) studied leguminous and nonleguminous crop residues, while McAuliffe *et al.* (Refs. 79, 80) applied similar techniques to a study of sheep manure. The availability of the phosphorus in the materials studied compared favorably with that in superphosphate. In considering these findings, perhaps it should be recognized that more than 50% of the phosphorus in the materials studied was in inorganic forms.

The half-life of P^{32} precludes studies which extend for a period exceeding 120 days. Thus a direct measurement of the residual value to succeeding crops of phosphatic fertilizers is not possible. However, the effects of previous fertilizer application can be studied through an extension of the technique involving radioactive-labeled fertilizer. In one study (Ref. 81) the effect of phosphorus applied 8 years previously was measured by comparing the percentage of phosphorus in the crop derived from it, with that obtained from P^{32}-labeled superphosphate applied to the soil. A highly significant decrease in the percentage of phosphorus derived from the fertilizer as a result of past phosphorus applications was observed, thus demonstrating the presence of residual phosphorus in available forms. More quantitative estimates of residual values are possible if *A* values (Ref. 37) are calculated and compared. This technique was used (Ref. 126) to study the residual effect of a number of fertilizer materials. Superphosphate and calcium metaphosphate applied at equal rates of P_2O_5 over a period of years appeared to have an equal effect upon the residual level of available soil phosphorus. Fused tricalcium phosphate compared favorably with these materials on an acid soil, but was less effective on a calcareous soil. Phosphate rock at equal or higher rates was less effective on all four of the soils studied. Others (Refs. 94, 102, 97, 121, 83) have also used P^{32}-labeled fertilizers to study the residual effect of previous phosphorus applications.

Fertilizer Utilization

Phosphorus. The mineral nutrition of crops leads to a consideration of the amounts and rates of absorption. Recent studies involving the use of P^{32} have supplied additional information regarding the influence of soil and fertilizer phosphorus on phosphorus absorption. A number of crops growing under a variety of conditions have been studied, including cereals, pasture species, potatoes, sugar beets, sugar cane, tobacco, and cotton (Refs. 15, 30, 67, 69, 61, 62, 82, 120, 123, 114, 115, 127, 133). With few exceptions, the results of these studies have followed a rather typical pattern. Phosphate fertilizer is usually applied at the time of planting and this application is localized in the proximity of the seed. In the early stages of growth, seedlings absorb phosphorus at a faster rate from the fertilizer than from the soil, but as the plant develops and the root system enlarges, the amount of soil phosphorus accessible increases. Thus the absorption of soil phosphorus increases as compared with fertilizer phosphorus. The net result is that young plants characteristically have a higher percentage of phosphorus in the crop derived from the fertilizer than mature plants. This pattern of decline in the percentage of phosphorus in the crop derived from the fertilizer with maturity varies with crop species. Relatively little decline has been observed for potatoes as compared with corn. This difference between crops is attributed to differences in root systems and their development.

Crops utilize phosphorus from applications of phosphatic fertilizers with low efficiency. The experiments which compare the phosphorus absorption by crops also supply information relative to the efficiency of fertilizer utilization. The percentage of the phosphorus in a given application which was absorbed and thus utilized by the crop grown during the first season was found to vary between the limits of 2% and 35%. With increasing rates of fertilizer application, the total amount of fertilizer phosphorus absorbed usually increases, but the efficiency of utilization usually decreases. The question is frequently asked: Does the application of fertilizer phosphorus influence the amount of soil phosphorus absorbed by plants? With many soils the addition of fertilizer phosphorus has little effect on the amount of soil phosphorus absorbed. Under some conditions of phosphorus deficiency where the phosphate fertilizer increases root growth, the addition of fertilizer phosphorus increases the amount of soil phosphorus absorbed. Other experiments have shown that adding fertilizer phosphorus actually decreases the amount of soil phosphorus absorbed. The exact mechanism involved is not understood.

Soil management practices such as fertilizer placement have been designed to enhance the efficiency of fertilizer use by crops. Many of the studies with P^{32}-labeled fertilizers included comparisons of methods of fertilizer application. Experiments with corn and cotton (Refs. 86, 117) and sugar beets (Refs. 88, 73) are cited as examples. The most pronounced effects of placement of superphosphate on the utilization of this fertilizer phosphorus were observed during the early stages of growth. The closer this material was placed to the seed, the greater its utilization. The pronounced differences observed during the early stages of growth largely disappeared as the crop matured. Broadcast applications are the most convenient method of applying fertilizers to pastures and other established stands of crops. Since it is known that phosphates penetrate soils only to a limited degree, this method has not been considered a very satisfactory means of applying phosphate fertilizers. In one study (Ref. 119) established meadows were top-dressed with P^{32}-labeled superphosphate. This method of application was found to be surprisingly effective. Subsequent experiments (Refs. 21, 74) with established stands of legume hays have confirmed these findings.

Phosphate fertilizer utilization has been shown to be affected by liming, nitrogen fertilization, and irrigation practices. In studies (Refs. 84, 101) of phosphorus utilization from soils of varying sesquioxide content as influenced by lime applications, liming above pH 6.5 reduced the fertilizer utilization regardless of the sesquioxide content of the soil. One study (Ref. 100) demonstrated that more phosphorus was utilized from a mixture of phosphate and nitrogen fertilizers applied to corn than when these same materials were applied in separate bands. Nitrogen fertilization has been shown to enhance the absorption of fertilizer phosphorus (Refs. 98, 109). The frequency of irrigation of sugar beets increases the uptake of fertilizer phosphorus (Ref. 46).

Other elements. The foregoing discussion on the use of radioisotopes to study fertilizer utilization has been limited exclusively to P^{32}; however, other isotopes have been used on several occasions. The average mixed fertilizer sold in the United States contains 17% CaO. Under conditions of low soil calcium supply, it seemed possible that crops might obtain substantial quantities of their calcium from the fertilizer applied. To test this premise, a field experiment with tobacco involving the use of Ca^{45} was carried out (Ref. 13). The results indicated that plants derived only 2% to 6% of their calcium from the fertilizer. In contrast, 25% to 36% of the phosphorus absorbed was derived from this fertilizer. Studies on the plant uptake of Zn^{65} from soils and fertilizers have been confined to greenhouse experiments (Refs. 107, 111, 132). The efficiency of utilization of average applications of zinc fertilizers was lower than that observed for average phosphate applications. Liming was shown to reduce the uptake of zinc applied to soils.

References for Chapter 15

1. S. E. Allen, R. J. Speer, and M. Maloney, "Phosphate fertilizers for the Texas blacklands: II. Utilization of phosphate as influenced by plant species and by placement and time of application," *Soil Sci.* 77:65–73 (1954).
2. D. K. Banerjee, R. H. Bray, and S. W. Melsted, "Some aspects of the chemistry of cobalt in soils," *Soil Sci.* 75:421–31 (1953).

3. G. Barbier and E. Tyszkiewiez, "Agitation permanente des ions phosphoriques retenus par une argile de sol," *C. R. Acad. Sci.* 235:1246–48 (1952).
4. G. Barbier and E. Tyszkiewiez, "Mobilité des ions phosphoriques 'fixes' dans le sol etudiée au moyen de P^{32}," *Trans. Int. Soc. Soil Sci. Comm. II and IV, 1952,* 2:79–82 (1953).
5. G. Barbier and E. Tyszkiewiez, "Etude par échange isotopique du calcium diffusable des sols calcaires," *C. R. Acad. Sci.* 236:2105–6 (1953).
6. G. Barbier and M. Lesaint, "Définition au moyen d'isotopes de P_2O_5 assimilable du sol et des engrais," *C. R. Acad. Sci.* 238:1532–4 (1954).
7. G. Barbier and E. Tyszkiewiez, "Sur la détermination, par échange isotopique des ions $(PO_4H_2)^-$ autodiffusibles du sol," *C. R. Acad. Sci.* 238:1733–5 (1954).
8. G. Barbier and E. Tyszkiewiez, "Extension, par desiccation, de l'auto-diffusion des ions phosphoriques dans le sol," *C. R. Acad. Sci.* 238:1908–10 (1954).
9. H. Behrens and C. E. Fernández, "Aplicaciones analíticas del fósforo radioactivo (P^{32}) al estudio de algunos problemas agricolas," *An. Edofol. Fisiol. Veg.* 12:675–93 (1953).
10. R. E. Blaser and C. McAuliffe, "Utilization of phosphorus from various fertilizer materials: I. Orchard grass and ladino clover in New York," *Soil Sci.* 68:145–50 (1949).
11. J. M. Blume, "Radioisotopes in soil fertilizer research," *Plant Food J.* 6:2–5, 13 (1952).
12. J. M. Blume, "Leaching of calcium in a fine sandy loam as indicated by Ca^{45}," *Soil Sci.* 73:383–9 (1952).
13. J. M. Blume and N. S. Hall, "Calcium uptake by tobacco from band applications of fertilizer materials," *Soil Sci.* 75:299–306 (1953).
14. J. M. Blume and D. H. Smith, "Determination of exchangeable calcium and cation-exchange capacity by equilibration with Ca^{45}," *Soil Sci.* 77:9–17 (1954).
15. J. A. Bonnet and A. Riera, "Absorption by sugar cane of phosphorus from tagged superphosphate added to a phosphorus-fixing latosol," *Soil Sci.* 76:355–9 (1953).
16. N. I. Borisova, "The entry plants of phosphorus from granulated and non-granulated phosphates applied simultaneously," *Izv. Akad. Nauk. Ser. Biol.* No. 1, 110–12 (1954).
17. J. W. Borland and R. F. Reitemeier, "Kinetic exchange studies on clays with radioactive calcium," *Soil Sci.* 69:251–60 (1950).
18. D. R. Bouldin and C. A. Black, "Phosphorus diffusion in soils," *Soil Sci. Soc. Amer. Proc.* 18:255–9 (1954).
19. M. F. Bureau, H. J. Mederski, and C. E. Evans, "The effect of phosphatic fertilizer material and soil phosphorus level on the yield and phosphorus uptake of soybeans," *Agron. J.* 45:150–4 (1953).
20. G. W. Burton, E. H. DeVane, and R. L. Carter, "Root penetration, distribution and activity in southern grasses measured by yields, drought symptoms and P^{32} uptake," *Agron. J.* 46:229–33 (1954).
21. A. C. Caldwell, A. Hustrulid, *et al.*, "Absorption of phosphorus from radioactive phosphate fertilizer applied to established meadows," *Soil Sci. Soc. Amer. Proc.* 18:440–3 (1954).
22. C. V. Cole, S. R. Olsen, and C. O. Scott, "The nature of phosphate sorption on calcium carbonate," *Soil Sci. Soc. Amer. Proc.* 17:352–6 (1953).
23. P. A. Collier, "The use of isotopes in agriculture," *Chem. & Ind.* 1122–24 (1951).
24. C. L. Comar and J. R. Neller, "Radioactive phosphorus procedures as applied to soil and plant research, *Plant Physiol.* 22:174–80 (1947).
25. D. E. Davis, W. H. MacIntire, *et al.*, "Use of Ca^{45}-labeled quenched calcium

silicate slag in determination of proportions of native and additive calcium in lysimeter leachings and in plant uptake," *Soil Sci.* 76:153–63 (1953).

26. L. A. Dean, W. L. Nelson, *et al.*, "Application of tracer technique to studies of phosphatic fertilizer utilization by crops: I. Greenhouse experiments," *Soil Sci. Soc. Amer. Proc.* 12:107–12 (1948).
27. L. A. Dean, "The evaluation of phosphorus fertility of soils through the medium of radiophosphorus," *Trans. Int. Soc. Soil Sci. Comm. II and IV, 1952,* 1:48–52 (1952).
28. L. A. Dean, "Yield-of-phosphorus curves," *Soil Sci. Soc. Amer. Proc.* 18:462–6 (1954).
29. M. Deribère, "Utilisation des radioisotopes dans l'étude des engrais et dans les techniques agricoles," *Industr. Chem.* 38:33–6, 65–7 (1951).
30. H. G. Dion, J. W. T. Spinks, and J. Mitchell, "Experiments with radioactive phosphorus on the uptake of phosphorus by wheat," *Sci. Agr.* 29:167–72 (1949).
31. H. G. Dion, J. E. Dehm, and J. W. T. Spinks, "Study of fertilizer uptake using radioactive phosphorus: IV. The availability of phosphate carriers in calcareous soils," *Sci. Agr.* 29:512–26 (1949).
32. H. G. Dion, J. E. Dehm, and J. W. T. Spinks, "Tracer studies with phosphate fertilizers," *Cand. Chem. Process Ind.* 34:905–9 (1950).
33. T. Egawa, K. Magai, and A. Sato, "Tracing the behavior of phosphorus added to water-logged soil: Preliminary experiment with radioactive phosphorus," *J. Sci. Soil Man.* (*Japan*) 22:61–3 (1951).
34. T. Egawa, A. Sato, and K. Sekiya, "The exchangeability of phosphate ions absorbed by the soil," *J. Sci. Soil* (*Tokyo*) 23:249–52 (1953).
35. J. G. Fiskell, G. A. DeLong, and W. F. Oliver, "The uptake by plants of labelled phosphate from neutron-irradiated calcium phosphates: III. Penetration into soil and uptake by pasture herbage," *Canad. J. Agr. Sci.* 33:559–65 (1953).
36. M. Fried and A. J. MacKenzie, "Rock phosphate studies with neutron-irradiated rock phosphate," *Soil Sci. Soc. Amer. Proc.* 14:226–31 (1949).
37. M. Fried and L. A. Dean, "A concept concerning the measurement of available soil nutrients," *Soil Sci.* 73:263–72 (1952).
38. M. Fried, "Progress in agronomic research through the use of radioisotopes," *Proc. 4th Ann. Oak Ridge Summer Symp.*, 1953, pp. 452–80.
39. M. Fried, "Quantitative evaluation of processed and natural phosphates," *Agri. & Food Chem.* 2:241–4 (1954).
40. M. Fried and L. A. Dean, "Phosphate retention by iron and aluminum cation exchange systems," *Soil Sci. Soc. Amer. Proc.* 19:143–7 (1955).
41. W. H. Fuller and L. A. Dean, "Utilization of phosphorus from green manures," *Soil Sci.* 68:197–202 (1949).
42. W. H. Fuller and R. N. Rogers, "Utilization of phosphorus from barley residues," *Soil Sci.* 74:373–82 (1952).
43. W. H. Fuller and R. N. Rogers, "Utilization of the phosphorus of algal cells as measured by the Neubauer technique," *Soil Sci.* 74:417–30 (1952).
44. W. H. Fuller, *Effect of Kind of Phosphate Fertilizer and Method of Placement on Phosphorus Absorption by Crops Grown on Arizona Calcareous Soil,* Arizona Agri. Expt. Sta. Tech. Bul. 128, 1953.
45. O. Gunnarsson and L. Fredriksson, *Méthode pour déterminer, au moyen de P^{32}, la teneur du sol en phosphore "assimilable par la plante,"* Bull. Document, Ass. Int. Fabr. Superph. No. 11, 16–20, 1952.
46. J. L. Haddock, "The influence of soil moisture condition on the uptake of phosphorus from calcareous soils by sugar beets," *Soil Sci. Soc. Amer. Proc.* 16:235–8 (1952).

47. N. S. Hall, W. L. Nelson, *et al.*, "Utilization of phosphorus from various fertilizer materials: II. Cotton and corn in North Carolina," *Soil Sci.* 68:151–6 (1949).
48. N. S. Hall, W. F. Chandler, *et al.*, *A Tracer Technique to Measure Growth and Activity of Plant Root Systems,* North Carolina Agri. Expt Sta. Tech. Bul. 101, 1953.
49. N. S. Hall, "The use of phosphorus-32 in plant root studies," *Proc. 4th Ann. Oak Ridge Summer Symp.*, 1953, pp. 435–51.
50. H. C. Harris, W. H. MacIntire, *et al.*, "Radioactive calcium in the study of additive and native supplies in the soil," *Science* 113:328–9 (1951).
51. H. C. Harris, W. H. MacIntire, *et al.*, "Use of Ca^{45}-labeled calcium carbonate in determining proportions of native and additive calcium in lysimeter leachings," *Soil Sci.* 73:288–98 (1952).
52. W. J. Henderson and U. S. Jones, "The use of radioactive elements for soil and fertilizer studies," *Soil. Sci.* 51:283–8 (1941).
53. S. B. Hendricks and L. A. Dean, "Radioisotopes in soils research and plant nutrition," *Ann. Rev. Nuclear Sci.* 1:597–610 (1952).
54. W. Herbst, "Radiophosphor (P^{32}) als Hilfsmittel zur Bestimmung der Auswaschung von Bodenphosphorsäure durch Niederschlage," *Z. Pfl. Ernahr. Dung.* 59:139–44 (1952).
55. Herbst, W., "Radiophosphor in der wirtschaftlichen Forschung," *Phosphorsäure* 13:232–256 (1953).
56. J. M. Heslep and C. A. Black, "Diffusion of fertilizer phosphorus in soils," *Soil Sci.* 78:389–401 (1954).
57. W. L. Hill, E. J. Fox, and J. F. Mullins, "Preparation of radioactive phosphate fertilizers," *Ind. & Eng. Chem.* 41:1328–34 (1949).
58. W. C. Hulburt and R. G. Menzel, "Soil mixing characteristics of tillage implements," *J. Amer. Soc. Agri. Eng.* 34:702–4 (1953).
59. A. S. Hunter and O. J. Kelley, "The extension of plant roots into dry soil," *Plant Physiol.* 21:445–51 (1946).
60. A. S. Hunter and O. J. Kelley, "A new technique for studying the absorption of moisture and nutrients from soil by plant roots," *Soil Sci.* 62:441–50 (1946).
61. W. C. Jacob, C. H. van Middlem, *et al.*, "Utilization of phosphorus by potatoes," *Soil Sci.* 68:113–20 (1949).
62. W. C. Jacob and L. A. Dean, "The utilization of phosphorus by two potato varieties on Long Island," *Amer. Potato J.* 27:439–45 (1950).
63. H. Jenny and R. Overstreet, "Contact effects between plant roots and soil colloids," *Proc. Nat. Acad. Sci. U.S.A.* 24:384–92 (1938).
64. H. Jenny, R. Overstreet, and A. D. Ayers, "Contact depletion of barley roots revealed by radioactive indicators," *Soil Sci.* 48:9–24 (1939).
65. W. B. Johnston, "Autoradiography of soil sections and its applications," *Soil Sci.* 78:247–55 (1954).
66. J. Jordan, C. Simpkins, *et al.*, "Uptake and movement of fertilizer phosphorus," *Soil Sci.* 73:305–13 (1952).
67. L. C. Kapp, J. R. Hervey, *et al.*, "Response of evergreen sweet clover and cotton to phosphorus applications on Houston black clay," *Soil Sci.* 75:109–18 (1953).
68. T. D. Koritskaya, "The use of radioactive phosphorus isotope for determining relative assimilability of various phosphates," *Izv. Akad. Nauk. Ser. Biol.*, No. 2: 111–15 (1954).
69. B. A. Krantz, W. L. Nelson, *et al.*, "A comparison of phosphorus utilization by crops," *Soil Sci.* 68:171–8 (1949).
70. C. Krishnamoorthy and R. Overstreet, "An experimental evaluation of ion-exchange relationships," *Soil Sci.* 69:41–53 (1950).

71. S. Larsen, "The use of P^{32} in studies on the uptake of phosphorus by plants," *Plant & Soil* 4:1–10 (1952).
72. K. Lawton and J. A. Vomocil, "The dissolution and migration of phosphorus from granular superphosphate in some Michigan soils," *Soil Sci. Soc. Amer. Proc.* 18:26–32 (1954).
73. K. Lawton, A. E. Erickson, and L. S. Robertson, "Utilization of phosphorus by sugar beets as affected by fertilizer placement," *Agron. J.* 46:262–4 (1954).
74. K. Lawton, M. B. Tesar, and B. Kawin, "Effect of rate and placement of superphosphate on the yield and absorption of legume hay," *Soil Sci. Soc. Amer. Proc.* 18:428–32 (1954).
75. A. Lecrenier, C. Corin, *et al.*, "Utilisation du radio-phosphore P^{32} pour l'étude de la dispersion des solutions salines dans le sol," *Bull. Inst. Agron. Gembloux* 21:89–112 (1953).
76. W. L. Lott, D. P. Satchell, and N. S. Hall, "A tracer-element technique in the study of root extension," *Amer. Soc. Hort. Sci. Proc.* 55:27–34 (1950).
77. A. J. Low, "The use of isotopes in agricultural research," II, *Chem. & Indus.* 1124–8 (1952).
78. C. D. McAuliffe, N. S. Hall, *et al.*, "Exchange reactions between phosphates and soils: Hydroxylic surfaces of soil minerals," *Soil Sci. Soc. Amer. Proc.* 12:119–23 1948.
79. C. McAuliffe and M. Peech, "Utilization by plants of phosphorus in farm manure: I. Labeling of phosphorus in sheep manure with P^{32}," *Soil Sci.* 68:179–84 (1949).
80. C. McAuliffe, M. Peech, and R. Bradfield, "Utilization by plants of phosphorus in farm manure: II. Availability to plants of organic and inorganic forms of phosphorus in sheep manure," *Soil Sci.* 68:185–96 (1949).
81. C. McAuliffe, G. Stanford and R. Bradfield, "Residual effects of phosphorus in soil at different pH levels as measured by yield and phosphorus uptake by oats," *Soil Sci.* 72:171–8 (1951).
82. J. Mitchell, H. G. Dion, *et al.*, "Crop and variety response to applied phosphate and uptake from soil and fertilizer," *Agron. J.* 45:6–11 (1953).
83. J. R. Neller and H. W. Lundy, "Availability of residual phosphorus of superphosphate and rock phosphate determined by phosphorus in crops from radioactive superphosphate," *Soil Sci.* 74:409–16 (1952).
84. J. R. Neller, "Effect of lime on availability of labeled phosphorus of phosphates in Rutledge fine sand and Marlboro and Carnegie fine sandy loams," *Soil Sci.* 75:103–8 (1952).
85. W. L. Nelson, B. A. Krantz, *et al.*, "Application of tracer technique to studies of phosphatic fertilizer utilization by crops: II. Field experiments," *Soil Sci. Soc. Amer. Proc.* 12:113–8 (1948).
86. W. L. Nelson, B. A. Krantz, *et al.*, "Utilization of phosphorus as affected by placement: II. Cotton and corn in North Carolina," *Soil Sci.* 68:137–44 (1949).
87. K. F. Nielsen, P. F. Pratt, and W. P. Martin, "Influence of alfalfa green manure on the availability of phosphorus to corn," *Soil Sci. Soc. Amer. Proc.* 17:46–9 (1953).
88. S. R. Olsen and R. Gardner, "Utilization of phosphorus from various fertilizer materials: IV. Sugar beets, wheat and barley in Colorado," *Soil Sci.* 68:163–70 (1949).
89. S. R. Olsen, W. R. Schmehl, *et al.*, *Utilization of Phosphorus by Various Crops as Affected by Source of Material and Placement*, Colorado Agri. Expt. Sta. Tech. Bul. 42, 1950.
90. S. R. Olsen, "Measurement of surface phosphate on hydroxyl-apatite and phosphate rock with radiophosphorus," *J. Phys. Chem.* 56:630–2 (1952).

91. S. R. Olsen, *Measurement of Phosphorus on the Surface of Soil Particles and Its Relation to Plant-Available Phosphorus* (Conf. on Use of Isotopes in Plant and Animal Research), U.S. Atomic Energy Comm. TID-5098, Kansas Agri. Expt. Sta., 1953.
92. S. R. Olsen and F. S. Watanabe, "Kinetic exchange studies on residual phosphates in calcareous soils with phosphorus-32," *Proc. 4th Ann. Oak Ridge Summer Symp.*, 1953, pp. 405–17.
93. S. R. Olsen, C. V. Cole, *et al.*, *Estimation of Available Phosphorus in Soils by Extraction with Sodium Bicarbonate*, U.S. Dept. Agri. Cir. 939, 1954.
94. S. R. Olsen, F. S. Watanabe, *et al.*, "Residual phosphorus availability in long-time rotations on calcareous soils," *Soil Sci.* 78:141–51 (1954).
95. R. A. Olson, M. B. Rhodes, and A. F. Dreier, "Available phosphorus status of Nebraska soils in relation to series, classification, time of sampling and method of measurement," *Agron. J.* 46:175–80 (1954).
96. F. W. Parker, "Radioactive materials in soil-fertilizer research," *Plant Food J.* 2:4–9, 33 (1948).
97. A. B. Prince, "Residual effects of superphosphate application on soil phosphorus level and growth of crimson clover as measured by yield and phosphorus uptake," *Soil Sci.* 75:51–8 (1953).
98. D. A. Rennie and J. Mitchell, "The effect of nitrogen additions on fertilizer phosphate availability," *Canad. J. Agri. Sci.* 34:353–63 (1954).
99. D. Ririe, S. J. Toth, and F. E. Bear, "Movement and effect of lime and gypsum in soil," *Soil Sci.* 73:23–5 (1952).
100. W. K. Robertson, P. M. Smith, *et al.*, "Phosphorus utilization by corn as affected by placement and nitrogen and potassium fertilization," *Soil Sci.* 77:219–26 (1954).
101. W. K. Robertson, J. R. Neller, and F. O. Bartlett, "Effect of lime on the availability of phosphorus in soils of high and low sesquioxide content," *Soil Sci. Soc. Amer. Proc.* 18:184–7 (1954).
102. E. J. Rubins, "Residual phosphorus of heavily fertilized acid soils," *Soil Sci.* 75:59–68 (1953).
103. R. S. Russell, J. B. Rickson, and S. N. Adams, "Isotopic equilibria between phosphates in soil and their significance in the assessment of fertility by tracer methods," *J. Soil Sci.* 5:85–105 (1954).
104. A. C. Schuffelen, "De betekenis van 'tracers' bij het bemestingsonderzoek," *Medel. Direct. Tuinb.* 15:610–24 (1952).
105. W. R. Schmehl and E. J. Brenes, "The availability of high-temperature process alkali (Rhenania-type) phosphates to crops when applied to calcareous soils," *Soil Sci. Soc. Amer. Proc.* 17:375–8 (1953).
106. L. F. Seatz, "Phosphate activity measurement in soils," *Soil Sci.* 77:43–51 (1954).
107. E. Shaw, R. G. Menzel, and L. A. Dean, "Greenhouse studies on plant uptake of zinc-65 from soils and fertilizers," *Soil Sci.* 77:205–14 (1954).
108. D. H. Smith, J. M. Blume, and C. W. Whittaker, "Radiochemical measurement of reaction rates of liming materials in soil," *J. Agri. & Food Chem.* 1:67–70 (1953).
109. J. C. Smith, J. F. Fudge, *et al.*, "Utilization of fertilizers and soil phosphorus by oats and crimson clover as affected by rates and ratios of added N and P_2O_5," *Soil Sci. Soc. Amer. Proc.* 15:209–12 (1951).
110. R. J. Speer, S. E. Allen, *et al.*, "Phosphate fertilizers for the Texas blacklands: I. Relative availability of various phosphate fertilizers," *Soil Sci.* 72:459–64 (1951).
111. R. J. Speer, S. E. Allen, *et al.*, "Plant utilization of zinc nutrients on Houston black clay," *Soil Sci.* 74:291–3 (1952).

112. W. F. Spencer and J. E. Gieseking, "Cobalt adsorption and release in cation-exchange systems," *Soil Sci.* 78:267–76 (1954).
113. J. W. T. Spinks and S. A. Barber, "Study of fertilizer uptake using radioactive phosphorus I," *Sci. Agri.* 27:145–56 (1947).
114. J. W. T. Spinks and S. A. Barber, "Study of fertilizer uptake using radioactive phosphorus II," *Sci. Agri.* 28:79--87 (1948).
115. J. W. T. Spinks, H. G. Dion, *et al.*, "Study of fertilizer uptake using radiophosphorus: 3," *Sci. Agri.* 28:309–14 (1948).
116. R. W. Starostka, J. H. Caro, and W. L. Hill, "Availability of phosphorus in granulated fertilizers," *Soil Sci. Soc. Amer. Proc.* 18:67–71 (1954).
117. G. Stanford and L. B. Nelson, "Utilization of phosphorus as affected by placement: I. Corn in Iowa," *Soil Sci.* 68:129–36 (1949).
118. G. Stanford and L. B. Nelson, "Utilization of phosphorus from various fertilizer materials: III. Oats and alfalfa in Iowa," *Soil Sci.* 68: 157–62 (1949).
119. G. Stanford, C. D. McAuliffe, and R. Bradfield, "The effectiveness of superphosphate top-dressed on established meadows," *Agron. J.* 42:423–6 (1950).
120. L. H. Stein and J. S. C. Marais, "Fertilizer studies with tobacco plants using radiophosphorus-labeled superphosphate, Part II," *South African Indus. Chem.*, June, 1952.
121. M. Stelly and H. D. Morris, "Residual effects of phosphorus and lime on cotton grown on Cecil soil as determined with radioactive phosphorus," *Soil Sci. Soc. Amer. Proc.* 17:267–9 (1953).
122. P. R. Stout, R. Overstreet, *et al.*, "The use of radioactive tracers in plant nutrition studies," *Soil Sci. Soc. Amer. Proc.* 12:91–7 (1948).
123. K. Strzemienski, "Phosphate uptake studies with radioactive phosphorus," *New Zealand J. Sci. Tech.* 34A:496–506 (1953).
124. L. F. Thompson and P. F. Pratt, "Solubility of phosphorus in chemical extractants as indexes of available phosphorus in Ohio soils," *Soil Sci. Soc. Amer. Proc.* 18:467–70 (1954).
125. A. Ulrich, L. Jacobson, and R. Overstreet, "Use of radioactive phosphorus in a study of the availability of phosphorus to grape vines under field conditions," *Soil Sci.* 64:17–28 (1947).
126. J. R. Webb and J. T. Pesek, "Evaluation of residual soil phosphorus in permanent fertility plots," *Soil Sci. Soc. Amer. Proc.* 18:449–53 (1954).
127. C. D. Welch, N. S. Hall, and W. L. Nelson, "Utilization of fertilizer and soil phosphorus by soybeans," *Soil Sci. Soc. Amer. Proc.* 14:231–5 (1949).
128. J. L. White, M. Fried, and A. J. Ohlrogge, "A study of the utilization of phosphorus in green manure crops by the succeeding crop, using radioactive phosphorus," *Agron. J.* 41:174–5 (1949).
129. L. Wiklander and J. E. Geiseking, "Exchangeability of adsorbed cations as influenced by the degree of saturation and the nature of the complementary ions with special reference to trace concentrations," *Soil Sci.* 66:377–84 (1948).
130. L. Wiklander, "Fixation of potassium by clays saturated with different cations," *Soil Sci.* 69:261–8 (1950).
131. L. Wiklander, "Kinetics of phosphate exchange in soils," *Ann. Roy. Col. of Sweden* 17:407–24 (1950).
132. S. Woltz, S. J. Toth, and F. E. Bear, "Zinc status of New Jersey soils," *Soil Sci.* 76:115–22 (1953).
133. W. G. Woltz, N. S. Hall, and W. E. Colwell, "Utilization of phosphorus by tobacco," *Soil Sci.* 68:121–8 (1949).

Part V

GENETIC AND BIOLOGICAL HAZARDS OF NUCLEAR RADIATION

Chapter 16

BIOLOGICAL DAMAGE BY IONIZING RADIATION *

Isotopes provide a unique tool for the investigation of the dynamic aspects of the metabolism of living organisms. Radioactive isotopes are in general easier to assay and may be followed to much higher dilutions than stable isotopes, but they must be used with caution. Since the radiations which they emit are probably damaging in some degree to all living cells, carelessness or insufficient awareness of the possible hazards may result in injury to the worker; failure to limit the amount of activity introduced into the system under investigation may vitiate an experiment by altering the natural course of metabolism, or in the case of a clinical investigation, may injure the patient. It is well known that some early field experiments concerned with phosphorus metabolism in plants led to erroneous conclusions because the amount of P^{32} used as tracer was sufficient to damage the sensitive root meristems (Ref. 1). These considerations not infrequently limit the use of tracers for clinical investigations. They are more rarely limiting in laboratory investigations.

Basic Physical Mechanisms

The production of biological damage by ionizing radiations starts with a physical act of absorption whereby individual atoms are ionized or excited. This leaves the molecules of which they form a part in a highly excited state, and in almost all cases these molecules will undergo some change of configuration; in many cases they will break up.

Dose is defined in terms of the energy absorbed per gram of tissue. The unit of dose is the rad, which is the same for all ionizing radiations and is equal to precisely 100 ergs per gm. The number of atoms which are primarily ionized or excited, though proportional to the absorbed dose, is probably not precisely the same for all ionizing radiations and has never been measured in living tissue or even in water. It is known, however, that there must be roughly

* This chapter is taken from Geneva Conference Paper 899, "Biological Damage Resulting from Exposure to Ionizing Radiation" by L. H. Gray of the United Kingdom.

two molecules ionized and rather more excited in each cubic micron of tissue exposed to a dose of one rad.

These ions are not distributed at random but are formed close to the geometrical path of the individual ionizing particles. The spacing of the ions along the track is an important parameter, which can influence the biological effectiveness of a given dose by as much as a factor of 20. Indeed, the relative biological efficiency of different types of radiation may be related empirically through this one parameter. Physically the spacing of the ions and excited molecules along the track is determined by the magnitude of the charge carried by the ionizing particle and by its speed. Protons, deuterons, and electrons traveling at equal speeds produce equal biological effects, as has been beautifully demonstrated by Tobias and his colleagues (Ref. 2), who showed that the behavior of two different varieties of yeast after exposure to 190 MeV deuterons was the same as after exposure to an equal dose of 200-kv X rays. In each case the ionized molecules would be formed irregularly along the tracks of the ionizing particles, sometimes as a simple pair of positive and negative ions, and sometimes in groups of two or three pairs of ions. The average spacing between the groups would be about 0.05 micron. The spacing increases with the speed of the particle, but never exceeds about 0.5 micron. At the other extreme, protons, at the peak of their ionizing efficiency, produce about 2000 ions per micron of track, and alpha particles about 4000 ions per micron. These figures represent comparatively large amounts of energy, having regard to the very small dimensions of the track, and give rise to high local concentrations of molecules which are in a very reactive state. Considering for example the hydroxyl radicals formed from ionized water molecules, it is apparent that the initial instantaneous concentration of these radicals along the tracks of the different ionizing particles will vary by nearly a factor of 1000 as between fast electrons and alpha particles, and that in the latter case the concentrations may reach the molar level. It is presumably for this reason that a single alpha particle may cause the ultimate disruption of relatively large formed components within the cell, as, for example, individual chromosomes. That chromosomes can be broken as an end result of energy dissipated in or near the chromosome by a single alpha particle has been shown experimentally by working with such low doses of alpha particles that most of the affected nuclei could have been traversed by no more than one alpha particle (Refs. 3, 4).

Biological effects which are produced in this way by single particles have the important characteristic that at low doses the proportion of cells affected increases strictly linearly with the dose. There is no threshold dose below which the radiation is without effect. Moreover, the effect of a succession of exposures to the same cell is in this case cumulative and independent both of the manner in which the exposures are spaced and the rate at which the energy is delivered during each exposure—provided, of course, that the cell is not developing in such a way as to change its sensitivity with time.

Genetic Damage

The production of mutations is an example of outstanding importance of this class of biological effect. Each mutation is believed to be brought about by a single particle. The vast majority of mutations are deleterious, and it is generally agreed that an increase in mutation rate in a human population is undesirable because it leads to an increase in the load of unfit persons which the community has to carry (Ref. 5). It is worth noting, however, that a minority of mutations may be desirable from a particular standpoint. Radiations have been used to produce varieties of mold giving a higher yield of penicillin more rapidly than these varieties would arise spontaneously in nature, and Gustafsson has produced new strains of cereals which give better yields of grain in particular habitats by means of radiation. He is now quite convinced that this is an economically important use of radiation (Ref. 6).*

Injury to the genetic material of cells of many different types in the form of structural damage to the chromosomes can be seen under the microscope when the cell enters its first division after irradiation. Some of these forms of chromosome structural damage are observed to increase in strict proportion to the dose. This is true of plants, insect embryos, mammalian tissues, and some kinds of cancer cells. In the case of roots, this form of damage to the meristem cells leads, at low dose levels, to a proportional temporary reduction in growth rate, which therefore also has no threshold. In the more sensitive roots 30 r may cause a 10% drop in growth rate. It is also true that certain criteria of abnormality may be set up, not necessarily related in any way to genetic damage, in terms of which a small proportion of the blood cells of persons exposed to low doses of radiation may be judged to be abnormal, and that this proportion increases approximately linearly with the dose, however small. Thus, both in regard to the use of radioactive tracers in biological investigations and to the assessment of permissible levels of exposure to personnel, it often cannot be said that the level of radiation has been such as to produce no biological damage, but only that such biological damage as may have been produced is not significant. As is well known, another long-term hazard associated with exposure to ionizing radiation is the induction of cancer. The cancer may arise at the site of an intense local irradiation, or a chronic local irradiation. Chronic whole-body exposure in man may give rise to leukemia. In experimental animals it has been shown that the total destruction of one organ by radiation may result in a hormonal imbalance which leads to hyperplasia and eventual tumor formation in another (Ref. 7). It has not yet been established whether these diverse aetiologies of radiation-induced cancer have any features in common at the cellular level.

* See also Part VII of this volume.

Cell Sensitivity

In passing from one functional state to another, a cell may undergo enormous changes in sensitivity. Reference was made above to the fact that 30 r appreciably reduces the growth rate of some roots for a period of days. A few hundred röntgens will cause such extensive damage to the meristem that the root dies. On the other hand, 200,000 röntgens delivered to a portion of the root only 1 cm from the apex causes no more than a slight temporary depression in growth rate. The cortical cells 1 cm from the apex were themselves meristematic cells 24 hr previously. Thus, in 24 hr the cells have matured in such a way as to have increased a thousandfold in resistance to the lethal effects of irradiation. Protoplasmic streaming is still active in some stamen hairs after 400,000 r. Unicellular organisms such as amoebae and paramecia likewise are not killed by doses of several hundred thousand r. It is not to be supposed that these cells are uninjured by such heavy irradiation, but only that the injuries sustained are not manifest. *Paramecium* provides interesting confirmation of this latter point of view, for, if organisms which have been exposed to less than one-tenth of the dose which would ordinarily be needed to kill them are induced to undergo the normal process of autogamy, whereby the macronucleus spontaneously disintegrates and is rebuilt from the haploid micronucleus, a large proportion die. It is fairly clear in this case that autogamy has uncovered extensive genetic damage to the micronucleus, which produced no visible deleterious effect in the presence of the macronucleus.

Similarly, in meristem or germinal tissue in plants and animals, the act of entry into division very frequently reveals an injury which was actually produced at the time of irradiation. The injury may appear within minutes, or it may be stored over long periods of time, according to the rate at which the cells are progressing through their cycle. In *Vicia* meristems in which the division cycle lasts 24 hr, damaged chromosomes are visible at metaphase within an hour of irradiation, and the number of damaged cells reaches its peak a few hours later. In lens epithelium the division cycle is much longer, and nuclear damage appears in the course of one or two weeks. In the thyroid gland, cells come into division so slowly that very few are seen in division at any one time. If, however, as long as twelve weeks after irradiation the cells are stimulated to divide, a considerable proportion are seen to be abnormal. The most extreme case is perhaps that of dormant seeds. Seeds may stand for several years after irradiation, but will nevertheless, when germinated, show abundant abnormalities among the cells of the root meristem when these enter their first division. Dormant seeds are rather resistant to radiation, but the other tissues referred to show extensive chromosome structural damage or nuclear pyknosis on entering division after doses in the range of 200 to 1000 r.

The time of entry into division is frequently the point at which radiation damage becomes manifest in meristematic and germinal tissues, but nuclear pyknosis may occur unassociated with mitosis. It does so in lymphocytes a few hours after exposure to comparatively small doses of radiation. Moreover, differentiated tissues have occasionally been found to show evidence of damage at very low dose levels. For example, Conard observed changes in intestinal rhythm in the rat a few minutes after doses as low as 100 r (Ref. 8). Many years ago Forssberg recorded a series of experiments in which he observed that the rate of elongation of the sporangia of *Phycomyces blakesleeanus* suffered a temporary check, which was followed by a compensatory acceleration, after doses of gamma radiation less than 1 r (Ref. 9). For the present, these are two isolated phenomena which bear no apparent relation to the main body of radiobiological knowledge. The phenomena appear to be well attested, and if followed up might give us an entry into virtually unexplored territory.

Metabolic Activity and Radiation Damage

Much is known about the primary physical acts of ionization and excitation, and more recently the chemical changes which follow upon the irradiation of water, aqueous solutions, polymers, and many macromolecules of biological importance have been closely studied; yet we are quite unable to trace the chain of events leading to any one form of cytological damage. It is generally difficult to detect any kind of metabolic derangement immediately after irradiation in the more sensitive types of cell, despite the fact that they will appear grossly or even lethally injured on entering division.

Respiration and aerobic glycolysis are rarely measurably affected below dose levels of the order of 30,000 r. Pirie has examined many enzymes extracted from rabbit lens exposed *in vivo* to 1200 r and found no evidence of impairment in any of them as late as 24 hr after irradiation (Ref. 10). Within a period after irradiation equal to their normal division cycle, bacteria show no drop in respiration after exposure to 60,000 r. Only about one in ten thousand of these bacteria will be able to continue to multiply indefinitely. The remainder will die, generally after passing through one or more divisions. Many of these bacteria, which are destined eventually to die, are nevertheless capable of supporting virus growth immediately after irradiation. In view of the fact that the virus relies on the enzymes of its host for most of the syntheses necessary for its own reproduction, this fact is impressive testimony of the essential integrity of the enzyme systems of the irradiated bacteria.

The synthesis of nucleic acid is at present outstanding as a metabolic activity which is grossly disturbed at dose levels in the range of 100 to 1000 r in a variety of plant and animal cells. The circumstances of this inhibition, and its implications as regards different forms of cytological damage, are being vigorously investigated. A disturbance in nucleic acid synthesis does not appear to be an important step in the production of chromosome structural damage by ionizing radiation, though it may be so in the case of some radiomimetic chemicals, for it has been demonstrated in both plants and animals that chromosomes may be broken as a result of exposure to radiation delivered late in the cell cycle after nucleic acid synthesis is complete.

Biochemical Effects of Radiation

While the metabolic pathways leading to cell death after irradiation remain for the present highly speculative, recent studies of radiation-induced chemical changes in comparatively simple systems such as ferrous sulfate appear to throw light on the earliest chemical links between the physical act of absorption and the metabolic chain. From the fact that the influence of radiation quality and anaerobiosis on many forms of biological damage closely resembles the influence of the same factors on radiochemical yield, it is generally believed that some of the chemical intermediates in the development of biological lesions may be identified. Three of these are the radicals OH, H, and HO_2. These radicals had all been postulated by chemists previously in other connections. It is postulated that in irradiated water and aqueous solutions the first two are derived directly by dissociation of the positively and negatively charged water molecules H_2O^+ and H_2O^-, and that the third is formed in the presence of dissolved oxygen by direct combination between the H radical and an oxygen molecule. There is thus an important distinction between the irradiation of systems in the aerobic and anaerobic states, the radicals OH and HO_2 being generated in the former, as compared with OH and H in the latter. The number of these radicals formed by moderate doses of radiation is very small, and being very reactive, they have an extremely short lifetime. For these reasons it has not been possible hitherto to characterize them by physical means in irradiated solutions; but the rapid advances that are now taking place in flash and microwave spectroscopy, when combined with pulsed radiation sources, should make such characterization possible. The hydrogen atom has been identified already in irradiated ice by microwave spectroscopy, and there is much indirect evidence in support of the existence of all three radicals. An experimental observation of some importance is that the chemical yields obtained when solutions are exposed to X and gamma radiation are often very different from those which result from exposure to neutron or alpha radiation, and indicate that in the case of the latter radiations most OH radicals are used up in the formation of H_2O_2 and most hydrogen atoms combine to form hydrogen gas. Fewer OH radicals are therefore available for reaction with solutes, and fewer HO_2 radicals are formed in the presence of oxygen, so that oxygen has little influence on the over-all yield.

These observations in the field of radiation chemistry are to be compared with the fact that many cells and organisms are much more sensitive to damage by X or gamma radiation under aerobic than under anaerobic conditions, whereas they show little or no dependence on the presence of dissolved oxygen as regards their sensitivity to alpha radiation. Furthermore, changes in X- or gamma-ray sensitivity are found to follow immediately upon changes in the concentration of dissolved oxygen in the fluid bathing the cells at the time of irradiation, and to be related to the molarity of the dissolved oxygen in a manner closely resembling some of the chemical reactions which have been studied.

Oxygen plays such an important role in biology that it would be remarkable if the differences between the aerobic and anaerobic sensitivities of cells were not in any way connected with the control exerted by oxygen tension on respiration and general metabolism. There are in fact some bacteria whose sensitivity to X radiation is no longer dependent on oxygen in the presence of certain enzyme poisons (Ref. 11). The same is also true of chromosome structural damage induced by the only chemical mutagen which has so far shown any dependence on oxygen tension (Ref. 12). In some of the biological effects of radiation for which the influence of oxygen tension has been most carefully investigated, this influence has been found not to be diminished when the respiratory enzymes are inactivated. On the basis of all the available evidence at hand at present, there appear to be good grounds for believing that the radicals revealed by studies in radiation chemistry play an important role in the development of some forms of radiobiological damage.

References for Chapter 16

1. R. Scott Russell and R. P. Martin, "Biological effects of radioactive isotopes in plant physiological studies," from Haddow, ed., *Biological Hazards of Atomic Energy,* Clarendon Press, Oxford, 1952.
2. C. A. Tobias, "The dependence of some biological effects of radiation on the rate of energy loss," from Nickson, ed., *Symposium on Radiobiology,* John Wiley & Sons, Inc., New York, 1952.
3. J. P. Kotval and L. H. Gray, "Structural changes produced in microspores of *Tradescantia* by α-radiation," *J. Genet.* 48:135–154 (1947).
4. L. H. Gray, "Some characteristics of biological damage induced by ionising radiations," *Rad. Res.* 1:189–213 (1954).
5. H. J. Muller, "Our load of mutations," *Amer. J. Hum. Genet.* 2:111–176 (1950).
6. A. Gustafsson, "Swedish mutation work in plants: Background and present organisation," *Acta Agricul. Scand.* 4:361–364 (1954).
7. J. Furth and E. Lorenz, "Carcinogenesis by ionising radiations," from Hollaender, ed., *Radiation Biology,* Maple Press Co., York, Pa., 1954, 1:i.
8. R. A. Conard, "The effects of X-radiation on intestinal motility of the Rat," *Amer. J. Physiol.* 165:375 (1951).
9. A. G. Forssberg, "Studien über einige biologische Wirkungen der Röntgen- und γ-Strahlen, insbesondere am *Phycomyces blakesleeanus,*" *Acta Rad.* Supp. 49 (1943).
10. R. van Heyningen, A. Pirie, and J. W. Boag, "Changes in lens during the formation of X-ray cataract in rabbits, 2," *Biochem. J.* 56:372–379 (1954).
11. H. Laser, "The oxygen-effect in ionising radiation," *Nature* 174:753 (1954).
12. B. Kihlman, "Studies on the effect of oxygen on chromosome breakage induced by 8-ethoxycaffeine," *Expl. Cell Res.* 8:404 (1955).

Chapter 17

EFFECTS OF DAILY LOW DOSES OF X RAYS ON SPERMATOGENESIS IN DOGS *

The prospect of radiation exposure of increasing numbers of people because of the greatly expanded usage of nuclear energy has made it advisable to obtain additional data on long-term biologic effects of low chronic doses of radiation. It is especially important to test experimentally, in a biologically suitable long-lived animal species, the effects of the present maximum permissible exposure level for humans for chronic X and gamma radiation (.3 r/week),† and of other dosage levels in this range, on extremely radiosensitive physiologic processes such as spermatogenesis and reproduction.

A life-term experiment on male beagle dogs was designed for the study of long-term effects of low daily doses of X rays on these processes, using previous work in this laboratory (Ref. 1) as a guide in selection of dose levels.

The design of the experiment, methods used, a morphologic description of the dog spermatozoon, and the normal ranges for all parameters of semen analysis have been presented in great detail in previous reports (Refs. 2, 3), and only the most important of these aspects are presented here briefly.

Experimental Design

There are four groups of pedigree male beagle dogs (see Table 17.1)—a non-irradiated control group and three irradiated groups receiving 0.06, 0.12, or 0.6 r of X rays daily on five consecutive days each week. Litter mates were separated from one another generally in their assignments to experimental dosage groups. X-ray exposure of the dogs began when they became about 86 weeks old. A total of 60 dogs are used in the experiment.

Each daily dose is given in 10 min by a General Electric 1000-kvp generator

* This chapter is taken from Geneva Conference Paper 257, "Effects of Daily Low Doses of X-rays on Spermatogenesis in Dogs" by G. W. Casarett and J. B. Hursh of the United States.

† This maximum permissible exposure has since been reduced. (Ed.)

having a 0.5-in. lead parabolic filter. Control dogs receive sham exposure. Doses are measured in air by Victoreen condenser r-chambers.

Semen is collected monthly by manual masturbation of dogs excited by the mounting of a bitch in "heat." Female dogs are kept in "heat" by stilbestrol for this purpose.

Semen Analysis

Volume of ejaculate. Ranges from 1 ml to 50 ml.

Motility differential count. Systematic count of active and inactive cells in an haemacytometer in a microscope stage incubator at 85° F.

pH of semen. Determined by fractional range pH test papers.

Concentration of spermatozoa. Systematic quantitative count of "fixed" and stained sperm in an haemacytometer.

Viability differential count. In smears of semen mixed with eosin solution, which differentiates living from dead sperm, 300 cells are evaluated (Ref. 4).

Morphology differential count. In smears of semen treated with an aniline blue-eosin-phenol solution (Ref. 5) for differential staining of structures of sperm, 300 cells are examined routinely for morphologic abnormalities.

Aberrational interrelationship studies. Using all above preparations and others, with ordinary and phase microscope, aberrational interrelationships are recorded and the data are being analyzed for functional implications.

Test mating. For correlation with semen analysis, a test mating consists of a single observed successful copulation at a standard time optimum for reproduction, in this instance the fifth day of receptivity of the female for the male.

Calculations and correlations. Absolute sperm counts are calculated from volume of ejaculate and sperm concentration. Concentrations or absolute numbers of sperm, dead, immotile, abnormal, or with any specific abnormality are calculated from sperm counts and percentages obtained in differential counts. When data on interrelationships of morphologic and physiologic aberrations have been obtained in sufficient amounts, these data will be used as a basis for calculation of "effective count," i.e., the number of apparently potentially effective spermatozoa in the ejaculate, obtained by subtracting from the absolute count the number of sperm probably or obviously ineffective for normal reproduction. Effective counts can then be correlated with mating results and thus provide as reliable a criterion or index of male fertility as is permitted by the extent of our knowledge of the influence of various aberrations of sperm on reproductive ability. Meanwhile, mating results are correlated with absolute sperm counts qualified by expressions of percentages of sperm with various morphologic and physiologic aberrations.

Experimental Results (To May 1, 1955)

TABLE 17.1. TOTAL DOSES

Dose, r/week	No. of Dogs	Exposure, weeks	Range of Total Doses, r
0.0	20	99–208	0.0
0.3	20	99–208	29–62
0.6	10	97–196	58–117
3.0	10	97–176	291–528

Criteria of effect. Disregarding its radiosensitivity, the stability of a parameter in control animals is a measure of its usefulness as a potential single criterion of slight effect.

Volume of ejaculate varies widely. Sperm concentration is a poor single criterion because it varies inversely with changes in volume when the volumes are in excess of the 2 or 3 ml which deliver most of the sperm in the first part of the ejaculate. The pH of semen, although fairly stable for individuals, varies with changes in volume, particularly with low volumes. Absolute sperm count is more reliable than sperm concentration, the general order of magnitude being fairly stable despite large fluctuations at times. Percentage of sperm alive is fairly stable usually, but erratic at times. Percentage of sperm motile may vary widely at times for consecutive samples, and although less useful on this basis, it is relatively sensitive as a physiologic process and is therefore employed chiefly in conjunction with other criteria.

Not all motile or morphologically normal sperm are capable of fertilization, while some sperm not showing motility or some sperm revealing certain morphologic abnormalities may actually be potentially effective under physiologic conditions. Some sperm proved dead by viability stain are motile if the structure primarily responsible for motility, the midpiece, is not dead along with the head.

Percentage of sperm morphologically normal is for most dogs one of the most stable values over long periods of time. Percentages of sperm with coiled tails, multiple tails, or cytoplasmic appendages are fairly stable in control dogs, while percentages of sperm amorphous or tailless, for examples, are more erratic. Coiled tails and separation of the head result in loss of effective motility. Amorphous sperm (head attached at an angle) have defective circular movement, and sperm with multiple heads or bodies also show defective motility. Coiled tails and cytoplasmic appendages are found frequently in semen of young dogs during sexual development and may be signs of maturation defect. There is preliminary evidence from test matings that in a semen ejaculate a high incidence of sperm with cytoplasmic appendages may be associated with a high incidence of microscopically undetected maturation defects among the other sperm which reduce the fertility of the semen.

Radiation effects. No effects of irradiation have been demonstrated conclusively to date in the 0.3 r per week and 0.6 r per week dogs. However, during the past 18 months or so, and beginning at various times, sustained increases in percentage of sperm with coiled tails have appeared in 80% of the dogs in the 0.6 r per week group, 45% of the dogs in the 0.3 r per week group, and in only 5% or possibly 10% of control dogs. The increase is in most cases about 5 to 10 percentage units (15 units in a few cases) and higher in a control sterilized by some undiagnosed disease. Although suggestive, this finding should not lead to definite conclusions regarding radiation effect without further study.

The dogs in the 3.0 r per week group, which had average pre-exposure sperm counts for individual dogs ranging from about 225 million to 770 million, revealed progressive decline in absolute sperm count, beginning generally after 20 to 30 weeks of exposure (60 r to 90 r). In 8 cases the count fell to less than 10% of average control values, generally within 40 to 60 weeks (120 r to 180 r) and remained depressed or declined further. There were concomitant moderate or marked increases in percentages of sperm immotile, dead, and showing morphologic abnormalities, especially coiled tails, separation of heads from the body, and cytoplasmic appendages. In Fig. 17.1 are graphs of some data

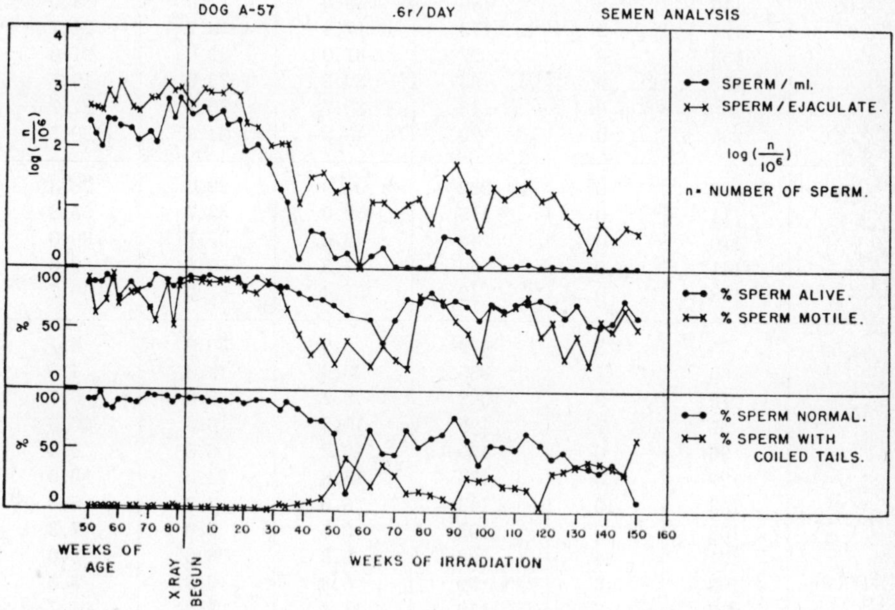

FIG. 17.1. Effects of daily low doses of X-rays on spermatogenesis in dogs.

from a dog in this group illustrating some of the principal effects of irradiation.

Six of these 8 dogs proved sterile by test mating (Table 17.2) and 2, not willing to mate, are presumed sterile on the basis of very low sperm counts.

The 2 remaining dogs (A1-59 and A1-88) are showing a much slower decline in sperm count, down to about 25% of control values at present, with transient tendencies toward recovery, and reveal little or no change in motility, viability, or morphology of sperm.

TABLE 17.2. TEST-MATING RESULTS FOR DOGS RECEIVING 3.0 r PER WEEK

Dog No.	Total Dose at Mating, r	Pups in Litter	Absolute Sperm Count,[a] millions	Sperm Qualities[a]		
				Nonmotile, %	Dead, %	Abnormal, %
A-51	153	0	31	96.7	49.5	63.9
	163	0	24	78.6	41.3	64.7
	183	0	14	93.6	21.3	54.0
A-57	154	0	19	80.0	31.0	37.0
	167	0	24	61.5	39.0	86.0
	180	0	0	—	—	—
A-98	116	8	156	25.0	7.0	4.0
	127	2	173	17.4	13.7	21.7
	153	8	27	37.0	22.0	21.6
	175	0	53	39.5	35.0	19.7
	190	0	13	30.7	30.2	21.7
	265	0	20	43.8	34.3	30.7
A-54	100	0	232	34.9	22.0	28.3
	111	0	115	39.0	53.7	55.3
	120	0	46	48.6	37.7	39.0
	141	0	69	47.5	50.3	59.3
	198	0	42	57.4	57.0	50.7
A1-59	55	3	479	6.9	36.0	5.7
	69	7	365	23.9	19.3	4.3
	75	9	303	8.1	5.0	6.7
	91	7	195	10.2	6.3	6.3
	96	0	195	10.2	6.3	6.3
	129	3	133	18.8	24.0	6.0
	153	10	143	6.6	7.7	5.3
	168	3	193	2.6	5.0	7.3
	177	4	92	6.3	3.3	5.0
	193	3	92	6.3	3.3	5.0
	277	7	156	11.4	5.3	8.7

TABLE 17.2. TEST-MATING RESULTS FOR DOGS RECEIVING 3.0 r PER WEEK (*Continued*)

Dog No.	Total Dose at Mating, r	Pups in Litter	Absolute Sperm Count,[a] millions	Sperm Qualities [a]		
				Nonmotile, %	Dead, %	Abnormal, %
A1-88	25	0	523	8.3	11.0	20.0
	57	3	1004	5.3	10.0	15.0
	67	1	481	9.4	12.0	22.0
	115	7	182	17.5	16.7	15.3
	133	5	105	14.7	15.0	10.8
	142	0	204	16.7	14.7	27.7
	159	0	195	7.7	2.3	26.3
	174	10	194	19.8	13.0	34.0
	256	8	632	6.3	9.0	26.7
B-5	30	8	1022	21.4	11.7	30.7
	55	0	848	45.3	18.0	34.7
	109	0	47	40.7	66.3	85.3
	129	0	23	92.3	92.0	98.5
	136	0	23	92.3	92.0	98.5
	158	0	36	75.0	91.7	95.7
B-6	76	2	89	37.8	20.0	27.3
	110	0	28	40.5	41.3	68.3
	138	0	11	50.0	44.3	84.9
	156	0	12	75.0	56.4	86.0
	168	0	8	80.6	73.0	83.0

[a] In general, a semen analysis from 6 days to 3 weeks prior to mating.

Tables 17.2 and 17.3 present results of completed mating tests correlated with sperm count data obtained generally from about 6 days to 3 weeks prior to mating. These counts may represent better or poorer quality of semen than that involved in the mating. Most male or female dogs may at one time or another in a single copulation perform in a physiologically defective or atypical manner in regard to reproductive processes. For these and other reasons the correlation is expected to be imperfect. Examination of the sperm count data and mating results may yield knowledge of the general order of magnitude of critical or minimal numbers of live, normal, moving sperm or apparently effective sperm on this basis necessary for reproduction.

Each of 11 matings of control and 0.3 r per week dogs has produced no fewer than 4 pups. In only 3 cases are the apparently effective sperm counts likely to be less than 200 million, perhaps between 100 million and 200 million.

TABLE 17.3. TEST-MATING RESULTS FOR DOGS RECEIVING 0, 0.3, AND 0.6 r PER WEEK

Group and Dog No.	Total Dose at Mating, r	Pups in Litter	Absolute Sperm Count,[a] millions	Sperm Qualities[a]		
				Nonmotile, %	Dead, %	Abnormal, %
Controls						
A-49	0	8	1086	13.8	31.3	5.3
A-49	0	7	937	12.1	29.7	9.7
A-58	0	5	218	14.6	18.7	17.0
A-100	0	5	598	20.2	20.7	20.3
A-100	0	9	598	20.2	25.3	12.0
A-100	0	6	741 [b]	10.0	23.3	8.0
98	0	9	214	54.3	59.7	35.0
0.3 r per week						
A-60	20	6	778	15.0	17.6	21.3
A-60	22	4	202	17.5	33.3	35.7
25	43	6	856	16.7	20.7	10.7
A-97	43	6	304	7.9	8.7	4.0
0.6 r per week						
116	82	10	870	4.2	9.0	13.0
A-50	69	8	310	18.8	16.0	12.0
A-50	94	0	383	3.1	8.0	10.7
A-59	43	7	385	17.0	39.0	7.7
A-59	46	5	941 [b]	11.6	23.7	18.7
A-59	59	8	313	3.5	7.0	13.0
A-59	75	5	553	6.5	13.0	33.0
A-67 [c]	59	0	25	27.3	4.7	18.0
A-67 [c]	73	0	234	23.8	16.7	21.7
A-67 [c]	90	5	45	18.4	12.3	16.0
A-99	62	0	208	7.7	4.3	5.0
A-99	67	2	100	20.7	10.3	11.0
A-99	85	8	96	10.0	11.3	17.7
A1-53	35	0	1067	19.1	21.0	20.3
A1-53	41	1	942	21.7	8.0	15.7
A1-53	56	2	603	29.1	20.7	37.7
A1-57	34	0	486	17.6	29.0	27.3
A1-57	54	0	377	23.0	8.7	30.0
A1-87	23	6	767	16.7	17.0	20.0
B-4	33	0	972	6.8	3.0	38.0
B-4	48	0	972	12.1	9.3	41.7

[a] In general, a semen analysis from 6 days to 3 weeks prior to mating.

[b] Sperm count a week or more after mating. Count prior to mating not typical.

[c] This dog exhibits unusually erratic sperm counts.

Examination of all productive matings in Tables 17.2 and 17.3 suggests poor correlation between sperm count and number of pups in a litter. Of the 27 unproductive matings of 3.0 r per week dogs, in 25 cases the apparently effective sperm counts seem to be below 200 million. Of the 21 productive matings at this dosage level, the correlated apparently effective sperm counts are probably well above 200 million in 7 cases, between 100 million and 200 million in 9 cases, and at or below 100 million in 5 cases.

These data suggest the preliminary generalization that for these dogs the minimum level of corrected absolute sperm count for successful reproduction by a single copulation may be roughly in the general region of 100 million to 200 million sperm, with some variability among individual dogs.

Of 9 unproductive matings of 0.6 r per week dogs (Table 17.3) only 3 instances are associated with apparently effective sperm counts which might fall below 200 million, and in 6 instances the subtraction of a reasonable number of defective sperm on the basis of tabulated sperm qualities would probably result in net sperm counts well above 200 million, ranging perhaps from 250 million to 850 million approximately. This might suggest the possibility of undetected defects in many of the spermatozoa of some of these dogs. However, the meaning of these data is not clear, and in view of the erratic spermatogenic behavior in some of these dogs, many more pertinent data are required before definite conclusions can be reached. Of the 12 productive matings in this group, the associated apparently effective sperm counts seem to be probably well above 200 million in 9 cases and between 40 million and 100 million in 3 cases.

Summary

In general, 0.6 r per day (3.0 r per week) of X rays produced in dogs progressive decline in absolute sperm count after the twentieth to the thirtieth weeks of exposure to levels less than 10% of control values within 40 to 60 weeks of exposure, with sustained levels or further depressions thereafter. There were concomitant moderate or marked increases in the percentages of sperm morphologically abnormal, immotile, and dead. Most of the dogs have been proved infertile by test mating.

Apparently effective sperm counts and critical sperm counts are discussed.

No effects of radiation have been demonstrated conclusively as yet in the 0.3 r per week and 0.6 r per week dogs after radiation exposures of from 2 to 4 years.

References for Chapter 17

1. R. D. Boche, "Effects of chronic exposure to X-radiation on growth and survival," in H. A. Blair, ed., *Biological Effects of External Radiation,* McGraw-Hill Book Co., Inc., New York, 1954, 1st ed., Chap. 10.

2. G. W. Casarett, "Effects of X-rays on spermatogenesis in dogs," *University of Rochester Atomic Energy Project Quarterly Technical Report* UR-127, 1950, pp. 14–23.
3. G. W. Casarett, "Effects of daily low doses of X-rays on spermatogenesis in dogs," *University of Rochester Atomic Energy Project Technical Report* UR-292, 1953.
4. E. Blom, "A one-minute live-dead sperm stain by means of eosin-nigrosin," *Fertility & Sterility* 1, No. 2:176–177 (1950).
5. G. W. Casarett, "A one-solution stain for spermatozoa," *Stain Technol.* 28, No. 3:125–127 (1953).

Chapter 18

EFFECTS OF WHOLE-BODY EXPOSURE TO IONIZING RADIATION ON LIFE SPAN AND LIFE EFFICIENCY *

All are concerned with consequences to life which may follow acute irradiations in operating plants releasing nuclear energy for utilitarian ends. Analyses of these problems on mice, as "stand-ins" for man, are under way in our laboratory. High-energy irradiations affect several body characteristics simultaneously. Many of these changes have been examined singly. Life span integrates all these effects in one unit, a unit which is of most interest to us. Concordant with its effects on life span, nuclear energy may affect the well-being of the exposed throughout their lives or through the lives of their progeny. Reduced vigor may be more serious than death. Of the several measures of vitality used in our work, two may be considered—reproductive efficiency of the exposed and necropsy examinations of the progenies for congenital defects or mutations.

Materials and Methods

Ten of our inbred strains of mice were utilized in these experiments. These strains have been bred brother by sister generation to generation from ten to forty years. Each strain has its unique natural characteristics, among the most important of which are differences in size, growth, disease resistance, reproductive efficiency, and longevity.

The number of mice, of two of these strains, that were exposed to radiations from nuclear detonations under field conditions was 655. The radiations were heavily filtered to increase the neutron component. The mice were exposed to 12 different dosages. Field conditions were such as to make it impossible to balance the numbers of mice in the individual experiments. Extreme difficulties in separating neutron and gamma irradiations in nuclear detonations of the range of energies involved cause real problems in measuring dosage with

* This chapter is taken from Geneva Conference Paper 256, "Effects of Whole-Body Exposure to Nuclear or X-ray Energy on Life Span and Life Efficiency" by J. W. Gowen of the United States.

physical instruments. Consideration of the rep effects should take this fact into account in evaluating the observations.

The actions of nuclear energy are contrasted with and paralleled by X-ray irradiations. X rays were generated by a Coolidge type tube having a half-layer value of 0.78 for al. when operating at 98 kvp, 2.5 ma at 36.5 cm from the tungsten anode with no filtration other than the glass of the tube. The dose rate was 45 roentgens per min. The 1046 male mice divided into 7 groups received 0, 20, 80, 160, 320, 640, and 960 roentgens in a balanced design. The X-ray treatments were all single exposures.

Although the numbers of mice in any one treatment group are small, the data will be analyzed by life-table methods rather than as some single statistic of the distribution, because in our experience this form of analysis is more meaningful.

The individual mice are mated monogamously following irradiation and housed in pairs throughout their lives. Successive progenies remain with their parents for 21 days. Results for experiments on nuclear energy are integrated with those for X rays and with simultaneous experiments on the fly *Drosophila*, bacteria *Salmonella* and *Phytomonas*, and viruses of the tobacco mosaic group. Mice were mated within their own inbred strain but never within litters. If a spouse died, it was replaced by an untreated mouse of comparable sex and age. Pairs were examined for deaths or young daily. Progenies of these matings were grown to weaning, killed, and examined for effects of the irradiations transmitted to progeny.

All mice were maintained on our diet. This diet has been in use for ten years in this colony with satisfactory results. The thoroughly mixed ingredients were compressed and fed in pellet form. Pellets, together with tap water, were before the animals at all times. The pellets are composed of the following ingredients in pounds-for-ton mixture: yellow corn meal (fine) 500, oats whole ground 200, wheat whole ground 300, meat and bone meal 280, fish meal 100, dried buttermilk 200, alfalfa meal dehyd. 17% 100, linseed oil meal 140, lard 60, dried brewer's yeast (debittered) 60, salt iodized 10, ground limestone 10, manganese sulfate 0.7, vitamin A 3000 I.U. per lb gm 400 D 10, vitamin D 2000 AOAC per lb gm 2, and riboflavin 500 microgram per gm 30.

The effect of the radiation on longevity of the exposed will be emphasized in this chapter. The material on general well-being, fecundity, fertility, physiological and pathological findings, and progeny characteristics will be covered insofar as space or discussion may be available.

Effects of Single Nuclear Irradiations on Life Span

Life tables were constructed for each dose. The data are presented as expected life spans of the groups of mice at specific ages following the irradiation treatments. These values were calculated by dividing the total number of days lived by the groups alive at a given age by the numbers alive at the specific ages. Fig. 18.1 presents the data for the combined sexes and strains. The ordinates give the mean life spans of those mice which are alive at the ages indicated as the abscissas of the chart.

The most "acute" effects of the irradiations are observed during the first 20 to 30 days immediately following exposure. Within this portion of the graph

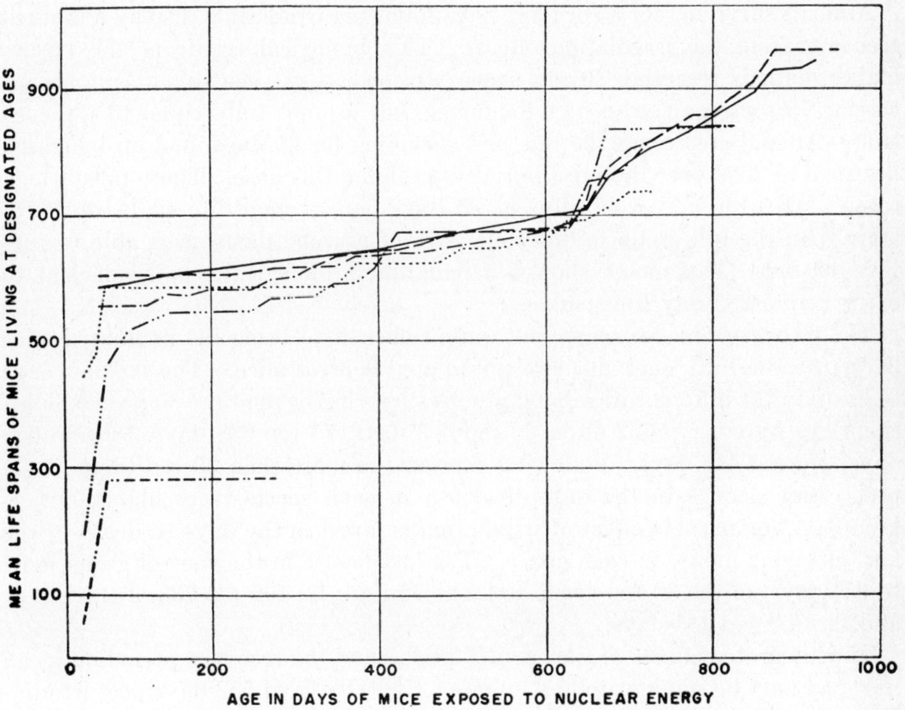

FIG. 18.1. Relation between expected life spans and ages of mice exposed to nuclear detonation when 50 ± days of age. The curves present date for the controls – – –, the 30 rep ———, the 75 rep —·—, the 166 rep ······, the 270 rep —···—, and 360 rep —··—.

the nuclear radiations have pronounced effects on mean life spans of mice receiving more than 75 rep. These effects become progressively greater as the rep dosage increases to the point where at 510 rep or above, all mice were dead within three days following exposure to the radiation. In this "acute" phase when certain individuals were unable to recover from the absorbed radiant energy, the doses from 0 to 75 rep were not lethal, since the observed survivals were control 100%, 30 rep 99%, and 75 rep 100%. Doses beyond 75 rep were progressively more lethal. The 166 rep had but 73%, the 270 rep 34%, and the 360 rep 3% of the mice survive. These data show evidence for an important threshold in radiation effects of nuclear detonations which could be related to "a safe dosage." Individuals receiving up to somewhat more than 75 rep show little or no immediate effects, whereas those receiving 75 rep have increasingly dangerous reactions as the energy absorbed becomes greater.

Recovery Phase

Animals surviving for 35 or more days following irradiation display a marked recovery from the irradiation effects. This biological repair of the tissues makes animals receiving 30 rep nearly as good as the controls. Animals receiving 75 rep repair most of the damage, but are not fully equal to the controls. Animals receiving 166 rep, and surviving for 35 days, had an increased life span of 25% over the mice initially receiving this dose. Those mice which received 270 rep and were alive at 35 days had average life spans of 150% more than the full group initially exposed. Only one mouse was able to survive 360 rep. This mouse showed a remarkable biological repair sufficient to carry it to a 285-day life span.

The recovery processes were not quite sufficient to bring the irradiated mice back to those fully equivalent to unradiated control mice. The median days to death in the different irradiated groups surviving beyond the 35 post-irradiation days were 0 rep 652 days, 30 rep 665 days, 75 rep 638 days, 166 rep 625 days, 270 rep 415 days. Following exposure to more than 30 rep there was a progressive decline in the days to death in each successively higher dosage group. A comparable effect of irradiation is noted in the days to death of the last surviving mouse in each group. The last mouse in the control group lived to 951 days, in the 30 rep group 941, the 75 rep 841, 166 rep 733, 270 rep 836, and for 360 rep 285 days.

The integrated effects of the "acute" period and the recovery period on mean life span may be summarized as follows. Controls lived 552 days post irradiation, 30 rep 542 days, 75 rep 509 days, 166 rep 398 days, 270 rep 147 days, 360 rep 9 days. Doses higher than 510 rep were lethal in less than one week. Significant curvatures are observed in the lines relating mean further life spans to ages of the mice, which must be taken into account in any analyses of these problems.

Effects of X Rays on Life Span of Mice

The pattern of the effects described for nuclear energy is confirmed through experiments on the more easily controlled X-ray energy. The mice treated were maintained under identically the same conditions as were the mice exposed to nuclear radiations. The treatments and illustrations of the data follow a like pattern. Fig. 18.2 presents the life spans of 1046 male mice treated at 40 ± 3 days old with 7 X-ray dosages. The ordinates give the

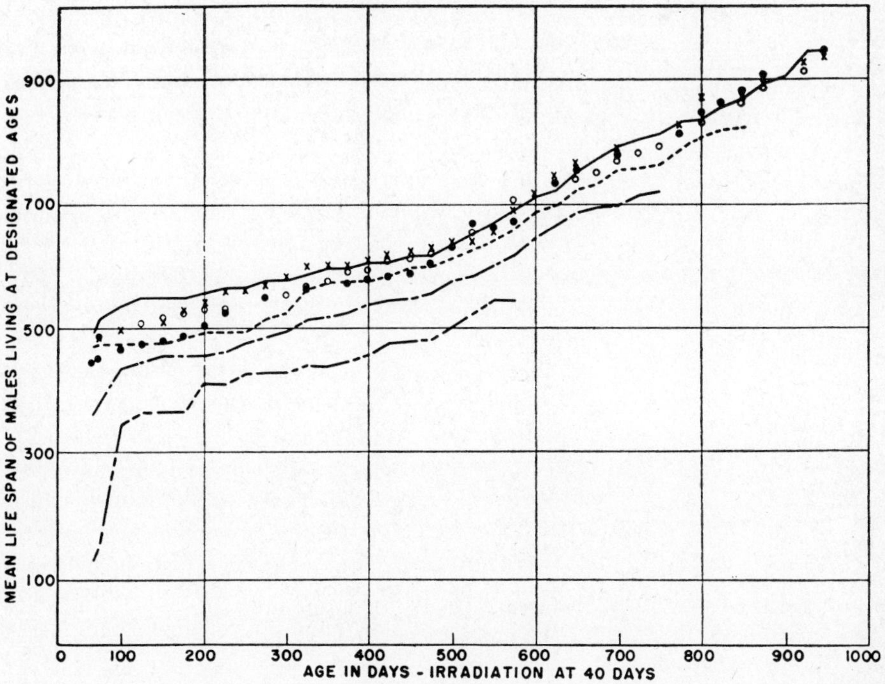

FIG. 18.2. Relation between expected life spans and ages of mice X-rayed at 40 days of age. The curves present data for the controls -----, the 320 r ———, the 640 r —·—, and the 960 r ———— irradiations. The data for the mice of the other treatments 20 r, 80 r, and 160 r are indicated as points.

mean life spans of those mice which are alive at the ages indicated as the abscissas of the chart. The X-ray irradiations show their most acute effects during the 35-day period immediately following exposure. Control mice of the 10 strains had a mean life span of 497 days. This life span was less than for the controls in the experiments on nuclear energy. Mice exposed to 20 r, 80 r, 160 r closely resemble the control mice in life span, with means of 485 days, 488 days, and 443 days respectively. The 320-roentgen, full-body irradiated mice have a similar mean expectancy of 474 days, but evidence for more lasting damage appears near 125 days of age. Mice exposed to 640 roentgens show a 27% reduction in life span—365 days as contrasted with the 497 for the controls. These effects were differentially expressed by the components of population as noted when expectancies for the individual strains are considered. The greatest effects on life span were observed when the mice were exposed to 960 roentgens whole-body radiation. For this dosage the mean life span was reduced to 131 days, or but 26% of that for the unexposed mice. Only ¼ of the original population of mice exposed to 960 r lived more than 35 days after irradiation. Life expectancy increased remarkably for the 75-

day-old survivors. Their mean life spans were more than double, 2.7 times, those of all mice in the original population immediately after exposure. The life curves are clearly separated into two different reacting systems; the period of "acute" effects with immediate deaths and the period following when biological recovery of the survivors has set in.

The average deficits in male life span for the intervals 100 to 500 days of age were significant and increasingly pronounced with increases in radiant energy absorbed—20-roentgen-exposed mice versus controls 15 days, 80-r mice 29 days, 160-r 46 days, 320-r 48 days, 640-r 74 days, and 960-r 153 days. Analyses of similar data on female mice show a similar trend, but one subject to more variation.

The age of the last mouse in any treatment group (Fig. 18.2) furnishes another measure of the treatment effect. For the controls the oldest mouse had a life span of 948 days, the 20-r 935, 80-r 914, 160-r 947, 320-r 827, 640-r 725 and 960-r 551 days. Life spans of mice receiving 320 r or more became progressively less as the radiation exposures increased, the oldest mouse in the 960-r group having a life span about 58% of that of the controls. These reductions in life span may be attributed to permanent X-ray damage which the animal is unable to repair.

The similarity in pattern of the irradiation effects of nuclear and X-ray energies on mice indicates a quantitatively similar biological mechanism for each. The noticeable features of difference lie in the equivalence of the rep and roentgen. One rep dose was 2 to 3 times as detrimental to life as one roentgen dose. Nuclear irradiations are proportionately more lethal to the mice in the region covered by the "acute" period. The data may be arranged in a different form useful in clarifying these differences.

Loss in Life Span of Those Irradiated

The mean further life spans of the control mice for each successive age group may be considered as 100 for each age group presented in the left column of Table 18.1. The ratio of life spans of the treated mice divided by the controls gives a measure of the loss or gain in longevity through irradiation.

In the age cohorts, improvements in life spans of the irradiated over controls are shown by values greater than 100. There are 4 such values in both nuclear and X-ray irradiations. None are significantly greater than 100. Three in each case appear in the lowest-dosage group. Slight depressions without statistical significance in life spans are seen for 75-rep and 80-r respectively. Depressions in the 166-, 270-, and 360-rep and 640- and 960-roentgen groups appear significant. The threshold effects were fairly consistent. Each threshold marks a change in the reacting system. The table has a further use in that from it may be derived rough estimates of compensations for irradiation exposure damages.

TABLE 18.1. LIFE EXPECTANCIES FOLLOWING IRRADIATION EXPOSURES TO NUCLEAR OR X-RAY ENERGY
(Unexposed Mice = 100)

Age of Mouse Cohort, days	Nuclear Energy, rep					X-Ray Energy, roentgens					
	30	75	166	270	360	20	80	160	320	640	960
40–74	98	92	78	40	12	94	94	87	93	72	23
75–124	98	90	98	77	34	91	90	84	85	79	61
125–224	101	92	97	84	23	95	92	87	83	75	57
225–324	104	96	94	84	8	101	89	91	84	74	53
325–574	104	95	88	97		102	97	91	88	66	33
575 and older	89	96	71	115		107	90	103	82	55	

Host Genetic Differences in Effects of Nuclear and X-ray Irradiations

The two strains studied in the nuclear irradiation tests were characterized by several well-known genetic differences. The S mice were of medium size and medium longevity and good in litter size, but the number of young raised was small. The Ba mice were long mice, somewhat larger than the S, longest lived of our strains, and high in litter size and raised most of their young. The Ba mice as controls had a longer life span. Their longevity was lowered by 30 rep and still further reduced by 75 rep. The S mice were little affected by the 30- and 75-rep doses. The Ba mice had their life spans reduced to 51%, 3%, and 1% by exposures to 166, 270, and 360 rep. The S mice life spans were 85%, 47%, and 3% of the control for the same respective doses.

Longevities under X-ray irradiations likewise differ according to genetic background. Thresholds of effect appear in the strains as they did in the combined data, but these thresholds may appear at slightly different ages. The detrimental action of the X rays are withstood better by the S mice than by the Ba. For exposures up to 320 r both strains are equally affected. Beyond 320 r the Ba mice become increasingly susceptible to more X rays, their longevity curve crossing that of the S at 600 r, even though they start with a 140-day advantage at 320 r. At 960 r the Ba is our most susceptible strain, and the S is our most resistant, the difference between the strains being 270 days in life span.

The nuclear and X-ray energies each lower life spans of both strains, especially at the higher dosages. The S strain is distinctly more resistant than the Ba. This difference is hereditary. Possibly the difference is partly due to the resistance of the S strain to the *Salmonella* type bacteria, but the full basis for resistance is likely to be more complex.

Effects of Radiant Energy on the Progeny of Those Exposed

Radiant energy affects the progeny through alteration in the ovaries or testes of those irradiated, leading to sterility, primary changes in the sperms or eggs (making them incapable of fertilization), deaths of the fertilized eggs or foetuses in uteri, and phenotypic differences in the adults or later-generation progenies. Nuclear energy displayed a differential action on the fertility of the sexes. Males that received as much as 270 rep have been fertile. Females receiving 75 and 166 rep were highly sterile. With X rays of 100 kvp 2.5 ma, males were fertile even though they had received 960 r. Female fertility was much reduced by irradiations of 320 r or above. Both types of radiation acted on fertility in a comparable manner, but to produce the same grade of effect, the dose in rep was less than that when the roentgen was employed. Litter size paralleled observations on fertility, except that the relative rep: roentgen effects were not so pronounced in the males. Exposure of the females to 75 rep reduced litter size noticeably. Males treated with X rays showed a progressive decrease in litter size, but the rate of decline was much less than that observed for the females. The decrease in numbers of young was roughly described by simple exponentials.

Careful examinations by Dr. Hollander were made of all young over the period of birth to three weeks of age. First-litter young reaching 3 weeks of age and many of those of subsequent litters were necropsied and the internal organs examined for any gross effects that could be attributed to the irradiations. Of the nuclear energy series 620 mice and 2171 in the X-ray series herein described were examined without notice of any changes attributable to the effects of the radiation. Besides these series some 17 or 18 thousand progeny from other X-ray-treated parents have been studied alive, necropsied, and searched for any changes from parental types. To date such changes due to radiation have yet to be found. The discovery of an easily overlooked gynandromorphic type that was naturally present in low frequency in one of the inbred strains is proof of the fact that the search was thorough (Ref. 4). The lack of observable genetic effects is not entirely unexpected. The changes observed in the F_1 generation would be due to visible dominant gene mutation and to chromosome aberrations having dominant effects. These changes are known to be rare in *Drosophila*. The frequency could be greater in mice, but the above data indicate the rate is not high. The large possible classes of genetic radiation effects, recessive visibles, lethals, and sterility genes would be missed in the experiments. The problem requires further extensive experiments some of which are now in progress in our laboratory.

Discussion

Integration of these results with the excellent data of a notable number of investigators (Ref. 1) working on similar problems must be omitted for lack of space. However, discussion of two outstanding papers (Refs. 2, 3) by W. P. Davey which preceded and exceeded most in the field and universally were overlooked require consideration. From his work Davey concluded that increasing dosages of X rays altered the life span through effects on different reacting systems important to life. Transfer of the X-ray action from one system to another created the thresholds observed when the curves for life spans were plotted against dosage. Working with a large population of flour weevils, *Tribolium confusum,* he observed that up to 200 ma-min at 25 cm (I am indebted to Dr. E. D. Trout for the estimate that 1 ma-min is of the order of 8½ to 10 roentgens) caused little change in the weevil's life span. Beginning at 200 ma-min increasing doses of X rays increased the deaths until at 500 ma-min all weevils were dead in 10 to 14 days. A second vital system which requires 500 ma-min or more to operate on it replaces the first at 500 to 4000 ma-min. A third vital system requiring 4000 ma-min or more X rays for it to react and extending to 16,000 ma-min later replaces both previous systems in importance to life. If the radiation dose is expressed as its logarithm the survival curves are straight lines but with different slopes breaking at the points 500 and 4000 ma-min.

The life spans of our mice exposed to nuclear irradiations from 0 to 360+ rep or X rays from 0 to 960+ roentgens kvp 100 ma 2.5 showed one threshold. In the range 0 to 75+ reps of nuclear irradiations or 0 to 320+ roentgens of X rays the animals had their life spans lowered, but the days lost were few. Between 75 and 166 rep, and between 320 and 640 roentgens, "acute" deaths appeared and increased in number as the radiation dosages were increased. This new cycle of effect killed the more susceptible mice. The survivors repaired a fair amount of the damage but not all. Other work has indicated other cycles at still higher dosages. The effects of radiation on the mouse are suggestive of those observed in *Tribolium.* Genetic differences between the mouse strains modify the severity of the effects.

Summary

Effects of exposures to nuclear and X-ray energies on the life spans and general well-being of 1700 mice have been discussed. F_1 progeny 2991 in number were examined for any visible genetic changes from the parental type either as external changes or as those evident on necropsy. Life spans were altered in a comparable manner by the two types of radiation, but the thresholds came at different dosages as measured by the rep and roentgen. The surfaces describing the effects of dosage, age, and mean further life spans show, within the ranges of energy used, two cycles of effects, a low-quality effect at the low irradiations followed by "acute" change developing between 75 and 166 rep and between 320 and 640 roentgens. The life spans were changed by genotypes of the individual exposed to each type of radiation. Progeny of exposed and of controls were similar in that no effects of the irradiations could be detected. This was not entirely unexpected, for the genetic effects subject to detection were known to be rather rare. Work now in progress is necessary for detection of the larger classes of radiation effects expected on the genotypes.

References for Chapter 18

1. A. M. Brues and G. A. Sacher, "Analysis of mammalian radiation injury and lethality," in James J. Nickson, ed., *Symposium on Radiobiology,* John Wiley & Sons, Inc., New York, 1952, Chap. 23.
2. W. P. Davey, "Effect of X-rays on the length of life of *Tribolium confusum,*" *General Electric Review* 20:174–182 (1917).
3. W. P. Davey, "Prolongation of life of *Tribolium confusum* apparently due to small doses of X-rays," *General Electric Review* 22:479–483 (1919).
4. W. F. Hollander, J. W. Gowen, and J. Stadler, "A study of 25 gynandromorphic mice of the Bagg albino strain," *Anatomical Record* 124:223–243 (1956).

Chapter 19

GENETIC EFFECTS OF RADIATION IN MICE AND THEIR BEARING ON THE ESTIMATION OF HUMAN HAZARDS *

Studies on the genetic effects of ionizing radiation in mice conducted over the past few years have made it necessary to reconsider various aspects of the estimation of the genetic hazards of radiation in man. In the space available here, only the general conclusions from these studies can be presented. Most of the work has been done with irradiated males. The limited results so far obtained from irradiation of females will not be discussed here.

In considering the genetic effects of irradiation it is important to distinguish the stage or stages in gametogenesis at which the gametes have been irradiated. This distinction can be made clearly in experimental work with mammals. Male mice exposed to an acute X-ray dose of a few hundred roentgens remain fertile for about four or five weeks. A period of sterility then sets in, after which fertility returns and is maintained. The temporary sterility is due to the depletion of spermatogonia, which are especially sensitive to killing by radiation (Ref. 1). Fertility returns when adequate repopulation of spermatogonia has occurred. Matings made during the presterile period utilize germ cells that were already in post-spermatogonial stages at the time of irradiation. Matings made after the sterile period utilize germ cells that were in spermatogonial stage at the time of irradiation. The genetic effects of irradiation in these two broad stages, spermatogonial and post-permatogonial, can therefore be measured separately.

It has long been known, from the pioneer investigations of Strandskov on guinea pigs and of Snell and Hertwig on mice (see review in Ref. 2), that when the male gametes are heavily X-irradiated in post-spermatogonial stages, the offspring show high incidences of dominant defects—such as prenatal lethality, partial sterility, and sterility—that result primarily from major chromosomal aberrations. Hertwig's work also showed that in the offspring that resulted from

* This chapter is taken from Geneva Conference Paper 235, "Genetic Effects of Radiation in Mice and Their Bearing on the Estimation of Human Hazards" by W. L. Russell of the United States.

male gametes that had been irradiated in spermatogonial stage, these particular defects were either absent or greatly reduced in frequency. This finding has been confirmed in extensive experiments in our laboratory.

In man the conditions are fortunately such that the radiation dose received by a male germ cell in its post-spermatogonial stages will usually be small compared to that received in spermatogonial stages. For example, even with continuous irradiation at the official maximum permissible level right up to the time of mating, only a small dose is accumulated in the five weeks or so required for development of the spermatozoa from spermatogonia. In exceptional cases, such as accidents or emergencies requiring the taking of unusual risks, the gonads of a man may be exposed to a considerable dose of radiation within a short time. In this event it is recommended that the individual abstain from procreation for the few weeks required for the disappearance of germ cells that were irradiated in post-spermatogonial stages. Thus, even in these cases, the risk of transmission of mutational changes can be reduced to that incurred from irradiation of spermatogonia. Unfortunately, as will be shown, even this risk is far from negligible.

Since the practical importance of mutation induction in spermatogonia far outweighs that in post-spermatogonial stages, determinations were made of the X-radiation-induced point mutation rates at seven specific autosomal loci for spermatogonia of mice (Ref. 3). The mean rate obtained from an experiment in which the males were exposed to a dose of 600 roentgens was $(25.0 \pm 3.7) \times 10^{-8}$ per roentgen per locus. Additional unpublished data from this experiment, and from another experiment, in which the males were exposed to 300 roentgens, give approximately the same mutation rate per roentgen. This rate is approximately one order of magnitude higher than that found in similar experiments with *Drosophila*. One of the *Drosophila* experiments was designed to parallel the mouse experiments as closely as possible, and was carried out in our laboratory by Alexander (Ref. 4). In this experiment the X-ray-induced mean mutation rate for eight specific autosomal loci in *Drosophila* spermatogonia was determined. The mouse induced mutation rate is 15 times higher than the *Drosophila* mutation rate obtained in this particular experiment. From a scientific point of view, it would be unwise to generalize from experiments involving a limited number of loci. However, since we are concerned with the immediate practical problems of protection in man, it would be risky to ignore the indication that calculations of human hazards based on *Drosophila* mutation rates may seriously underestimate the damage.

It has sometimes been suggested that the mutation frequency in the offspring of an irradiated individual might decline as the interval between irradiation and conception increased. It has already been shown that there is a decline in frequency of transmitted chromosomal aberrations when the father has exhausted the gametes that were irradiated in post-spermatogonial stages. After the father has reached this state—that is, when all future offspring will come from germ cells that were irradiated as spermatogonia—does the frequency of point mutations transmitted to the offspring decline as the time after irradiation increases? No information from experiments on irradiated *Drosophila* spermatozoa could be considered an answer to this question about spermatogonia. Furthermore, it was desirable to have this particular problem investigated on an organism with a life span considerably longer than that of *Drosophila* and with a testis histologically similar to that of man. Analysis of the data now accumulated on mutations induced at specific loci in spermatogonia of the mouse shows no significant change in mutation rate with time after irradiation. It may be concluded that these data indicate that in man, offspring conceived long after exposure of the father to radiation are just as likely to inherit induced mutations as those conceived a few weeks after exposure. Putting this in practical terms, the results of the mouse work indicate that although postponement of procreation for a few weeks following exposure to radiation would reduce the total risk of transmission of mutational changes, by excluding those induced in post-spermatogonial stages, further postponement would not give any additional reduction in risk.

In addition to the data on mutation rate, other information obtained from the specific loci experiments on the mouse has a bearing on the estimation of human hazards. One type of information that is important in this respect is the nature of the mutations. More than one-half of the observed mutations induced in mouse spermatogonia have proved to be lethal in homozygous condition. From *Drosophila* results, it might be expected that these would on the average have deleterious effects in heterozygous condition. Clear-cut deleterious effects of some of the individual radiation-induced mouse mutations in heterozygous condition have been found. This increases the importance of investigating whether significant damage can be found in the first-generation descendants of male mice whose gametes have been irradiated in spermatogonial stage.

The specific loci experiments were not designed for measuring over-all damage in the offspring. However, one possible indication from these experiments that such damage may not be negligible, even when the gametes receive the radiation in spermatogonial stage, is provided by the data on litter size. This was recorded when the litters reached approximately three weeks of age, the age at which they were examined for mutations. In all experiments, the litters sired by irradiated males showed a slightly lower mean litter size than those sired by control males. In the 300-r experiment the reduction was between 3% and 4%. It should again be emphasized that all these data were obtained from gametes irradiated in spermatogonial stage. If the effect on litter size really represents a reduction in viability of the offspring, it is quite possible that appreciable effects will be found in experiments now being conducted to measure over-all damage in the first generation descendants of mice whose gametes were irradiated as spermatogonia. In any case, first-generation effects, even from irradiated spermatogonia, appear to be important enough to warrant further investigation. Adequate protection against the genetic hazards of peaceful uses of atomic energy may require a limitation not only of the average dose of radiation received by the population as a whole but also of the dose accumulated by individuals. The magnitude of the first-generation effects already observed, with the doses used in these preliminary experiments with mice, indicates that it is quite possible that if the present permissible weekly dose is to be kept, a total accumulated dose limit may have to be established to protect the individual from incurring too great a risk of damage to his own offspring.

References for Chapter 19

1. E. F. Oakberg, "Sensitivity and time of degeneration of spermatogenic cells irradiated in various stages of maturation in the mouse," *Radiation Research* 2:369–391 (June 1955).
2. W. L. Russell, "Genetic effects of radiation in mammals," in A. Hollaender, ed., *Radiation Biology*, McGraw-Hill Book Co., Inc., New York, 1954, Vol. 1, Chap. 12, pp. 825–859.
3. W. L. Russell, "X-ray-induced mutations in mice," *Cold Spring Harbor Symposia Quant. Biol.* 16:327–336 (1951).
4. M. L. Alexander, "Mutation rates at specific autosomal loci in the mature and immature germ cells of *Drosophila melanogaster*," *Genetics*, 39:409–428 (1954).

Chapter 20

GENETIC STRUCTURE OF MENDELIAN POPULATIONS AND ITS BEARING ON RADIATION PROBLEMS *

Of the many problems accompanying the extensive use of atomic energy, the most insidious and potentially the most dangerous is that resulting from genetic damage. Each of man's technological advances, each of his inventions has exacted its toll of human health and lives. The utilization of atomic machines will inevitably result in routine industrial accidents; there is no reason to suspect that atomic accident rates will be lower than those of today's industry. Atomic radiation possesses in addition, however, the unique ability to induce gene mutations. In their passage from generation to generation, these mutations will impose an additional increment of human suffering. The concern of this chapter is limited to these genetic effects; this limitation in no way implies that non-genetic injuries are unimportant.

The evaluation of the cumulative effect that radiations will have on future generations requires three types of information—(1) an estimation of the amount of radiation to which persons will be exposed, (2) information concerning the genetic effects of these radiations, and (3) knowledge of the genetic structure of populations. The study of the genetic effects, *radiation genetics,* has since 1927 become an impressive branch of genetics. The vast literature that has developed in this field includes some of the most precise data in biology. For the moment we avail ourselves of this information for but one purpose—to assert that we have no doubt that irradiation induces gene mutations in man. This and the additional assumption that these mutations enter the gene pool of the human population are made without discussion. Arguments supporting these assumptions are to be found elsewhere (Refs. 12, 15). The acceptance of these assumptions leads to the admission that our population is confronted with a radiation problem. Our discussion of this problem will differ from most: the sections that follow will review briefly

* This chapter is taken from Geneva Conference Paper 238, "The Genetic Structure of Mendelian Populations and Its Bearing on Radiation Problems" by B. Wallace of the United States.

concepts of population genetics; the concluding section will deal with the bearing these concepts have on the probable effects of artificially induced mutations.

Individual Variation

Modern taxonomy is outstanding in its recognition of variation within populations. Individuals are considered not mere replications of one another but members of a population exhibiting a spectrum of real differences. The type specimen was once regarded as the exemplar of a given species; individuals differing sufficiently from the type were classified as separate species. Today the type specimen has lost much of its biological importance. Well-studied groups are represented by samples of many individuals from which estimates of both means and variances can be obtained. The type specimen has been relegated to an administrative role; it is simply the physical counterpart of the species name.

An appreciation of variability is essential to an understanding of change within populations. A proper approach to the problems of evolution was impossible under the view that each individual was genetically identical with every other individual of its kind. Many who fail to comprehend the extent and nature of this variability are mystified today by the development of DDT resistance by insect populations. The same confusion is found, too, in connection with the development of resistance to antibiotics by bacteria.

Genetic Variation

The change in taxonomic philosophy was based on the development of the science of population genetics. An important corollary of Mendel's laws is the equilibrium (Hardy-Weinberg equilibrium) that exists between individuals of different genotypes within a population. If the genes A and a have frequencies p and q respectively, the frequencies of AA, Aa, and aa individuals in a large, panmictic population are p^2, $2pq$, and q^2. The proportion (q^2) of individuals in a population exhibiting a rare recessive trait is always much smaller than the frequency (q) of the recessive gene itself. Many "normal" individuals carry recessive genes in the heterozygous condition. Since the number of loci at which mutations are known is extremely large in well-studied groups, it is apparent that virtually every individual must carry at least one recessive mutation. This reasoning underlies the claim that aside from identical twins, no two individuals are genetically alike. This reasoning, too, brought about the modification of taxonomic thought.

In a sense, however, the geneticist retained the "type specimen" philosophy of the taxonomist. The type of the geneticist was the "wild type" gene, the allele symbolized above by A. The normal individual, according to this view, carried at least one wild-type allele at every locus; the non-existent "ideal" individual would be homozygous for these alleles at all loci. The wild-type allele was considered virtually constant for a species; the allele present in every normal individual was identical with that in all others. However, in the late 1930's even this concept of constancy was weakened by the identification of wild-type iso-alleles—alleles distinguishable only through the use of special genetic tests. The existence of differences of this order increased once more the amount of demonstrable genetic variation in populations (see, for instance, Ref. 14).

The revelation by geneticists of the existence of this genetic variation in populations represented a tremendous advance for population genetics. The dominance of wild-type alleles allowed recessive mutations to accumulate in a population without imposing an undue strain upon its fitness. The necessity for possessing this concealed store of recessives was the need for adapting to novel demands imposed by a changing environment. Rates of spontaneous mutation were assumed to be established by natural selection at a level which endowed the species with an optimal amount of genetic variability. At any given time, however, the necessity for keeping these mutant genes resulted in a burden on the population.

Balanced Polymorphism

The concept of genetic variability in the form of concealed recessive mutations failed to explain certain instances of polymorphism in nature. Polymorphism is the existence of several phenotypically distinct types of individuals in one population; balanced polymorphism is the existence of these types in fairly constant proportions over a long period of time. Fisher (Ref. 9) proposed very early that the stability of a balanced polymorphic system based on a single gene pair is assured if heterozygous individuals are favored by natural selection over both homozygotes. Mutant alleles in cases of balanced polymorphism are tolerated, but not because they are concealed; they are actively selected for within populations through the adaptive superiority of heterozygous individuals.

Our knowledge of the extent and importance of balanced polymorphism was increased by the discovery of cryptic polymorphism—polymorphism for chromosome structure—in *Drosophila*. Cytological examination of giant salivary chromosomes of *Drosophila* larvae of numerous species revealed a wealth of naturally occurring inversions. Wright and Dobzhansky (Ref. 25) demonstrated that in artificial populations of *D. pseudoobscura,* individuals carrying chromosomes of two different arrangements were favored by "natural" selection over individuals homozygous for either one of the two inversion types. Dobzhansky and Levene (Ref. 5) were able to show later that evidence for the selective superiority of heterozygous individuals was present in data that had previously been misinterpreted (Ref. 4).

At the present time it is most profitable to assume that selection is responsible for all cases of balanced polymorphism in which the rarer gene has a considerable frequency. This view has been strengthened in human populations by evidence that heterozygosis for the gene causing sickle-cell anemia confers resistance to malaria (Ref. 1) and that various blood group genes are associated differentially with a number of human ailments (Refs. 2, 13).

Coadaptation and Genetic Homeostasis

The early studies on inversions in populations of *D. pseudoobscura* showed that these inversions constituted cryptic, balanced polymorphic systems within these populations. Later studies (Ref. 3) indicated that a stable equilibrium was obtained in experimental populations only if chromosomes of the two arrangements came initially from the same geographical region. If they came from different localities, equilibria were not always established, and furthermore, erratic and unpredictable changes in relative frequencies occurred in different experimental populations (Ref. 6). Chromosomes of different arrangements were said to be coadapted by natural selection within local populations.

The concept of coadaptation has been extended to include cytologically invisible as well as visible factors. If selection favors individuals heterozygous at many gene loci and if combinations of alleles at each of these loci interact with those at other loci, natural selection must coadapt the elements of any local population into an integrated gene pool.

The concept of integrated gene pools based upon selection of heterozygous individuals of superior adaptive value is scarcely more than a working hypothesis. As a hypothesis it has proved to be a rewarding one. In our own work with irradiated populations of *D. melanogaster* the hypothesis was suggested initially by the non-correspondence between frequencies of obviously deleterious chromosomes in populations and the viabilities of individuals carrying random combinations of these same chromosomes. The explanation seemed most likely to lie in the indifference of populations to properties exhibited by chromosomes when homozygous; within populations these chromosomes occurred only in random, heterozygous pairs (Refs. 21, 22). The hypothesis was tested further by an analysis of variation arising through gene recombination. Selection for homozygosis should curtail the variety of novel gene combinations that can be formed by recombination of chromosomes from the same population. Crossing-over between chromosomes from different populations can produce novel combinations even if selection in each population favors homozygous individuals; any combination of chromosomes from the two populations should yield, however, the same array of cross-over products. Experimental data obtained through an analysis of 100 cross-over chromosomes obtained from each of the 55 combinations of 10 chromosomes—5 from each of two experimental populations—supported the hypothesis that selection favors heterozygotes (Ref. 23).

Evidence for the integration of local gene pools has also been obtained from natural populations. Wallace (Refs. 18, 19) interpreted the geographical distribution of gene arrangements of *D. pseudoobscura* as evidence for the existence of integrated gene complexes. Combinations of inversion types incapable of maintaining the integrity of gene combinations are seldom found within the same population. Vetukhiv (Refs. 16, 17) studied the survival under semistarvation conditions of interpopulation F_1 and F_2 hybrids and of intrapopulation, non-hybrid individuals. He found that a higher percentage of F_1 hybrids than of non-hybrids survive; F_2 hybrids, however, have a much lower survival rate than either of the other two classes. This has been demonstrated for three species: *D. pseudoobscura, D. willistoni,* and *D. paulistorum.* The results of these experiments have been paralleled recently by King (Ref. 10) in an analysis of DDT resistance in experimental populations. In this study the F_2 hybrids of two resistant populations of *D. melanogaster* possess less resistance than do either of the two original populations or the F_1 hybrids. Finally, recent experiments in our own laboratory (Ref. 20) using *D. melanogaster* have indicated that only the interpopulation F_1 hybrids have higher viabilities than the non-hybrid members of local populations. By studying three-way and double-cross hybrids it was possible to show that flies genetically as heterozygous as the F_1 hybrids frequently have much lower survival rates. These studies demonstrate the integration of the gene pools of local populations. They also demonstrate the importance of heterozygosity for selected gene complexes; this demonstration supports the earlier conclusion based on the geographical distributions of gene arrangements.

Evidence indicating why selection might favor heterozygous individuals has also been obtained. Wallace and Madden (Ref. 24) and Dobzhansky and Spassky (Ref. 7) have re-examined the problem of the frequency of subvitals (chromosomes with slightly deleterious effects on viability) in populations. In the course of these analyses it was necessary to determine the various components of the total variance observed between tests of different chromosomes: (1) the component due to the number of flies counted in making the tests, (2) the component representing differences between replicated cultures, and (3) the "true" variance between different chromosomes from a population. As reported (Ref. 8), replicated tests of homozygous individuals possessed a large between-culture variance, while similar replicated tests of heterozygous individuals did not. Heterozygous individuals appear to be better buffered against the effects of these different environments. Selection may decrease total phenotypic variance by increasing genotypic variance.

It is necessary to mention in this connection the work of Lerner (Ref. 11). Working with entirely different material and using a different approach, Lerner has reached a series of conclusions regarding homeostasis and population structure that virtually coincide with those presented above.

Mutations in Natural Populations

One's concept of the genetic structure of populations determines one's concept of the role mutations play in these populations. The presence of genetic variation was originally unsuspected. Next, the Hardy-Weinberg equilibrium and its modifications allowed accurate analyses to be made of the relation between mutation rates and frequencies of recessive mutations in populations. These mutations were regarded as a burden tolerated by natural selection because they enabled the population to adapt to fluctuations of the environment. Finally, we have a model that implies that mutations are retained in populations largely by virtue of their adaptive values in heterozygous individuals.

The fate of a recent mutation under this model might be as follows: Provided that it is not lost by chance from the population immediately, the new mutation may exist in a small number of individuals. This is a rare gene, so it exists only in heterozygous individuals. These individuals carry background gene combinations as well as the mutation under discussion, and so they will suffer different fates. In certain instances their destinies will not depend upon the new mutation. Among others, however, there will be some individuals favored by selection because of the over-all genotypic characteristics determined by interactions between the mutation and alleles at other loci; similarly, there will be other individuals eliminated by selection because of corresponding interactions. Consequently, the mutation in question may be preserved, and to the extent that gene recombination allows, so are the background combinations in which it was favorable. This is the course of natural selection at a low gene-frequency level. The proportion of new mutations that possess selective advantages in the heterozygous condition is unimportant: those that have an advantage, regardless of their initial proportions, will be preserved and will become the gene pool of the population. Similarly, the behavior of a favored mutation when it is homozygous is also unimportant; the selection described is based upon properties of heterozygous individuals. The duration of the selective superiority of a given mutation does not determine its selection at a given moment; the genetic background is composed of a continuously shifting complex of mutant alleles each of which participates in a selective process like the one described.

Our "load of mutations" (Ref. 12) becomes, then, a complex burden consisting of a spectrum of mutations. At one end are those whose favorable action in any genotypic background is highly unlikely; these mutations are maintained in the population at low frequencies simply by mutation pressure. At the other end are mutations—perhaps grotesquely aberrant when homozygous—that are retained in populations through their advantageous behavior in heterozygous individuals. Perhaps the bulk of all mutations includes those alleles that are favorable in some genetic combinations and unfavorable in others; depending upon the course of future selection, these may have their position shifted towards either end of the spectrum.

Our understanding of the problems confronting populations under this model is still inadequate. We can see that coadaptation in the case of large blocks of genes leads to exceptionally stable examples of balanced polymorphism; this was demonstrated in the case of chromosomal rearrangements in *D. pseudoobscura*. In looking beyond present populations by means of the geographical distribution of the gene arrangements, we can discern faintly just how stable these balanced heterozygous systems are; not only do some of them appear to have antedated species but it also appears that in some instances speciation has occurred during the preservation of these balanced systems (Ref. 19). In attempting to understand some of the present systems, we lack the insight necessary to evaluate the price that different populations can pay for their genetic structures. Populations of *D. tropicalis* inhabiting Central America, for instance, possess a balanced-lethal inversion complex; for some unknown reason the populations of this single area find it advantageous to preserve this system even though half of the zygotes formed each generation are lost (Dobzhansky, oral communication).

Radiation Damage

In the introduction two assumptions were made which led to the conclusion that human populations are confronted with a radiation problem. This problem arises not only from the utilization of atomic energy but also from the widespread use of X radiation for medical and industrial purposes. A proper evaluation of the nature of this problem depends not only upon an understanding of radiation genetics but also upon a knowledge of population genetics. What bearing does coadaptation within Mendelian populations have on the question of radiation damage?

Under a system of coadaptation, as under any other system, gene mutations whose average effect is to lower the viability or "fitness" of individuals must be eliminated by natural selection. In human populations this elimination must be recognized in terms of human suffering and individual misery. Since the preponderance of newly induced mutations are probably deleterious in virtually all genotypes, these mutations will harm individuals and handicap their families. We cannot wantonly bestow pain and misery upon individuals of succeeding generations; there must be positive reasons for exposing persons to irradiation at any time. Unnecessary or irresponsible exposure cannot be justified by the negative arguments that radiations have no effects or that, at most, they induce just a few mutations.

From the point of view of populations, the concept of coadaptation indicates that genetic events move more rapidly than they do under the model of "concealed recessives." One of the fundamental aspects of the model presented above is the importance placed upon differential selection among heterozygous individuals. Negative selection for heterozygous individuals will rapidly eliminate the responsible genes from a population. If selection is positive, these genes are quickly coadapted to the rest of the gene pool. Since the incorporation of "favorable" mutations occurs regardless of the rarity of such mutations, we might expect that some of the radiation-induced mutations will be taken up by the genetic fabric of our population.

It seems unlikely—barring wholesale exposure of populations to near-lethal doses of irradiation—that any reasonable level of exposure will result in the extinction of human populations. The uncertainty in this matter lies in the definition and adherence to a "reasonable" level of exposure; it is difficult to see how reason will prevail in the frenzy of technological advances that is surely approaching. I personally doubt whether atomic disasters will burden populations for thousands or even hundreds of generations with deleterious mutations; natural selection seems to act much too rapidly in eliminating semidominant, deleterious mutations. As we saw above (Refs. 21, 22) the supply of genetic material passed on from generation to generation in irradiated *Drosophila* populations gives rise to remarkably normal individuals.

In discussing radiation effects, geneticists make a useful distinction between somatic and genetic damage. Somatic damage does not affect the germ cells, and hence is not transmitted from generation to generation. Regardless of the extent of this type of damage at any time, the next generation is spared its effects. Some types of genetic damage follow precisely the pattern of somatic damage; dominant lethals and dominant sterility factors are completely eliminated from populations in the generation following the cessation of their induction. The model proposed for coadapted gene pools, through its emphasis on semidominance, has the effect of blurring still more the transmission patterns of somatic and genetic effects. Perhaps by rendering the distinction between these two types of effects less clear, this model will allow geneticists to shift some of their concern to individuals of our own generation.

References for Chapter 20

1. A. C. Allison, "Protection afforded by sickle cell trait against subtertian malarial infection," *Brit. Med. J.* 1:290–294 (1954).
2. A. C. Allison, "Aspects of polymorphism in man," *Cold Spring Harbor Symp. Quant. Biol.* 20 (1955).
3. Th. Dobzhansky, "Genetics of natural populations: XIX. Origin of heterosis through natural selection in populations of *Drosophila pseudoobscura*," *Genetics* 35:288–302 (1950).
4. Th. Dobzhansky and C. Epling, *Taxonomy, Geographical Distribution and Ecology*

of Drosophila pseudoobscura and Its Relatives, Carnegie Institute, Washington, Publ. 554 (1944).

5. Th. Dobzhansky and H. Levene, "Genetics of natural populations: XVII. Proof of the operation of natural selection in wild populations of *Drosophila pseudoobscura,*" *Genetics* 33:537–547 (1948).
6. Th. Dobzhansky and O. Pavlovsky, "Indeterminate outcome of certain experiments on *Drosophila* populations," *Evolution* 7:198–210 (1953).
7. Th. Dobzhansky and B. Spassky, "Genetics of natural populations: XXI. Concealed variability in two sympatric species of *Drosophila,*" *Genetics* 38:471–484 (1953).
8. Th. Dobzhansky and B. Wallace, "The genetics of homeostasis in *Drosophila,*" *Proc. Nat. Acad. Sci.* 39:162–171 (1953).
9. R. A. Fisher, *The Genetical Theory of Natural Selection,* Clarendon Press, Oxford, 1930.
10. J. C. King, "Integration of the gene pool as demonstrated by resistance to DDT," *Amer. Nat.* 89:39–46 (1955).
11. I. M. Lerner, *Genetic Homeostasis,* Oliver and Boyd, Edinburgh and London, 1954.
12. H. J. Muller, "Our load of mutations," *Amer. J. Human Genetics* 2:111–176 (1950).
13. P. M. Sheppard, Synthesis of the section "Genetic variability and polymorphism," *Cold Spring Harbor Symp. Quant. Biol.* 20 (1955).
14. C. Stern and E. W. Schaeffer, "On wild-type iso-alleles in *Drosophila melanogaster,*" *Proc. Nat. Acad. Sci.* 29:361–367 (1943).
15. A. H. Sturtevant, "The genetic effects of high-energy irradiation of human populations," *Eng. and Sci. Monthly,* Calif. Inst. Tech., Pasadena, 1955.
16. M. Vetukhiv, "Viability of hybrids between local populations of *Drosophila pseudoobscura,*" *Proc. Nat. Acad. Sci.* 39:30–34 (1953).
17. M. Vetukhiv, "Integration of the genotype in local populations of three species of *Drosophila,*" *Evolution* 8:241–251 (1954).
18. B. Wallace, "On coadaptation in *Drosophila,*" *Amer. Nat.* 87:343–358 (1953).
19. B. Wallace, "Coadaptation and the gene arrangements of *Drosophila pseudoobscura*" (I. U. B. S. Symposium on Genetics of Population Structure) *I. U. B. S.,* Ser. B, No. 15:67–94 (1954).
20. B. Wallace, "Inter-population hybrids in *Drosophila melanogaster,*" *Evolution* 8 (1955).
21. B. Wallace and J. C. King, "Genetic changes in populations under irradiation," *Amer. Nat.* 85:209–222 (1951).
22. B. Wallace and J. C. King, "A genetic analysis of the adaptive values of populations," *Proc. Nat. Acad. Sci.* 38:706–715 (1952).
23. B. Wallace, J. C. King, C. V. Madden, B. Kaufmann, and E. C. McGunnigle, "An analysis of variability arising through recombination," *Genetics* 38:272–307 (1953).
24. B. Wallace and C. V. Madden, "The frequencies of sub- and supervitals in experimental populations of *Drosophila melanogaster,*" *Genetics* 38:456–470 (1953).
25. S. Wright and Th. Dobzhansky, "Genetics of natural populations: XII. Experimental reproduction of some of the changes caused by natural selection in certain populations of *Drosophila pseudoobscura,*" *Genetics* 31:125–156 (1946).

Chapter 21

THE GENETIC PROBLEM OF IRRADIATED HUMAN POPULATIONS *

In a civilized community the thesis is accepted that it is permissible to do harm to a few individuals when this is an unavoidable by-product of doing good to many. Airline passengers are still occasionally killed or injured, but nobody suggests that air travel should be banned. All technically possible steps are taken to minimize the harm and to minimize the few to whom it is done; then it is weighed, consciously or unconsciously, against the good to the many; and provided that the harm is only slight, or the few are very few, the public conscience is unoffended and the situation accepted. This procedure rests on the implicit acceptance of two premisses: first, that it is possible for each individual at risk to evaluate his chances of being among the unlucky few; and second, that he has a free choice to expose or refuse to expose himself to the risk. We know how many fatalities there are per hundred million airline passenger-miles; and if we think it too high, we can always walk or stay at home.

During the last year or two the public at large has become aware that even the peacetime use of nuclear power, with its benefit to the many, may entail danger to the few; and the public has become seriously disquieted by this realization. The reason, I think, is that the usual procedures for determining the acceptable level of harm cannot be applied in this instance. This is partly because there is no agreement, even among the informed, about the exact nature and magnitude of the genetic danger from ionizing radiations; and partly because there is no possibility of choice. The unlucky individual who suffers the genetic damage is not the one who exposed himself to the risk.

The awakening of the public conscience in this matter is due mainly to the writings of geneticists, notably H. J. Muller and A. H. Sturtevant. They pointed out that there is no threshold for the induction of genetic effects by ionizing radiations and that any exposure, however slight, implies the induction of some mutations; that when a population is in genetic equilibrium every new

* This chapter is taken from Geneva Conference Paper 449, "The Genetic Problem of Irradiated Human Populations" by T. C. Carter of the United Kingdom.

mutation must be balanced, on the average, by the loss of a mutant gene through selection; and that this loss—"genetic death," to use Muller's term—often means suffering for the individual or his family (Refs. 1, 2).

It was a natural consequence that there should be a demand for some quantitative assessment of the probable magnitude of the genetic dangers of ionizing radiations. But the geneticist found himself faced with such enormous gaps in our knowledge of the genetics of human populations and of the results of exposing them to radiations that even a rough quantitative assessment was impossible. As Sturtevant put it, "No scientist interested in exact quantitative results would touch the subject, were it not that its social significance leaves us no alternative" (Ref. 3). *Faute de mieux* it was necessary to use quantitative results obtained from experiments on other organisms, notably *Drosophila* and the mouse, and feed them into mathematical models intended to represent human populations. Perhaps the most surprising result of these operations has been the measure of agreement among geneticists; almost all who have written on the subject agree that to expose a human population to continuous irradiation with an accumulated dose as high as 25 r per generation would be very undesirable.

The object of his chapter is to re-examine some of the foundations on which assessments of genetic dangers to man rest, and to outline some of the research which will be needed to plug the larger gaps in our knowledge. I want to emphasize that the opinions expressed represent only the results of my own crystal gazing and must not be interpreted as a statement of official British policy.

In its simplest form, the theoretical model of a closed population consists essentially of a pool of genes, the genes carried by the individuals of which the population is composed. Most of them are "wild type" alleles, that is to say those alleles which have become characteristic of the population through the action of natural selection in the past; it is almost a matter of definition, therefore, that wild-type alleles confer an advantage on the individuals carrying them. Conversely, mutant alleles confer a disadvantage, relative to wild-type. In the gene pool wild-type genes are continually changing, by mutation, into their mutant alleles, so that the proportion of mutant genes tends to rise. This tendency is counteracted by the loss of mutant genes from the pool through the disadvantage which they confer on their carriers; this may express itself as reduced viability or reduced fertility or both. Thus a genetic equilibrium tends to be set up, at which the tendency of mutant genes to increase through mutation is exactly counterbalanced by their loss through selection. From this it follows that if equilibrium is to be maintained, for every mutant allele arising by mutation there must be another eliminated from the population by the early death or reduced fertility of some individual. At times of disequilibrium mutant alleles may arise more or less rapidly than they are eliminated, but on the average, in the long run, the rates will be equal.

Attempts to make quantitative estimates of genetic damage to human populations by ionizing radiations depend upon the concept of genetic equilibrium. They usually run on the following general lines. First, an estimate is made of the proportion of all deaths in the unirradiated population which are "genetic deaths" due to spontaneous mutation. The average mutation rate per locus, m, is estimated and multiplied by an estimate of n, the number of loci in the human genome; the product is an estimate of mn, the average number of newly arisen mutant genes per gamete; $2mn$ is the average number per zygote. It is then assumed that an individual suffers "genetic death" whenever he carries z mutant genes more than the average; thus $2mnp$ mutant genes would be eliminated through the genetic death of $2mnp/z$ individuals. But $2mnp$ is the number of mutant genes arising by new mutation in p individuals. In other words, among p deaths $2mnp/z$ will be genetic deaths; or a proportion $2mn/z$ of all deaths will represent genetic deaths. The next step is to estimate the factor, y, by which the mutation rate is increased under the given irradiation conditions. When a new genetic equilibrium has been reached, the fraction of all deaths which are genetic deaths becomes $2mny/z$. The radiation-induced genetic deaths are then a fraction $2mn(y - 1)/z$ of all deaths.

When attempts are made to put numerical values into this expression it becomes apparent that any present-day quantitative estimate of radiation-induced genetic damage to man has extremely tenuous foundations. Numerical estimates vary from 10^{-5} to 10^{-7} mutations per locus for m and from 10,000 to 100,000 loci for n; and there are no human data on which to base estimates of y and z. For several loci in the mouse, y probably has a value of about 2 for an accumulated radiation dose of 50 r. In the mouse also it is possible to obtain viable and fertile individuals homozygous for 8 major recessive mutants, i.e., carrying 16 major mutant genes, presumably over and above the minor mutants which are present in the average individual; it therefore seems quite possible that in this species z may be greater than 10. If these figures were applicable also to man, it would follow that exposure to chronic irradiation at a dose-rate of 50 r per generation would result in the induction of genetic deaths amounting to somewhere between 0.2 and 0.0002 of all deaths. Thus the range of uncertainty in such calculations is colossal; and I feel by no means convinced that the truth necessarily lies within even this enormous range.

Granted that any increase in radiation exposure will ultimately cause some genetic deaths, the form which they will take should be considered. In a very few cases it is possible to specify the form and make numerical estimates. Thus achondroplasia has a spontaneous mutation rate of about 4×10^{-5} per gamete; therefore in the United Kingdom, with about 793,000 live births per year (Ref. 5), the number of achondroplastics occurring annually through new mutation is about 60. Of these about 80% die in their first year (Ref. 4). If a radiation dose of 50 r doubles the human mutation rate, exposure of the whole population of the United Kingdom to 50 r per generation (i.e., about 2 r per annum) would lead to the induction and subsequent genetic death of a further 60 achondroplastics annually. Genetically simple, easily recognized conditions like achondroplasia are, however, the exception rather than the rule; probably the great majority of mutant genes have much less obvious effects. Furthermore, it does not necessarily follow that they are all harmful to the population, even though they may cause the genetic death of the individual. The existence of a negative correlation between fertility and intelligence, and the fact that intelligence is at least partly genetically determined, implies that an unduly high proportion of the most intelligent suffer genetic death. Thumbing over the pages of an encyclopedia is sufficient reminder that many who made the greatest contributions to the heritage of our civilization contributed nothing to the generation which followed them. If genetic death is the price of being a Beethoven or an Isaac Newton, it does not necessarily follow that all genetic deaths are undesirable.

The main lesson to be learned from the sort of calculation made above is, I think, that we are profoundly ignorant of all quantitative aspects of this subject; the corollary is that a massive research effort is needed. What form should this take?

Our final objective is to make accurate quantitative predictions about the effects on human populations of exposing them to chronic dosage with ionizing radiations. It seems to me that a research program with this end naturally falls into three main parts: research on mutation, since radiations induce mutations; research on animal populations, since man is a social animal; and research on human populations, since one cannot justify the application to one species of quantitative measurements made on another.

Research on mutation is in progress in genetical institutes throughout the world. It entails a great deal of counting heads, and for that reason much of it is done on virus, bacteria, and other microorganisms; enormous numbers of these can be cultured in a small space, and techniques have been developed whereby the identification of mutants becomes a simple routine. Furthermore, in these organisms the final, phenotypic effect of the gene is probably causally close to the primary gene action, a fact which makes it less difficult to investigate the processes involved. The disadvantage of work with microorganisms lies in the fact that the reproductive nucleus always occurs in the same type of cell; and we know that with higher organisms the mutational properties of a gene may vary with the type of cell in which it is carried. More studies of mutation are needed, but they must be spread over many types of cell in many different organisms, including vertebrates. In particular, we need studies of the effects of ionizing radiations when administered at very low dose rates over a long period, to determine whether the linear accumulative law is really strictly valid under these conditions.

Experimental research on population genetics has lagged far behind population genetic theory; and anyone who reads the literature in this field cannot avoid a disquieting feeling of doubt about the extent to which theoretical models of populations may bear any relationship to actual populations as they exist in nature. Heterosis, epistasis, linkage, chromosome mutation, and assortative mating play a small part, if any, in most population genetic calculations today; yet studies of populations, both in the wild and in the laboratory, suggest that these may be essential elements in population structure. It seems to me that we need a large program of research on animal population genetics, with the object of determining the factors involved, assessing their relative importance and developing a more adequate population genetic theory. In particular, we need more research on the effects of assortative mating and the extent to which it occurs in wild populations; there can be little doubt that assortative mating with respect to mental characteristics occurs extensively in man. Until we have a clearer picture of what occurs in animal populations, both free-living and experimental, we have little hope of making predictions about human populations.

The third main field of research should be human genetics, because we cannot make predictions about the changes which will be induced in human populations by additional radiation exposure until we know something about the changes which may occur in its absence. The first thing we need to know is the extent of genetic death in civilized populations and the form it takes. Does it occur through prenatal inviability, postnatal inviability, susceptibility to infection or accident, inability to attract a mate, organic sterility, disinclinaton to reproduce? Is it associated with high intelligence or the ability to make significant contributions to civilization? Is it associated especially with mental defect? To what extent does it lead to suffering by the individual concerned, or to his family? To what extent does it place a material load on the rest of the population—e.g., by the upkeep of institutions, hospitals and prisons? To answer these questions we need a study of the mental, physical, and demographic characteristics of normal human populations as they exist today and in the absence of additional radiation exposure.

Next we need to know the relationship, in man, between radiation exposure and mutation rate. Here two approaches are possible. One is to study small populations which have been exposed to high radiation doses, as at Hiroshima and Nagasaki; the other is to study large populations which are exposed to low radiation doses, e.g., the population of a large area in which the natural radiation background is high. My own view is that the latter is likely to be the more fruitful approach. There are granite areas in Scotland, England, Wales, and Scandinavia in which the radiation background is appreciably higher than in the granite-free areas of the same countries; and I think it may be within the bounds of what is technically possible today to make estimates of the mutation rates of a few clear-cut autosomal dominant or sex-linked recessive conditions (e.g., achondroplasia, epiloia, haemophilia) with sufficient accuracy to enable us to set useful fiducial limits to the mutation-rate-doubling radiation dose in man. Such a project would probably have to be on an international scale and would certainly take ten years or more; but I think the results would be well worth the effort.

To sum up: In my opinion we cannot today make any useful quantitative assessment of the genetic consequences of exposure of human populations to ionizing radiations at low dosage rates; we know far too little about human population structure and the induction of mutations in man. But we know enough to be apprehensive about the genetic dangers. We now need a research program with three main parts: fundamental studies of mutation, studies of animal populations, and studies of human populations. Such a program would have to be on a very lavish scale and parts of it would almost certainly require international cooperation.

References for Chapter 21

1. H. J. Muller, "Our load of mutations," *Amer. J. Human Genetics* 2:111 (1950).
2. H. J. Muller, "The manner of dependence of the 'permissible dose' of radiation on the amount of genetic damage," *Acta Radiologica* 41:5 (1954).
3. A. H. Sturtevant, "The genetic effects of high-energy irradiation of human populations." Address given to the California Institute of Technology (1955).
4. E. T. Mørch, "Chondrodystrophic dwarfs in Denmark," *Opera ex Domo Biol. Hered. Human Univ. Hafniensis* 3 (1941).
5. United Nations. *Demographic Year Book,* New York (1953).

Part VI

GENETIC ERADICATION OF INSECT PESTS

Chapter 22

ERADICATION OF THE SCREW-WORM FLY *

During 1954 screw-worms, *Callitroga hominivorax* (Cqrl.), were eradicated from the island of Curaçao, Netherlands Antilles, by means of a new tool in biological control—sterilized insects. Male flies, reared in the laboratory and sterilized with gamma rays from cobalt-60, were released in the field to compete for mates with normal males. It was possible to release sterile insects in greater numbers than existed in nature. As the females mate only once, those that mated with sterile males were incapable of reproduction. The sterile males competed so effectively that screw-worms were eradicated. This chapter is an account of the research conducted on this method of controlling the screw-worm.

Effects of Irradiation on Insects

It has been known for almost forty years that insects can be sterilized by radiations. Runner reported in 1916 that cigarette beetles, *Lasioderma serricorne* (F.), laid infertile eggs after exposure to X rays (Ref. 9). Extensive studies on the effect of radiations on insects followed the discovery by Muller (Ref. 13) that X rays induced mutations in *Drosophila melanogaster* Meig. In 1928 Muller reported that when untreated *Drosophila* females were mated to heavily treated males, they laid eggs that failed to hatch.

Following Muller's discovery that chromosomal changes can be induced by X-irradiation, other geneticists and cytologists tested its effects on various insects. White (Refs. 20, 21) treated four species of grasshoppers with X rays, and Carlson (Ref. 4) X-rayed still another species of grasshopper. Whiting (Ref. 22) irradiated *Bracon hebetor* Say. Reynolds (Ref. 18) reported on the effects of X-irradiation on *Bradysia ocellaris* (O.S.), and Crouse (Ref. 6) treated another species of fungus gnat, *B. coprophila* (Lint.). Koller and Ahmed (Ref. 9) compared the effects of X-irradiation on *D. melanogaster* and *D. pseudoobscura*. Thus, in addition to the extensive literature on *D. melano-*

* This chapter is taken from Geneva Conference Paper 114, "Eradication of the Screw-worm Fly by Releasing Gamma-Ray-Sterilized Males Among the Natural Population" by R. C. Bushland, E. F. Knipling, and A. W. Lindquist of the United States.

gaster, biologists had observed the effects of irradiation on still another species of fruit fly, a parasitic wasp, a beetle, two species of fungus gnats, and five species of grasshoppers before our work on the screw-worm fly.

Muller (Refs. 15, 16), Lea (Ref. 10), and Catcheside (Ref. 5) have reviewed the literature of genetics and cytology dealing with the effects of radiations on insects. There seems to be general agreement that X rays and gamma rays cause similar effects. If the germ cells of an insect are irradiated, chromosomal changes result, and the extent of the changes depends upon the amount of ionizing radiation. Extreme changes cause the cells to degenerate. Less extensive mutations may not prevent the sperm from fertilizing an egg, but the zygote may be incapable of maturing and usually dies in the embryonic stage. Geneticists call mutations that prevent the survival of the fertilized egg dominant lethal mutations.

Our work with the screw-worm fly has been to determine the dose of irradiation sufficient to cause dominant lethal mutations in all the germ cells. Technically speaking, the males so irradiated are not truly sterile, because they are still capable of producing sperm which fertilizes eggs; but for practical purposes they are called sterile because their progeny die as embryos.

Biology of the Screw-worm

The screw-worm fly (Fig. 22.1) is a calliphorid fly which is parasitic in the larval stage. The female deposits her eggs in a mass of about 200 on the edge

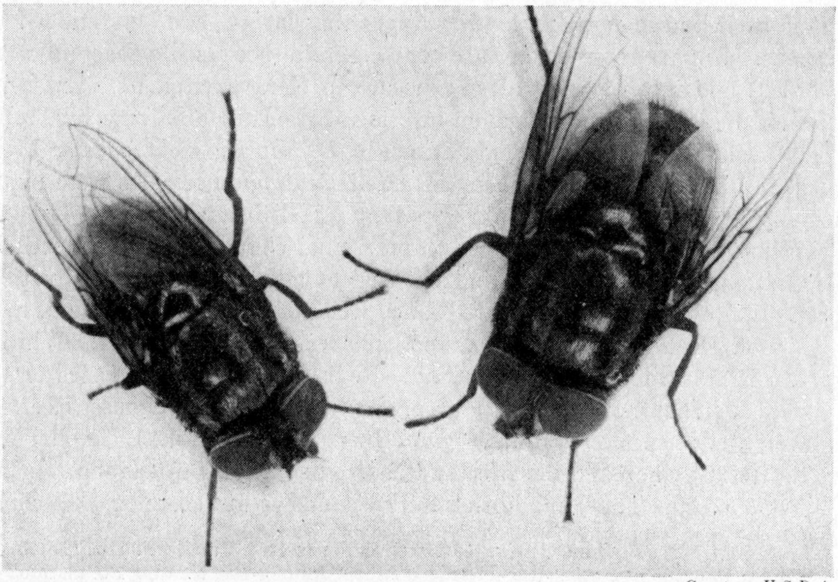

Courtesy U.S.D.A.

Fig. 22.1. Screw-worm flies. Male fly on left (note large eyes). Female on right (note space between eyes).

of an abrasion in the skin of a warm-blooded animal. Flies oviposit on cuts, scratches, the navels of newborn animals, and even such inconspicuous wounds as tick bites. The eggs hatch in 12 to 24 hr, and the larvae feed on the living muscle tissue. The insects feed as a colony, eating a hole in the flesh. Infested wounds attract other flies to oviposit, and untended animals may be literally eaten alive. Almost invariably infested animals die if not properly treated with a larvicide. The larvae (Fig. 22.2) complete feeding in 5 to 6 days and

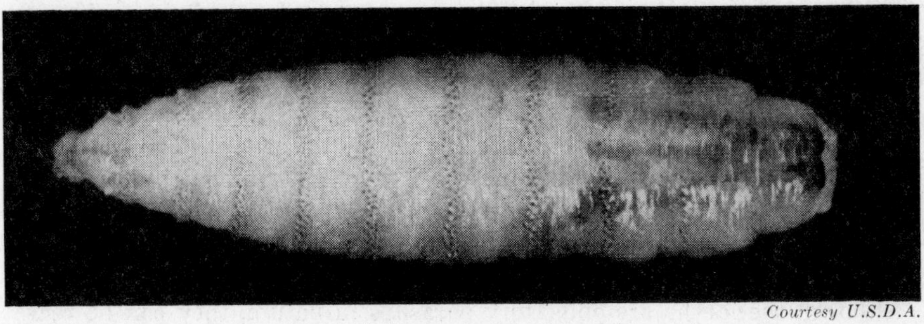

Courtesy U.S.D.A.

FIG. 22.2. Full-grown screw-worm larvae.

then crawl from the wound and burrow into the soil to pupate. The pupal stage (Fig. 22.3) lasts about 8 days in warm weather. Adult flies emerge, work their way to the surface of the soil, and crawl up on the nearest vegetation to expand their wings and fly away. They mate when 2 to 4 days old, and at

Courtesy U.S.D.A.

FIG. 22.3. Screw-worm pupae.

the age of 6 days the female is ready to lay eggs. The adults usually die of old age in about 3 weeks.

In nature the screw-worm is an obligatory parasite on warm-blooded animals. The flies oviposit only on living animals, and if the host animal dies before the larvae have completed two-thirds of their development, they cannot survive. The population is therefore limited, and the insects never reach such numbers as are attained by flies that breed in carrion. Domestic animals that are properly cared for may become infested, but proper wound treatment with a larvicide prevents the survival of any maggots. In order to survive the species must increase in the wounds of neglected domestic animals and wild animals.

Screw-worms are easily reared in the laboratory. The flies are maintained in cages provided with water and are fed chiefly on honey. The adults also feed readily on exudate from ground meat during the preoviposition period. The females will readily oviposit when confined in shell vials containing pieces of fresh beef. The egg masses are transferred from the vials to moist filter paper and held in a covered petri dish for hatching.

Although the larvae are obligatory parasites in nature, they can be reared in the laboratory by a modification of the procedure described by Melvin and Bushland (Ref. 12). A mixture of ground lean meat, citrated beef blood, and water with formaldehyde to retard putrefaction serves as the rearing medium. The larvae are reared at a temperature of about 95° F and grow almost as well as they do in their natural hosts. The larvae crawl from the rearing medium and drop into sand trays to pupate. The pupae are sifted from the sand and transferred to screen cylinders, which are held in a cabinet at 80° F and 100% relative humidity. The flies emerge after 7½ days.

The screw-worm is tropical and semitropical in its distributing, being found the year around only in those parts of the Americas where the winters are mild. In the United States screw-worms used to be limited in their winter distribution to regions bordering on Mexico. In the summertime they migrated up the Pacific coast area about as far as the Sacramento Valley, and in the Midwest they traveled as far north as Kansas before being killed by cold weather. The summer migration of the midwestern infestation extended northward rather than eastward. Screw-worms seldom were found very far east of the Mississippi River, and the population that migrated to Mississippi and Louisiana did not survive the winter.

In 1933 screw-worms were accidentally introduced into Georgia through shipment of infested cattle from Texas. The Georgia infestation spread into Florida, where the insects found a winter climate suitable for survival. During the summer months screw-worms spread over the Southeastern States, doing millions of dollars of damage each year. In average winters the cold kills screw-worms north of peninsular Florida. If the Florida overwintering population could be destroyed, screw-worms might be eliminated from the Southeastern States. It is believed that with proper precautions to prevent importation of infested livestock, the area might be kept free of the parasite.

Laboratory Tests with Irradiated Flies

The Entomology Research Branch's studies with irradiated flies have been directed toward the practical problem of eradicating the overwintering Florida population. When screw-worms first became established in Florida, the U.S. Department of Agriculture, in cooperation with the State, made every effort to control them through good animal-husbandry practices and the use of larvicides to treat infested wounds. However, enough insects survived in wild animals and neglected domestic animals that eradication was impossible. Some other supplementary procedure is necessary if screw-worms are to be eradicated.

In 1950 we began experiments with irradiated flies. The first tests (Ref. 2) were with X rays. Both pupae and adults were irradiated. When flies less than 2 days old were given a dose of 5000 roentgens, the males were sterilized, but the females were not affected. Pupae were more susceptible to the sterilizing effects of the irradiation. When they were irradiated 2 days before flies were due to emerge, a dose of 2500 roentgens caused sterility in the males, but 5000 roentgens were required to sterilize females. The sterilizing doses of irradiation did not appear harmful to the pupae, since as many adults emerged from irradiated groups as from untreated controls. About the only adverse effect seemed to be that flies from irradiated pupae did not live quite as long as controls.

Mating experiments with caged flies established that males mated repeatedly if mates were available, but females mated only once. If a female mated with a sterilized male, she laid only infertile eggs. When mixed populations of sterile and normal insects were caged together, the irradiated males seemed to compete for mates about equally with normal males. The proportion of infertile egg masses was similar to the ratio of sterile to normal males in a caged population.

X-ray equipment was adequate to treat small numbers of insects, but it would be too expensive for large-scale irradiation of screw-worms. Therefore, experiments were made with gamma rays from cobalt-60. Laboratory tests were made in which screw-worms treated with 200-kv X rays were compared with others treated with gamma rays from cobalt-60. As in the preceding work, pupae were X-rayed at Brooke Army Hospital, Fort Sam Houston, Texas. The gamma rays were from an 11-curie distributed source loaned us by the Biology Division of Oak Ridge National Laboratory. The two types of irradiation (Ref. 3) showed very similar effects. It was established that the minimum sterilizing dose was 5000 roentgens, which was best given to insects that had been in the pupal stage for 5 days at 80° F. Except for a somewhat reduced adult longevity, the sterilizing radiation seemed to be well tolerated by the insects.

Gamma-ray source for sterilizing insects. As the cobalt-60 source loaned by the Biology Division appeared satisfactory for screw-worm sterilization, arrangements were made for the Oak Ridge National Laboratory to construct a source especially for this purpose, described in Ref. 7. It consists of 70 curies of cobalt-60 in a lattice arrangement. The cobalt slugs are spaced in the wall of a brass cylinder 3½ in. in diameter and 12 in. long, which is held in a lead-filled steel shell. An aluminum irradiation chamber of about 900 cc capacity is suspended from a steel plug and is lowered into the brass cylinder. At the time the source was constructed, in 1953, the activity within the irradiation chamber was 235 roentgens per min, with a maximum variation of $\pm 13\%$.

The source is set up on an elevated base in a locked concrete structure separated from the main laboratory buildings at Orlando, Florida. Entomological technicians do not need to approach closer than 3 ft from the unit to sterilize insects, and the source is considered entirely safe for skilled biological technicians, since the radiation level within the room is negligible except for the region directly over the shield. The operator, using a chain and pulley to lower the irradiation chamber within the source, has practically zero exposure to irradiation.

Florida Field Tests with Irradiated Flies

From 1951 to 1953 a group consisting of A. H. Baumhover, A. J. Graham, D. E. Hopkins, F. H. Dudley, and W. D. New conducted field studies in Florida. Techniques were developed for estimating fly populations in the field and appraising the effects of releasing sterilized insects. Laboratory-reared screw-worms labeled with P^{32} according to the technique described in Ref. 17 were released and the numbers compared with natural populations. These field observations confirmed our earlier opinion that screw-worm populations, even where most abundant in Florida, did not exceed more than a few hundred insects per square mile.

Screw-worm flies could be caught in traps baited with decomposing liver, but the most reliable index to screw-worm abundance was obtained by collecting egg masses from artificially wounded and infested goats exposed to oviposition by flies of a natural population. Nearly all the females of the normal population were fertilized, as eggs from wounded goats hatched practically 100%.

When males sterilized with gamma rays were released on Sanibel Island at the rate of 100 per square mile per week, the fertility of egg masses collected from wounded goats declined to less than 25%, and in 8 weeks the natural population appeared to have been eradicated. However, after 12 weeks normal flies were again active. As Sanibel Island is only 2 miles off the coast of

Florida and is part of a chain of islands, the test area was not sufficiently isolated to prevent migration of flies from adjacent untreated areas. Therefore it was necessary to seek a truly isolated area to conduct a valid eradication experiment.

Eradication of Screw-worms from Curaçao

Late in 1953 the Veterinary Service of the Netherlands Antilles government requested advice from the U.S. Department of Agriculture on methods of screw-worm control. The insects were extremely abundant on the island of Curaçao, about 40 miles off the coast of Venezuela, and did heavy damage to domestic animals, particularly goats. The small size of the island (170 square miles), the high population of screw-worms, and the isolation made Curaçao ideal for a field experiment to test whether screw-worms could be eradicated through release of sterilized flies. Details of this experiment are described in Ref. 1.

A. H. Baumhover and W. D. New of the Entomology Research Branch worked with B. A. Bitter of the Curaçao Veterinary Service in releasing flies and making observations on the island. Screw-worms for release were reared and sterilized at Orlando, Florida, by A. J. Graham, D. E. Hopkins, and F. H. Dudley. When pupae were five days old they were irradiated, packed with a little excelsior at the rate of 130 per kraft-paper sack, and shipped by air freight to Curaçao. Upon arrival in Curaçao the sacks were unfolded to make room for the flies, which emerged within a few hours. When emergence was complete, the adults were distributed by airplane over the island. An entomologist riding with the pilot tore open the sacks and dropped them at desired intervals as they flew over the island in flight lanes 1 mile apart. From the 130 pupae packed in each sack about 100 active flies emerged. Thus it was a simple matter to compute the rate of fly release. Releases were made twice weekly, the flight lanes being shifted ½ mile each time. The practical effect was that insects were scattered over the island in ½ mile flight lanes, as many per mile as were required to attain the desired release rate.

The screw-worm activity and the efficiency of the release of sterilized flies were measured by observations on egg masses collected from wounded goats. Eleven goat pens, each stocked with eight goats, were established at suitable locations over the island. Each week two goats in each pen were infested by making a small incision in the skin near the shoulder and implanting 100 newly hatched larvae. When the larvae were three days old they were killed with benzol. The wounds were attractive to flies from the time of infestation until several days after the larvae were removed. Thus, there were always at least two wounds at each goat pen to attract ovipositing flies.

In March 1954, prior to the release of sterilized flies, native screw-worms deposited about 15 egg masses per goat pen per week. All the egg masses were fertile.

Sterilized flies were released at the rate of 100 males per square mile per week for 8 weeks. The fertility of the egg masses collected from wounded goats declined from the 100% recorded prior to release to 85%, but this reduction was not sufficient to affect the fly population, which actually increased during a period of favorable weather.

An experiment was then made in which one-half of the island was treated with sterilized flies at the rate of approximately 400 males per square mile per week, while the 100 rate was continued on the other half.

Female flies sterilized with 5000 roentgens could not oviposit normal egg masses, but some individuals deposited a few eggs that did not hatch. Those scattered eggs might have been confused with small egg masses deposited by normal flies. Females irradiated with 7500 roentgens were incapable of producing any eggs. To avoid any doubt as to source of screw-worm eggs, the sterilizing dose of 7500 roentgens was used during the eradication experiment. Any eggs found were laid by females of the normal population.

The rate of 400 males per square mile per week seemed to be so effective that the whole island was treated at that rate in an effort to attain eradication. This heavy release rate was started on August 8, and within 8 weeks normal fly activity had almost ceased. Observations on egg masses collected during this period are shown in Table 22.1.

TABLE 22.1. EGG-MASS RECORDS IN 11 GOAT PENS DURING RELEASE OF STERILIZED FLIES OVER ENTIRE ISLAND OF CURAÇAO

Week	Males Released, per sq mi	Number of Egg Masses		Percentage of Sterile Egg Masses
		Fertile	Sterile	
Aug. 8–15	491	15	34	69
16–22	224	17	38	69
23–29	175	17	36	68
30–Sept. 5	381	10	37	79
Sept. 6–12	451	7	42	86
13–19	701	3	23	88
20–26	450	0	10	100
27–Oct. 3	607	0	12	100

During the first four weeks the releases caused about 70% sterility. This high sterility caused a marked depression in the number of insects in the subsequent generation. Since the release of sterilized insects was maintained at a high rate, there was an even greater percentage of sterility, which was reflected in the hatching records on the reduced number of egg masses collected.

Only two egg masses were collected after October 3, and both of them failed to hatch. They were taken on November 4 and November 11. The goat pens were maintained through January 6, 1955, and fly releases were continued at the rate of approximately 400 males per square mile per week, but there was no further evidence of normal fly activity on the island of Curaçao.

The United States representatives came home in January, but the Curaçao Veterinary Service has continued to watch for evidence of renewed screw-worm activity. Since there have been no cases of myiasis, it is concluded that screw-worms were eradicated from the island of Curaçao through release of sterilized flies.

Screw-worm Eradication in the Southeastern United States

We believe that the Curaçao experiment demonstrated that it may be practical to use sterilized flies in an attempt to eradicate screw-worms from the Southeastern States (Ref. 11). In ordinary winters screw-worms survive in Florida over an area of approximately 50,000 square miles. An eradication campaign in Florida would involve rearing, sterilizing, and releasing about 300 times as many insects as were required for the Curaçao experiment. We are now engaged in investigations to establish procedures for rearing, sterilizing, and distributing 50 million flies per week. We hope that our techniques for mass production of sterilized flies can be perfected so that within two years we can undertake an eradication campaign with reasonable confidence of success.

The Sterilization Method on Other Insects

In considering whether this method will be effective on other insects, one should take cognizance of some of the requirements for success (Ref. 8). A low natural population must exist or the population must be reduced by other means so that it is possible to release an excess of sterilized males. The insect must be easily reared in mass numbers in the laboratory. The mating behavior of the males must not be adversely affected by sterilization, and native females must be willing to accept sterilized males. Preferably the females should mate only once, but on theoretical grounds multiple matings should result in the production of infertile eggs in a ratio similar to that of released sterilized to native males.

Methods of measuring the insect populations per unit area are necessary, so that the numbers required for release purposes can be accurately estimated. Obviously the area in which eradication is attempted should be isolated or protected by quarantine or other measures against reinfestation.

Each insect presents numerous problems, and a large amount of research on the effects of irradiation, habits of the insect, population trends of the species, rate and extent of migration, and other problems is necessary before the feasibility of the sterilization methods as a means of control can be determined.

References for Chapter 22

1. A. H. Baumhover, A. J. Graham, B. A. Bitter, D. E. Hopkins, W. D. New, F. H. Dudley, and R. C. Bushland, "Screw-worm control through release of sterilized flies," *J. Econ. Ent.* (1955).
2. R. C. Bushland and D. E. Hopkins, "Experiments with screw-worm flies sterilized by X-rays," *J. Econ. Ent.* 44:725–31 (1951).
3. R. C. Bushland and D. E. Hopkins, "Sterilization of screw-worm flies with X-rays and gamma-rays," *J. Econ. Ent.* 46:648–56 (1953).
4. J. Gordon Carlson, "Effects of X-radiation on grasshopper chromosomes," *Cold Spring Harbor Symposia Quant. Biol.* 9:104–11 (1941).
5. D. G. Catcheside, "Genetic effects of radiations." Adv. in *Genet.* 2:271–358 (1948).
6. Helen V. Crouse, "The differential response of male and female germ cells of *Sciara coprophila* (Diptera) to irradiation," *Amer. Nat.* 84:195–202 (1950).
7. E. B. Darden, Jr., E. Maeyens, and R. C. Bushland, "A gamma-ray source for sterilizing insects," *Nucleonics* 12(10):60–2 (1954).
8. E. F. Knipling, "Possibilities of insect control or eradication through the use of sexually sterile males," *J. Econ. Ent.* (1955).
9. P. C. Koller and I. A. R. S. Ahmed, "X-ray induced structural changes in chromosomes of *Drosophila pseudo-obscura*," *J. Genet.* 44:53–71 (1942).
10. D. E. Lea, *Actions of Radiations on Living Cells*, The Macmillan Co., New York, 1947.
11. A. W. Lindquist, "The use of gamma radiation for control or eradication of the screw-worm," *J. Econ. Ent.* (1955).
12. Roy Melvin and R. C. Bushland, "The nutritional requirements of screw-worm larvae," *J. Econ. Ent.* 33:850–2 (1941).
13. H. J. Muller, "Artificial transmutation of the gene," *Science* 66:84–7 (1927).
14. H. J. Muller, "The problem of genetic modification" (Proc. Fifth International Genetics Congress, 1927), *Z. induktive Abstam. u. Vererbungslehre,* Sup. Bd. I:234–60 (1928).
15. H. J. Muller, "An analysis of the process of structural change in the chromosomes of *Drosophila, J. Genet.* 40:1–66 (1940).
16. H. J. Muller, "Résumé and perspectives of the Symposium on Genes and Chromosomes," *Cold Spring Harbor Symposia Quant. Biol.* 9:290–308 (1941).
17. R. D. Radeleff, R. C. Bushland, and D. E. Hopkins, "Phosphorus-32 labeling of the screw-worm fly," *J. Econ. Ent.* 45:509 (1952).
18. J. Paul Reynolds, "X-ray-induced chromosome rearrangements in the females of *Sciara*," *Natl. Acad. Sci. Proc.* 27:204–8 (1941).
19. G. A. Runner, "Effect of Röntgen rays on the tobacco or cigarette beetle and the results of experiments with a new form of Röntgen tube," *J. Agr. Res.* 6:383–8 (1916).
20. M. J. D. White, "Effects of X-rays on mitosis in the spermatogonial divisions of *Locusta migratoria,* L.," *Roy. Soc. London Proc.* Ser. B 119:61–84 (1935).
21. M. J. D. White, "Effects of X-rays on first meiotic division in three species of orthoptera," *Roy. Soc. London Proc.* Ser. B 124:183–96 (1937).
22. P. W. Whiting, "Decrease in biparental males by X-raying sperm in *Habrobracon*," *Pa. Acad. Sci. Proc.* 12:74–6 (1938).

Part VII

CROP IMPROVEMENT

Chapter 23

PRODUCTION OF BENEFICIAL HEREDITARY TRAITS WITH IONIZING RADIATION *

Radiation-Induced Mutations

Ionizing radiations, irrespective of mode of production and of properties, readily induce hereditary changes (mutations) which are stable and lead to new traits manifested also in following generations. Truly induced changes of genes and chromosomes were first described by H. J. Muller in 1927. This discovery has influenced genetic research immensely. For a long period of time induced changes were considered by most workers to lead to a breakdown of the hereditary material, the induced mutations thus being exclusively "harmful" in the sense that the individual mutants manifesting the new property would be, if not entirely monstrous or lethalized, at least distinctly inferior to the original type as regards viability. It is a fact, indeed, that most of the induced mutations decrease viability in the homozygous state, i.e., when the mutated gene is present in the double dose, and that many mutations lethal when homozygous show a more or less detrimental effect even in the heterozygous state. But such a type of behavior prevails in spontaneous mutations, too.

Nevertheless, recent studies have shown that the induction of mutations is an important tool in plant breeding. For several hundred "harmful" hereditary changes there are one or two that raise production in the homozygous state or in some other respect than production imply a "positive" alteration. (We here define a *positive mutation* of an agricultural species as a hereditary change in some way beneficial to man's interests and well adapted to his very peculiar type of "ecological niche." The spontaneous mutation rate is so low that time, labor, and experimental area required for the collection of positive mutations naturally arising would be unreasonably great. Since by the application of ionizing radiations the mutation rate can be raised several thousand

* This chapter is taken from Geneva Conference Paper 793, "The Production of Beneficial New Hereditary Traits by Means of Ionizing Radiation" by L. Ehrenberg, I. Granhall, and A. Gustafsson of Sweden.

times, such a collection is now made possible and can be adopted as part of the breeding routine, without any special expansion of ordinary experimental stations.

In this connection it ought to be pointed out that many mutations which from the traditional point of view should be regarded as deleterious, since they cause lethal effects when homozygous, will increase viability when heterozygous. In organisms like barley or *Antirrhinum* these "beneficial lethals" are not uncommon, but their occurrence has been described also in *Drosophila,* even for the X chromosome. In fact, Gustafsson concluded that lethal factors play an important role in the population dynamics of cross-fertilizing organisms. This does not contradict the views, eloquently vindicated by Muller, about the eugenic hazards of an uncontrolled use or rather misuse of ionizing radiations in human activities. Chromosome rearrangements of various kinds, as well as long deficiencies and duplications of the chromosome material, are seen in relatively greater abundance when applying ionizing radiations than in nature, and many gross chromosome alterations show conspicuous effects of a deleterious kind even when heterozygous.

Induction of Mutations in Barley

The principles of the induction of positive mutations have now been worked out in a large number of agricultural species. The Swedish group treating the subject has concentrated chiefly on barley, which is a diploid self-fertilizing organism where the mutations are easily identified, and on wheat, which is a polyploid but still self-fertilizing organism suitable for mutation analysis. In barley, in striking contrast to wheat, there is a wide range of morphological drasticity, from small scarcely perceptible changes to most profound alterations in type, even breaking the morphological frame of the variety or species. In spite of their greatly altered appearance many mutants are surprisingly productive. This is for instance the case with the so-called *erectoid* mutants in barley, about 75 cases of which are now being studied. The erectoid mutation takes place, as far as we know, by changes in 16 to 20 different gene loci. Many of these mutants display features valuable from an agricultural point of view. The grain (and straw) production approaches or equals that of the mother strain; in some cases the mutant is even superior by a few per cent. At the same time erectoid mutants distinctly surpass the mother as regards stiffness of straw. Some are earlier, possess higher protein contents, etc. In addition, the erectoids are, more than the mother line, able to utilize very high nitrogen dressings, a desired property in today's farming. Other mutants, e.g., the so-called *bright-greens,* behave in a quite contrary way in this respect: they thrive better in a meager soil. On the other hand, they appear to be more resistant towards certain fungi than the respective mother strains.

The changed response to nitrogen dressing is one example of an altered reaction norm, involving the sudden registration of new ecological requirements by induced mutation. A new "ecotype" or race has arisen. As previously mentioned, ionizing radiations induce frequent chromosomal rearrangements (sometimes these are beneficial when homozygous, detrimental when heterozygous, in contrast to lethals) leading to deviating chromosome configurations detectable in cell division. Now and then the erectoid mutations originate simultaneously with such chromosome rearrangements. This causes the formation of more or less pronounced sterility barriers in the crosses between mutants and mother strain. In fact, by one stroke and simultaneously, we may induce all the essential characters distinguishing species in nature: (1) the origin of a sterility barrier, (2) a drastic change in morphology and anatomy, (3) an altered ecological response, and (4) a new karyotype detectable microscopically (Ref. 1).

Character of Positive Mutations

It has been definitely shown, in barley and other experimental plants, that induced mutation can increase the yielding capacity of a variety, or leave this capacity intact and improve upon special characters of importance in agriculture—not only stiffness of straw or response to an increased nitrogen dressing, but also earliness, protein or oil content, baking quality, malting properties, fiber strength, and grain size, in cereals as well as in peas, lupines, flax, mustard, tomatoes, etc.

A few examples of interest will be given. In Sweden, where the yellow sweet lupine (*Lupinus luteus*) is cultivated north of its center of adaptation, it has been considered especially important to breed for an increased earliness. This was found to be difficult with the traditional methods of breeding (hybridization and selection). After X-irradiation of seeds a greater variation in the actual property was obtained, and selection for earliness became successful, without any significant decrease of the yielding capacity (Ref. 7). A few X-ray varieties are at present released for the market: the Primex white mustard of Svalöf (Ref. 1), which by an increase of grain yield as well as oil content surpasses the mother strain in oil yields per area about 7%; the "Strålärt" (ray pea) of Weibullsholm, and the "Schäfers Universal" beans (*Phaseolus*). Numerous rather promising mutants in barley, wheat, peas, and lupine are being tested in large-scale experimentation; some are in fact included in the official Swedish state trials.

The majority of high-productive mutants, especially of species represented by very high-bred and specialized varieties like barley and wheat, are not suited for a direct marketing. The basic use of induced mutations therefore consists in the building up of a new variability, similar to that occurring in the old-time varieties of the species, but based on the highest-yielding modern varieties. The subsequent crossing of these mutants with one another or with other varieties will contribute to the gradual improvement of cultivated plants and may effect this with a considerable gain of time.

Somatic Mutations

In fruit trees spontaneous somatic mutations (bud sports), which can be propagated vegetatively, have played a great role in the production of new market varieties. Out of 143 apple varieties with known parentage marketed in the United States and Canada during the last thirty years, no fewer than one-fourth have originated as bud sports (Ref. 4); and in pears, peaches, plums, and cherries, about 10% have originated in such a manner. At the Balsgård Fruit Breeding Institute, Sweden, techniques have therefore been worked out to treat scions or seedlings with different ionizing radiations in order to induce bud sports. The investigations carried out have presented valuable information, since mutations in color and shape of fruits as well as in ripening time have been obtained (Ref. 4).

Effects of Different Ionizing Radiations

In barley, the effects of different ionizing radiations are extensively being compared, partly as a first step in investigating how to direct and control the mutation process, i.e., to influence the distribution of mutation types in a special direction. These studies have so far revealed that when seeds are irradiated, the total mutation frequency increases more rapidly with linear energy transfer (ion density *) than does the lethalizing action of the radiation (Ref. 3). The lethalizing action of fast neutrons—produced by (d, n) reactions in cyclotrons (ion density: 400–800 ion pairs per micron) or in the pile center (1000–3000 ion pairs per micron)—is about 20 times that of 175 kv X rays ($\sim$100 ion pairs per micron) or Co^{60} gamma rays (8 ion pairs per micron). Since, however, their mutagenic efficiency is about 40 times greater, about twice as many mutations are obtained at the point of 50% survival.

As to the distribution of various mutation types it has been shown (Ref. 3) that lethal mutants deficient in chlorophyll and other plastid-borne pigments are relatively more common after neutron irradiation, and that consequently the vital mutants, i.e, those of practical interest, are commoner after irradiation at a low ion density: twice as many mutants, relatively seen, being vital after X-ray treatment as after neutron treatment. With a corresponding survival of the irradiated generation, the proportion of mutants of practical inter-

* Expressed as ion pairs per micron, i.e., average along the linear particle path in tissue of unit density and assuming the absorption of 32.5 ev to correspond to the formation of one ion pair.

est will therefore be about the same after treatment with X rays and gamma rays on the one hand and fast neutrons on the other.

Although the material is still rather scant, it can be stated with a high degree of probability that within the group of vital mutants in barley, the relative frequency of *erectoid* mutations increases with increasing linear energy transfer of the radiation, the frequency in the 1953 experiments being about 25% after irradiation with X rays, 50% for cyclotron-produced neutrons, and nearly 100% after irradiation in the pile. Data from the 1954 experiments further evidence this difference in relative efficiency.

Within the group of chlorophyll-deficient mutations, several lethal types can be recognized, viz., a colorless *albina*, a light-green or yellow-green *viridis*, and a yellow *xantha*, which are the most common ones, and in addition several rare types. Each group of mutants is realized by changes in a great many genes. The relative frequencies of these types are seemingly identical whether X rays with 100 ion pairs per micron or neutrons ionizing at 400–800 ion pairs per micron are applied. When ion density is varied within a wider range, however, it becomes evident that the relative frequency of *rare chlorophyll mutations*, treated as a group, increases with increasing ion density.

In addition it has been shown that a variation of the irradiation conditions other than ion density may provoke displacements of the types of chlorophyll-deficient mutations. The *viridis* type, which occurs in combination with a high sterility of the irradiated generation, is thus more common under conditions favoring a high sterility (Ref. 2). About 22% *viridis* are obtained when moist seeds are irradiated with low doses ($\leq$5000 r), whereas the frequency amounts to about 62% when dry seeds are irradiated with high doses. The latter very high frequency approaches what is obtained when chlorophyll mutations are induced by means of the chemicals mustard gas and nitrogen mustard (Ref. 6).

Summary

(1) The induction of mutations by means of ionizing radiations has given and can give hereditary changes of high production capacity in agricultural plants. If the mutants, although high-productive, are not suited for a direct marketing, they form a source of variation valuable for the continued breeding along traditional lines.

(2) It has been proved, in principle, that the spectrum of mutation types can be intentionally displaced in different directions. This is achieved, for instance, by a variation of the ion density, from the minimum obtained with Co^{60} gamma rays or P^{32} beta rays, via X rays and neutrons of different origin to densely ionizing alpha rays, but also by a variation of the irradiation conditions or by the application of mutagenically active chemicals. In this way also the origin of positive, directly valuable mutations can be controlled.

These results form a first step in our attempts to learn how to influence selectively the mutation of individual genes.

Most agricultural species, even high-bred ones like barley, wheat, or corn are still rather old-fashioned as to morphology, anatomy, and karyology, in spite of fifty years or more of intense breeding work. They need to be reconstructed in agreement with the requirements of modern agriculture, facilitating a high mechanization and an intense fertilizing of the soil, in this way increasing yield and quality but decreasing labor and costs of cultivation. Recent studies by Wettstein indicate that the construction of the straw of wheat and barley ought to be fundamentally repatterned, with regard to internode number and internode length as well as cross-section area of the nodes and internodes, before the varieties can be considered fully adapted to high-nitrogen dressings and an intense mechanization. With regard to internode structure, barley should be changed towards wheat; with regard to cross-section area, wheat in its turn should be remade to resemble barley. These changes can be effected by induced mutation, possibly more easily than by hybridization of different varieties and subsequent selection. Similarly, in corn the mutual development of vegetative and generative parts ought possibly to be readjusted in such a way as to decrease the vegetative system and to increase, relatively, grain production, without an unnecessary waste of nutrients and photosynthetic materials. Finally, the karyotype of some species like barley seems to be rather old-fashioned, remaining on an ancient status. The modernization of the karyotype may render possible a better coordination of genes and mutations in the continued breeding program of the future.

References for Chapter 23

A full account of the mutation studies carried out by the Swedish group of research workers is given in a series of articles, "Mutation Research in Plants" in *Acta Agriculturae Scandinavica* 4:3 (1954). Further work on the mechanism of action of ionizing radiations in plant seeds is summarized by Ehrenberg in *Svensk Kemisk Tidskrift* 67:207–222 (1955).

1. G. Andersson and G. Olsson, *Acta Agr. Scand.* 4:3, 574–577 (1954).
2. L. Ehrenberg, *Svensk Kemisk Tidskrift* 67:207–222 (1955)
3. L. Ehrenberg and N. Nybom, *Acta Agr. Scand.* 4:3, 396–418 (1954).
4. I. Granhall, *Acta Agr. Scand.* 4:3, 594–600 (1954).
5. A. Gustafsson, *Acta Agr. Scand.* 4:3, 361–364; 601–632 (1954).
6. S. MacKey, *Acta Agr. Scand.* 4:3, 419–429 (1954).
7. O. Tedin, *Acta Agr. Scand.* 4:3, 568–573 (1954).

Chapter 24

IONIZING RADIATIONS IN PLANT BREEDING *

The most readily available and effective mutagenic agents are high-energy radiations which initiate changes in biological systems either directly or indirectly through ionizations. Prior to the advent of atomic fission most of the radiations used in biological studies were electromagnetic. However, there is presently a great deal of interest in the biological effects of high-energy particle radiations. Some of this interest has been directed toward determining the relative mutagenic effectiveness of radiations with different specific ionizations.

For maximum utilization of the various sources of radiation in plant breeding, basic information must be catalogued on the relative kinds and frequencies of genetic and physiological phenomena that result from irradiation. To collate such information adequately, all of the factors known to modify the sensitivity of one species of radiation must be investigated to determine their influence on the other kinds of radiations. At present it seems that only through such studies will it be possible to obtain detailed information on the spectrum of genetic events that may prove useful to the plant breeder.

The principal objective of the present studies has been to determine what, if any, biological differences that are related to ion density could be detected in the growth and cytology of the barley plant. Space limitations have placed severe restrictions on the presentation of all the data pertinent to this subject. It is expected, however, that detailed reports on most of the unpublished data cited herein soon will appear in the literature.

Because of their availability, X rays, 2-mev electrons, thermal neutrons, fast neutrons, and to a lesser extent, cobalt-60 gamma rays were employed as radiations sources. The X rays were generated with a G.E. Maxitron apparatus operated at 250 kvp and 30 ma, and were filtered through 1 mm of aluminum and the beryllium window of the tube. The 2-mev electrons were obtained from a Van de Graaff electrostatic generator.

* This chapter is taken from Geneva Conference Paper 101, "Ionizing Radiations as a Tool for Plant Breeders" by R. S. Caldecott of the United States.

Thermal neutron bombardments were conducted in the thermal column of the Brookhaven National Laboratory nuclear reactor. The fast neutrons were obtained from the fission of U^{235}. Details of the irradiation procedures, radiation purity, and the spectrum of energies from each radiation source have been presented elsewhere (Ref. 1). It is significant that in our test organism, the barley seed, 98% of the radiation from thermal neutrons resulted from protons and alpha particles through capture reactions with nitrogen and boron.

It is recognized that as we used them in these studies, a distribution of ion densities occurred with all these radiations. However, with X rays, 2-mev electrons, and gamma rays a relatively large fraction of the energy loss occurred at a low specific ionization. On the other hand, with the neutron sources of radiation, a relatively large fraction occurred at a high specific ionization. These studies, then, are concerned with comparing the biological effectiveness of radiations which have for the most part a high specific ionization with radiations which have a low specific ionization.

Physiological and Cytogenetic Effects of Radiations

General effects of ion density. *Seedling studies.* Earlier studies have shown that when barley seeds were subjected to a dose of X rays sufficient to reduce seedling growth materially, the seeds were not uniformly injured and the distribution of seedling heights about the mean was skewed. It was of considerable interest, therefore, when we found that seedlings from seeds treated with any dose of thermal neutrons gave a normal distribution of heights about the mean, irrespective of any reduction in average seedling height resulting from the treatment (Ref. 2).

All efforts to explain the lack of uniformity with X rays, either on the geometrical distribution of the X-ray beam and/or of the seed, failed. It was decided that this lack of uniformity must result from the physical nature of the radiation. From this inference, and from the fact that through capture in the barley seed, thermal neutrons indirectly have a much higher specific ionization than X rays, it seemed reasonable to assume that the difference may result from the spatial distribution of ion pairs in the irradiated material. In order to check this reasoning and to preclude the possibility that the observed phenomena resulted from some experimental idiosyncrasy, it was decided that the results should be compared with two different sources of radiation which had specific ionizations comparable to X rays and thermal neutrons. For this purpose, 2-mev electrons and fast neutrons were selected.

Dormant seeds were subjected to X rays, 2-mev electrons, thermal neutrons, and fast neutrons and then analyzed to determine the distribution of seedling heights from a number of doses. To elucidate the difference between the two types of radiations (those with a high as compared to a low specific ionization), doses of each radiation which caused a comparable average inhibition of seedling growth were graphed (Fig. 24.1). This figure lucidly illustrates the similarity between the curves obtained with thermal and fast neutrons on the one hand and between X rays and 2-mev electrons on the other. It seemed logical to conclude from these data that the differences observed were related to the physical characteristics of the radiations, and probably to their specific ionization (Refs. 4, 5).

Cytogenetic studies. In one study which compared the effects of X rays and thermal neutrons, treated barley seeds were planted in the field and permitted to produce mature plants. Cytogenetic specimens were taken at the pollen mother cell stage and the frequency of chromosomal interchanges as a function of dose was determined. In addition, counts of plant survival were made for all treatments before the plants were harvested for seedling mutation studies (Ref. 2).

By plotting the interchange frequency as a function of survival it was clearly demonstrated that more genetic interchanges could be obtained per unit survival with thermal neutrons than with X rays (Fig. 24.2). The same was later shown to be true for seedling mutations. Preliminary studies conducted in the greenhouse indicated that differences of the same order of magnitude were obtained when seeds treated with gamma rays from cobalt-60 were compared with seeds treated with fast neutrons (Refs. 4, 5).

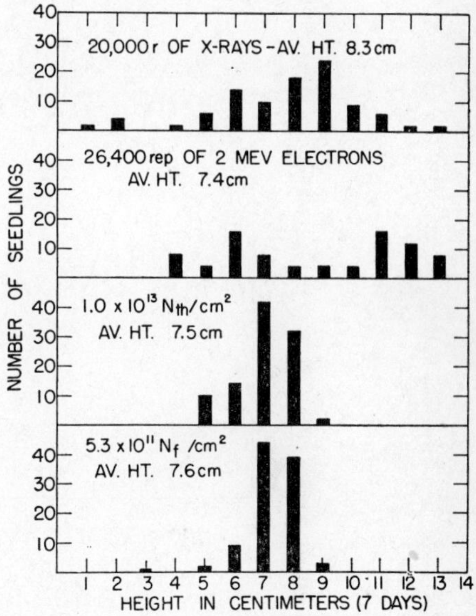

FIG. 24.1. Frequency distribution of seedling heights from a dose of X-rays, 2 MEV electrons, thermal neutrons and fast neutrons that reduce the average heights to a comparable degree.

Detailed comparisons of the kinds and relative frequencies of mutations induced with X rays and thermal neutrons have failed to demonstrate any major differences (Table 24.1), even though a higher frequency of any given type of change per unit survival is obtained with thermal neutrons than with X rays. The same was shown to be true for the types and frequencies of interchanges. This indicates that densely ionizing radiations should be of more value to plant breeders than sparsely ionizing radiations because they give him a better probability of inducing the genetic changes which he seeks.

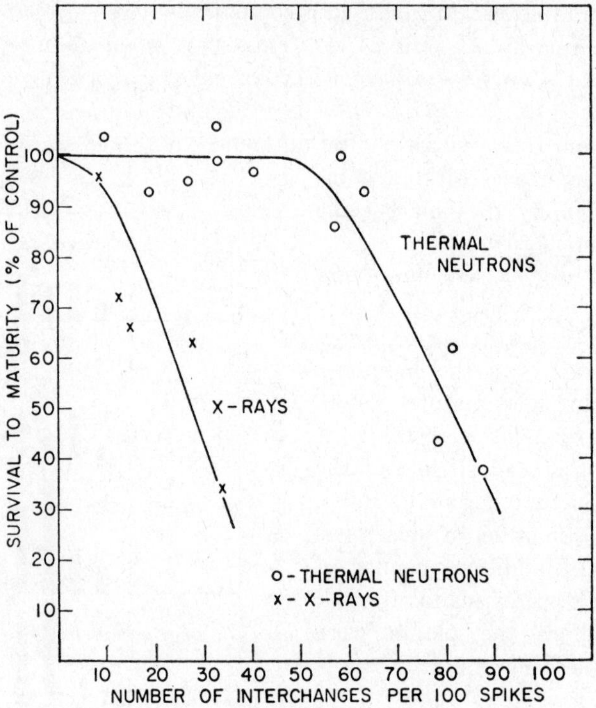

FIG. 24.2. Relation of survival to interchange frequency in seeds of barley treated with X-rays and thermal neutrons.

TABLE 24.1. FREQUENCIES OF TYPES OF SEEDLING MUTANTS INDUCED IN DORMANT SEEDS OF BARLEY IRRADIATED WITH X RAYS AND THERMAL NEUTRONS

Mutant Types	Treatment				Total	
	X Rays		Thermal Neutrons			
	No.	%	No.	%	No.	%
White	387	56.0	779	56.8	1166	56.5
Yellow	102	14.8	140	10.2	242	11.7
Yellow-green	114	16.5	237	17.3	351	17.0
Virescent	10	1.4	42	3.1	52	2.5
Striped white and yellow	17	2.5	36	2.6	53	2.6
Banded shrivel	20	2.9	52	3.8	72	3.5
Others	41	5.9	86	6.3	127	6.2
Total	691		1372		2063	

Influence of Hydration

Through a combination of storing seeds in atmospheres with different vapor pressures and actually steeping them in water, a wide range of embryo water contents was obtained. By subjecting seeds with different water-content levels to different types of ionizing radiations it was possible to demonstrate pronounced differences in sensitivity modification. These differences are of particular significance to the plant breeder because they give him an indication of the variables that must be controlled when seeds are irradiated if he is to be assured of a mature plant population to study.

Effect of vapor pressure. *Seedling studies.* Dormant barley seeds were stored over solutions with different vapor pressures (Table 24.2) until they

TABLE 24.2. THE RELATION OF THE WATER CONTENT OF EMBRYOS AND ENDOSPERMS OF BARLEY SEEDS TO THE WATER CONTENT OF THE ATMOSPHERE IN WHICH THEY ARE STORED

Seeds Stored in Desiccator at 20° C Over—	Water Content of Air, mg/liter	Approximate Number of Seeds	Water Content of Embryos and Endosperms at Equilibrium			
			Embryo		Endosperm	
			Wt., gm	% H_2O	Wt., gm	% H_2O
Dry P_2O_5	0.02×10^{-3}	440	0.8430	4.0	24.9801	5.0
Dry $CaCl_2$	0.20	440	0.8274	5.0	25.3545	6.0
Sat $CaCl_2$	5.52	440	0.8813	6.0	25.7708	8.0
Sat $NaHSO_4$	8.89	440	0.9347	8.0	26.7485	11.0
Sat $NaClO_3$	12.82	440	0.9102	11.0	27.7838	14.0
Sat $NH_4H_2PO_4$	15.92	440	1.0146	16.0	30.1678	19.0

reached weight equilibrium. At this time the water content of the embryo ranged from about 4% to approximately 16% of their dry weight. Seeds representing each water content level were subjected to one of a range of doses of X rays and then germinated and grown for 7 days in petri dishes, after which time seedling height measurements were made (Fig. 24.3). The data demonstrate that the sensitivity of the seeds decreased as the water content of the embryo increased from 4% to about 8%. At this upper point a plateau was reached, and added increments of water between about 8% and about 16% apparently gave little further modification of radiosensitivity. Efforts to obtain similar results with both fast and thermal neutrons were essentially

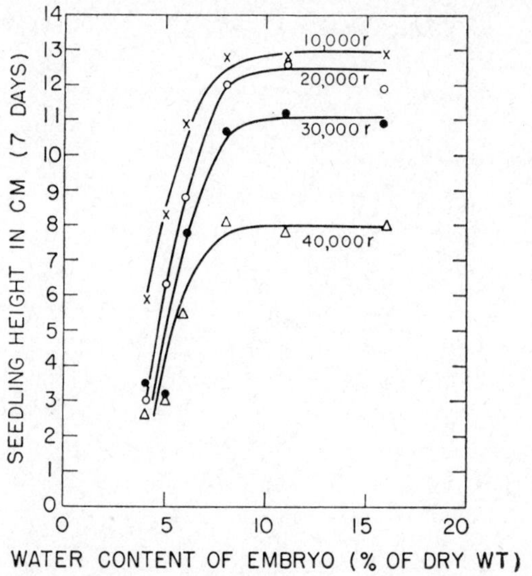

FIG. 24.3. Relation between the water content of the embryos of dormant barley seeds and their sensitivity to different doses of X-rays.

negative, although in some experiments there was some indication of a very slight sensitivity modification (unpublished).

These studies point to the fact that less rigid controls over seed storage conditions would be required to predict a given response when seeds are subjected to neutrons than when they are treated with X rays. This is of particular consequence to plant breeders because they often experience difficulty in predicting biologically effective X-ray doses from year to year, and a readily available neutron source for treating their material would largely alleviate this difficulty.

Cytogenetic studies. A sample of the seeds having the embryo water contents listed in Table 24.2 were subjected to 20,000 r of X rays and then sacrificed for cytological analyses by removing the shoot tip of the germinating seed during the first cycle of cell divisions.

Cells at anaphase from embryos of all water content levels were scored to determine the frequencies of normal cells and those that had chromosomal bridges (Fig. 24.4). As the water content of the embryo increased from about 4% to about 8%, the frequency of detectable aberrancies decreased. At the upper level of hydration, a plateau was reached which apparently was not subject to modification by further addition of water to the embryo up to 16%.

These data mean to the plant breeder that he can increase the X-ray sensitivity of the genetic material of dormant seeds simply by desiccating them over P_2O_5 prior to irradiation. This would be of greatest value where X-ray facilities were limited or expensive.

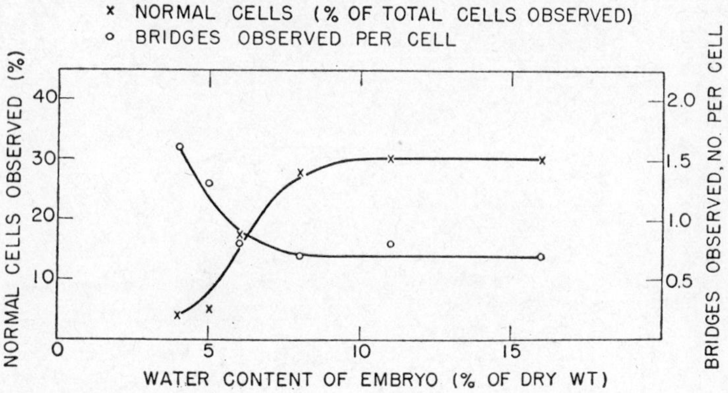

FIG. 24.4. The relation between the water content of the embryos of dormant barley seeds and their sensitivity to 20,000 r of X-rays as determined by cytogenetic analyses.

Effect of soaking. Two of the factors known to limit the germination and growth of seeds are water and temperature. Water provides the working medium for the biochemical reactions incident to growth. The temperature at which growth takes place determines the rate of these reactions. If both the water content and the growth of seeds modify their radiosensitivity, it should be possible to increase or decrease the efficiency, per unit energy absorbed, of any given radiation in producing physiological and genetic changes.

In this connection, seeds originally desiccated over P_2O_5 were soaked at 22 ° C and 0° C for various periods of time prior to X-radiation. Following irradiation the seeds were germinated and grown in petri dishes for 7 days, at the end of which time seedling height measurements were made. These studies showed that after 1 hr steeping at 22° C the seeds were much less sensitive to X rays than controls that had been stored over P_2O_5 and not steeped. However, steeping the seeds at this temperature in excess of 2 hr resulted in a striking increase in their X-ray sensitivity, so that after 6 hr they were more sensitive than dry controls. On the other hand, when soaked at 0° C the seeds just reached maximum resistance after about 4 hr steeping, and they remained at this level of tolerance for at least 8 hr before they underwent a gradual increase in sensitivity. After 24 hr of soaking at 0° C the seeds were about as sensitive to X rays as the dry controls (Fig. 24.5).

Further studies on the nature of the sensitivity changes that accompany soaking at 22° C have demonstrated that seeds soaked for about 8 hr are about 10 times as sensitive to X rays as dry controls that have been stored over P_2O_5 prior to irradiation. This increased sensitivity resulting from soaking for 8 hr is reflected by both a reduction in seedling growth rate and an increase in the frequency of chromosomal aberrations (unpublished). These studies indicate that soaking seeds at about 20° C for 8 hr or more gives the maximum increase

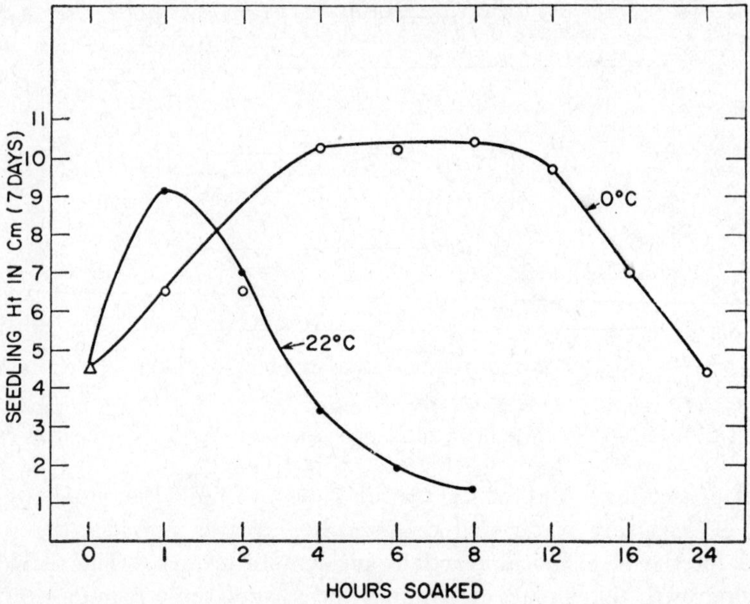

FIG. 24.5. Effects of soaking barley seeds at 22° C and 0° C on their sensitivity to 15,000 r of X-rays.

in sensitivity that can be obtained with seeds under the conditions we have tested.

Studies on the effects of increasing the water content of seeds by steeping prior to subjecting them to thermal neutrons have shown that the treatment reduces their sensitivity for about 16 hr at 22° C and for at least 24 hr at 0° C. Other cytogenetic studies on seeds soaked at 22° C suggest that the period at which the increased sensitivity occurs with thermal neutrons is associated with the time at which chromosome doubling takes place in the cells of the seed (unpublished).

The reduction in sensitivity to thermal neutrons that accompanies the initial stages of soaking seems to be best explained by the fact that in the unsoaked seed over 90% of the ionization results from protons and alpha particles from capture of thermal neutrons by nitrogen and boron respectively. Thus, when the water content is increased, the ratio of nitrogen and boron to the total elemental constituents of the seeds is decreased. From this it follows that relatively fewer thermal neutrons would be captured by nitrogen and boron, and thus the biological efficiency per neutron would decrease.

In contrast to the observations with thermal neutrons, when seeds are soaked at either 0° C or 22° C and then subjected to fast neutrons their sensitivity increases with increased time of steeping. The rate of increase is greater at the higher temperature (unpublished). Undoubtedly, some of the sensitivity increase is associated with cellular activity, but it seems likely that a considerable fraction results from the greater abundance of hydrogen for reaction with fast neutrons.

Influence of the atmosphere. It has been well established that a number of biological materials X-rayed in the presence of free oxygen are more sensitive than when X-rayed under anaerobic conditions (Ref. 3). There is some indication that the magnitude of this "oxygen effect" is dependent on specific ionization. Therefore, to help clarify the role of ion density in the oxygen effect, dormant barley seeds with about 10% water in the embryo were placed in atmospheres of air, oxygen, carbon dioxide, and nitrogen for 24 hr and then irradiated with either X rays, fast neutrons, or thermal neutrons. Following treatment, the seeds were germinated and height measurements were taken after 7 days (Table 24.3).

TABLE 24.3. THE INFLUENCE OF ATMOSPHERE DURING IRRADIATION ON THE SENSITIVITY OF BARLEY SEEDS TO X RAYS, THERMAL NEUTRONS, AND FAST NEUTRONS AS MEASURED BY SEEDLING HEIGHT AT 7 DAYS [a]

Treatment	X Rays		Thermal Neutrons		Fast Neutrons	
	No. of Seeds	Av. Ht., cm	No. of Seeds	Av. Ht., cm	No. of Seeds	Av. Ht., cm
Control (no treatment)	80	12.1	147	11.6	147	11.4
Air	40	6.7	142	3.8	147	6.5
Oxygen	40	4.0	147	3.9	147	6.5
Carbon dioxide	40	7.3	149	4.0	—	—
Nitrogen	40	7.9	146	3.9	147	6.9

[a] Doses of radiation used: X rays, 20,000 r; thermal neutrons, $1.7 \times 10^{13}/cm^2$; fast neutrons, $6.3 \times 10^{11}/cm^2$.

The atmospheres had little or no effect on susceptibility of the seeds to either neutron source of radiation. However, the seeds treated with X rays were distinctly affected by the atmosphere in which the irradiation was given. X-radiation in the presence of either oxygen or air reduced seedling height by about one-half and one-eighth respectively over that of the seeds treated in nitrogen and carbon dioxide.

Preliminary cytogenetic studies suggested that the oxygen effect produced when seeds were subjected to X rays was also manifested in an increase in the frequency of chromosomal aberrations (Refs. 3, 6).

These observations are further evidence of the manner in which radiosensitivity can be modified by environment. However, it is still too soon to say conclusively whether or not it will be possible to alter the kinds and frequencies of genetic events obtained from irradiations by modifying the conditions under which materials are exposed.

From the plant breeder's standpoint the most significant feature of this study appears to be the fact that it presents him with a means of increasing the efficiency of the X-ray facility at his disposal.

Summary

Optimum usefulness of ionizing radiations in plant-breeding programs can be obtained only if plant breeders have adequate basic information which can be utilized in designing their experiments. In an effort to collate such information we have irradiated barley seeds, subjected to different degrees of hydration and oxygenation, with some of the following radiations: 250 kvp X rays, cobalt-60 gamma rays, 2-mev electrons, thermal neutrons, and fast neutrons. The studies have shown that the neutron sources of radiation, which in seeds have a high specific ionization, produce from two to four times as many genetic events per unit survival as the more sparsely ionizing radiations.

In other studies it has been shown that the efficiency of X rays (per r unit) in producing genetic changes and lethality can be increased by about a factor of 10 when seeds are presoaked for 8 hr at about 22° C. Such a treatment reduces the effectiveness of the thermal neutron by about 25% and increases the effectiveness of the fast neutron by a comparable amount. In the same way, pretreatment with oxygen markedly increases the sensitivity of seeds to X rays, but has considerably less influence on their sensitivity to either neutron source of radiation.

It should be emphasized that while the effectiveness of X rays per r unit can be strikingly modified by pretreatments, it has not yet been demonstrated that the degree of genetic damage per unit survival can be altered.

Detailed cytogenetic analyses and seedling mutation studies have failed to indicate any appreciable difference in the kinds of genetic changes induced by X rays, gamma rays, fast neutrons, and thermal neutrons. As mentioned above, because neutron sources of radiation produce more genetic events per unit survival than more sparsely ionizing radiations, it appears that for plant-breeding purposes densely ionizing radiations should be more useful than sparsely ionizing radiations.

These studies will have particular significance if pile radiations can be made readily available to agricultural scientists.

REFERENCES FOR CHAPTER 24

1. R. S. Caldecott, "The effects of X rays, 2-mev electrons, thermal neutrons and fast neutrons on dormant seeds of barley," *Ann. New York Academy of Sci.* 59:514–535 (1955).
2. R. S. Caldecott, B. H. Beard, and C. O. Gardner, "Cytogenetic effects of X ray and thermal neutron irradiation on seeds of barley," *Genetics* 39:240–259 (1954).
3. B. Hayden and L. Smith, "The relation of atmosphere to biological effects of X rays," *Genetics* 34:26–43 (1949).
4. J. MacKey, "Neutron and X-ray experiments in barley," *Hereditas* 37:421–464 (1951).
5. J. MacKey, "The biological action of X rays and fast neutrons on barley and wheat," *Arkiv. Botanik.* 1(16):545–556 (1952).
6. R. A. Nilan, "Relation of carbon dioxide, oxygen and low temperature to the injury and cytogenetic effects of X rays in barley," *Genetics* 39:943–953 (1954).

Chapter 25

GENETIC EFFECTS OF CHRONIC GAMMA RADIATION ON GROWING PLANTS *

Most studies on biological effects of ionizing radiations have been made after acute exposures of radiation over short periods of time. Treatments of seeds or pollen have been the methods most frequently used to induce genetic effects in plants. A great number of reports of radiation-induced mutations in plants have been issued since the early work of Stadler (Refs. 16, 17).

More recently other methods have also been demonstrated to produce genetic effects in plants (Ref. 15). At Brookhaven National Laboratory in the United States, they installed a Co^{60} source in a field, and plants were grown around it and exposed to continuous or chronic gamma radiation during growth and development. Experiments with chronic gamma irradiation have been started in Sweden also (Ref. 1).

Although Stadler (Ref. 17) claimed that radiation-induced mutations are deleterious to plants, some optimism about a possible application of artificial mutations as a method in plant breeding has appeared. A considerable amount of work is being carried out and some positive results have been obtained. A special issue of *Acta Agriculturae Scandinavica,* Volume IV, 3, is devoted to the subject of mutation research in plants, and the most important work in the field is reviewed and discussed here. For further references we refer to that publication.

Although some advantageous mutations are known, the majority of radiation-induced mutations are deleterious. Thus, for plant-breeding purposes it is of great importance to work out methods by which not only the total mutation frequency but also the frequency of advantageous mutations can be increased.

This problem has been given much attention in our radiation studies at the Institute of Genetics and Plant Breeding at the Agricultural College of Norway since Professor H. Wexelsen started the radiation program in 1948. A great number of mutants are produced after different acute treatments of ionizing

* This chapter is taken from Geneva Conference Paper 890, "Studies of Genetic Effects in Plants of Chronic Gamma Radiation" by K. Mikaelsen of Norway.

radiations. Since 1953 experiments with chronic radiations have been possible, and new methods are being tried in our efforts to find out if ionizing radiation can be a useful tool in plant breeding. Some results from our preliminary experiments will be presented.

The Radiation Field

The arrangement of our radiation field is in principle similar to the radiation field at Brookhaven National Laboratory (Ref. 15). Our radiation source is located in a metal tube mounted on the end of a lead cylinder 20 cm long. The lead cylinder slides inside an aluminum pipe. By a cable attached to the lead cylinder and a windlass 60 meters from the source, it is possible to move the source up and down. The source can be lowered 1 m below surface. In this location the source is shielded by the lead cylinder and a concrete block so that only negligible radiations can be measured on the surface. This safety precaution makes it possible to enter the field for planting, cultivating, taking notes, collecting material, and the like, without exposing oneself to radiation. When the source is in operation it is elevated to 70 cm above surface. The plants are grown in concentric circles or arcs around the source. The radiation doses vary with the distances from the source. In 1953 a 12-curie Ir^{192} isotope was operating in the field. Because of a relatively short half-life of 74.5 days, Ir^{192} was inconvenient as a radiation source and consequently was considered as a preliminary arrangement. In 1954 a 25-curie Co^{60} was substituted. In the relatively small radiation field of 1953, more thorough studies were made only in barley and *Tradescantia paludosa,* while other species like oats, wheat, lupines, onions, and cabbage were grown only for observation.

In 1954 the field was considerably extended and included barley, oats, wheat, *Tradescantia,* potatoes, sweet corn, blackberries, tomatoes, beans, strawberries, roses, and other horticultural plants.

Effects of Chronic Gamma Radiations

The tolerance of different species to chronic gamma radiation varies, as is also indicated in Ref. 15. In our material, *Tradescantia paludosa* proves to be very radiosensitive, while barley seems to be one of the most radioresistant species tested. Physiological effects of radiation on growth and fertility are usually very pronounced at heavier doses or dose rates (Mikaelsen and Aastveit, to be published) and can be easily demonstrated in the field. Genetic effects can usually be verified only by more elaborate studies and analysis.

Effects on chromosomes. The cytological effects of chronic gamma irradiation which are presented in Ref. 15 indicate that the percentage of micronuclei in microspores of *Tradescantia paludosa* is linearly increasing with dose rates (r/day) and total doses given. The results also indicate that the type of aberrations represented by micronuclei are more dependent on the amount of radiation received per day than upon the total doses. The effects on chromosomes of chronic gamma irradiation is also studied by Mikaelsen (Refs. 7, 8) in anaphase of root tip cells of *Tradescantia paludosa.* He found that both frequencies of acentric fragments and bridges are linearly related to dose rates (r/day) and to total doses. The experiments seem also to indicate that dose rate (r/day) was a more important factor in the yield of aberrations than the total dose given. This finding seems to deviate from the usual concepts of dose relationship, and a comparison between an acute and a more chronic radiation exposure would be of interest.

The following experiments have been carried out. Cuttings of *Tradescantia paludosa* were used as experimental material. Part of the material was exposed to 12.5, 25, 50, 100, 200 r in a few minutes as an acute treatment, while other cuttings were exposed to the same doses over 24- and 48-hr periods as chronic exposure during the procedure of mitosis. Root tips are fixed in alcohol-acetic acid (3:1) 24 hr and 48 hr after the start of the experiment. Slides were prepared using acetocarmine squash technique. Both acentric fragments and bridges were scored at anaphase, and the results are summarized in Tables 25.1 and 25.2. The figures are based on analyses of 300 to 400 anaphase cells, except for the heavier doses of 200 r, since fewer anaphase cells were found in all acute and the 48-hr chronic series.

The results indicate that the yield of chromosome aberrations after doses from 12.5 to 100 r are greater when the radiation dose is given over 24- or 48-hr exposure periods than when the exposure time is only a few minutes. At a dose of 200 r, where the data are based on approximately a hundred cells only, no difference can be demonstrated.

An exact comparison between those two radiation treatments is difficult because the sensitivity of chromosomes varies during mitosis. During chronic

TABLE 25.1. NUMBER OF ACENTRIC FRAGMENTS PER 100 CELLS AFTER ACUTE AND CHRONIC EXPOSURES TO GAMMA RADIATION OF ROOT TIPS OF *Tradescantia paludosa*

Total Dose, r	After 24 Hr		After 48 Hr		Acute, from 2 to 168 Hr After Irradiation
	Acute	Chronic	Acute	Chronic	
12.5	1.6	6.8	1.0	—	2.3
25	4.1	15.7	2.4	9.1	5.0
50	3.2	29.3	12.1	16.9	9.9
100	50.0	59.7	45.7	35.3	30.0
200	117.2	—	50.0	76.1	93.7

TABLE 25.2. NUMBER OF BRIDGES PER 100 ANAPHASE CELLS AFTER ACUTE AND CHRONIC GAMMA IRRADIATION IN ROOT TIPS OF *Tradescantia paludosa*

Total Dose, r	After 24 Hr		After 48 Hr		Acute, from 2 to 168 Hr After Irradiation
	Acute	Chronic	Acute	Chronic	
12.5	0	1.3	3.2	—	0.9
25	0.6	2.3	1.4	1.9	1.8
50	0	5.3	0.8	3.1	3.0
100	4.6	5.1	0	5.3	5.6
200	13.8	—	14.3	12.0	24.2

radiation over the period used in these experiments, all stages of mitosis were exposed to radiation, while analyzed cells in the acute series were all in a particular stage during the radiation exposure. In the right-hand column of Tables 25.1 and 25.2 the sum of all analyses from the following periods after the acute exposures are listed: 2, 6, 12, 18, 24, 36, 48, 72, 96, 120, 144, and 168 hr. Thus, cells in all stages of mitosis and delayed dividing cells were included in the analyses, and the total radiation damage of the acute exposures should have been analyzed.

The frequencies are still lower than after chronic treatments at the moderate doses of gamma irradiation. The same tendency can be demonstrated if the data from 2 to 24 hr and from 2 to 48 hr are added together.

Effects on chlorophyll mutations in barley. The most striking effect of ionizing radiation in plants is the segregation of chlorophyll mutants in the progenies of the irradiated plants (Refs. 4, 5, 16, 17, and many others). Since chlorophyll mutations are the most frequent type of mutations in irradiated barley plants, analyses of chlorophyll mutants will be important means for evaluating the genetic effects of radiation treatments.

After radiation exposures of seeds, the segregations of chlorophyll mutants appear in the second generation (the progenies from the irradiated plants). After chronic radiation of growing plants, the segregation of chlorophyll mutants appears in the third generation. The second generation of plants from chronic radiation treatments and the first generation of plants from acute treatments of seeds have many characteristics in common. The appearance of chlorophyll mutants is as rare as the spontaneous rate by which those mutations occur. During late fall 1954 and winter 1955, analysis of chlorophyll mutants in the third generation of barley material grown in the gamma field 1953 has been carried out. A wide range of types of chlorophyll mutations have been defined and described (Ref. 5).

Table 25.3 gives the results of the analysis of chlorophyll mutations in the Norwegian barley variety, Domen. It appears that the relatively radioresistent barley has a considerable chlorophyll mutation frequency of 0.8 per 100 spikes at a dose rate of approximately 5 r/day and hence a total dose of approximately 280 r. The mutation rate increases proportional with dose. The increase is slower from 5 to 25 r/day than between 25 and 56 r/day. No conclusions about mutation-dose relationship can be reached on the basis of these data, but it is tempting to point out that the results seem to be in good

TABLE 25.3. FREQUENCIES OF CHLOROPHYLL MUTATIONS IN PROGENIES FROM BARLEY PLANTS TREATED WITH GAMMA IRRADIATION FOR 56 DAYS DURING GROWTH

Dose Rate, r/day	Total Dose	No. of Spike Progenies	No. of Mutations	No. of Mutations per 100 Spikes
±5	±280	1186	10	0.8
±13	±700	1121	15	1.3
±20	±1100	1100	9	1.6
±25	±1400	315	8	2.5
±31	±1700	376	18	4.8
±41	±2300	83	6	7.2
±56	±3700	84	8	9.5

agreement with the results obtained at Brookhaven (Ref. 15) on mutation in some particular loci in maize exposed to chronic gamma radiation. A gradual increase was observed between 5 and 32–57 r/day. Above 57 r/day a more marked increase was obtained.

It is surprising that these relatively small total doses of gamma radiation have produced such large numbers of chlorophyll mutations. Acute treatments of gamma rays on seeds would certaintly not have produced these mutation rates at those doses. No convincing gamma-irradiated materials are available for comparison. X-ray material of Domen variety from two independent experiments may be used as a comparison and as a measure of the effectiveness of the chronic irradiation. The X-ray doses are given to dormant seeds.

In Table 25.4 the chlorophyll mutation frequencies are tabulated from those two experiments, in 1951 and 1952 respectively.

From the seeds treated with 20,000 r and 25,000 r, only very few plants survived and the mutation rates of those doses are not reliable. It is striking that a considerably larger dose of X rays given to the dormant seeds is necessary to produce the same effects as the more sparsely ionizing gamma rays given to the growing plants.

TABLE 25.4. FREQUENCIES OF CHLOROPHYLL MUTATIONS IN BARLEY AFTER ACUTE X IRRADIATION OF SEEDS

Dose, r	1951	1952
5,000	—	1.2
7,500	—	4.5
10,000	2.7	4.0
15,000	6.7	6.8
20,000	5.6	34.8
25,000	25.0	—

Effects on somatic mutations. In vegetative propagating plants such as many fruit trees and some horticultural plants, bud mutations have played an important role in the production of new varieties and of genotypic variation. Such changes appearing in somatic tissue may be considered somatic mutations. Thus, the effects of chronic radiation on the production of somatic mutations should be of great interest. Sparrow (Refs. 13, 14) studied such somatic changes caused by mutation after acute and chronic irradiations in a number of flowering plants where the genotypic constitution of the plants is known.

Since *Tradescantia paludosa* seems to be relatively radiosensitive and can easily be propagated vegetatively, this material was found suitable for the study of somatic mutations. *Tradescantia paludosa* (Clone B 2-2) was grown in the gamma field for 56 days in 1953, and the plants were exposed to chronic gamma radiation at dose rates from 5.6 r/day to 76.4 r/day. Studies on the effects of radiation were confined to changes in flower morphology and color. By a thorough study of the flowers of the normal non-irradiated plants it appeared that among 465 flowers studied, 12.7% carried one or more abnormal flower characters (Table 25.5). At an average dose rate of 5.6 r/day the percentage of abnormal flowers had increased to 72.5. At dose rates between 5.6 and 18.2 r/day the percentage of abnormal flowers varied between 72.5 and 94.9.

The changes in flower morphology are changes in number and shape of sepals, petals, stamens, and pistil. Besides flowers with the normal three sepals, flowers with two and four sepals are found. Changes in petal color appear as white chimeras in the blue petals. At a higher dose of 87.2 r/day, flower development is almost entirely inhibited. The inflorescence is after a while overgrown by clusters of modified leaves.

Plants receiving dose rates of 54.1 and 76.4 r/day were completely stopped in development and finally died.

These drastic changes in flower morphology are quite different from the changes Sparrow (Refs. 13, 14) noticed in several plant species and defined as

TABLE 25.5. EFFECTS ON FLOWER STRUCTURES IN *Tradescantia paludosa* AFTER EXPOSURES TO CHRONIC GAMMA IRRADIATION FROM Ir^{192} FOR 56 DAYS

Average Dose Rate, r/day	Total No. of Flowers	Normal Flowers	Abnormal Flowers	Abnormal Flowers, %
0	465	406	59	12.7
5.6	51	14	37	72.5
6.8	92	21	71	77.2
7.7	126	9	117	92.9
8.9	93	20	73	78.5
10.3	84	16	68	81.0
12.1	39	2	37	94.9
14.4	78	12	66	84.6
18.2	43	3	40	93.0
37.2	0			
54.1	0			
76.4	0			

somatic mutations. They are of the same kind as we have found in carnations, where white spots appear in the petals on irradiated plants. Gunckel *et al.* (Refs. 2, 3), who have done a thorough study of the variations in vegetative and floral morphology of *Tradescantia paludosa,* claim that the changes they describe cannot be attributed to cytogenetic effects, but are considered physiological effects, because cuttings from those plants revert to normal growth pattern. We have obtained in our material, however, cuttings of the irradiated plants which have not reverted to normal growth so far. But eventually they may do so. Before the true nature of the changes in flower characters in *Tradescantia paludosa* is worked out by further studies, it is doubtful whether the changes observed can be defined as somatic mutations. The white spots or chimeras in carnations, however, may certainly be defined as somatic mutations.

Summary

Ir^{192} and Co^{60} have been installed in a field where different plant species are exposed to chronic gamma irradiation during growth and development. The installation and method of operation of sources of approximately 12 and 25 curies are described.

The effects of chronic irradiation on chromosome aberrations are studied in root tips of *Tradescantia paludosa* over a period of 24 and 48 hr. The results are compared with the results of acute exposure of the same doses given during a few minutes. The yield of chromosome aberrations was larger in the chronic series than in the acute at doses from 12.5 to 100 r.

The genetic effect of chronic gamma irradiation is studied on chlorophyll mutations in barley. The mutation rates increased with radiation dose. Compared with X-irradiation of barley seeds, relatively small doses of chronic gamma radiation during growth produced pronounced frequencies of chlorophyll mutations.

Variability in flower characters and abnormal development of flower parts was noted in 12.7% of the inflorescences on non-irradiated plants of *Tradescantia paludosa*. At a dose rate of 5.6 r/day of chronic gamma irradiation for 56 days, the percentage of abnormal flowers increased to 72.5. At dose rates between 5.6 r/day and 18.2 r/day the percentage of abnormal flowers varied from 72.5 to 94.5. At a dose rate of 37.2 r/day development of flowers is suppressed and the inflorescences are overgrown by clusters of modified leaves. At dose rates of 54.1 r/day and 76.4 r/day, development of flower heads is completely inhibited and no growth whatsoever is observed.

References for Chapter 25

1. I. Granhall, L. Ehrenberg, and S. Borenius, "Experiments with chronic gamma irradiation on growing plants," *Bot. Not.* 155–162 (1953).
2. J. E. Gunckel, A. H. Sparrow, I. B. Morrow, and E. Christensen, "Vegetative and floral morphology of irradiated and non-irradiated plants of *Tradescantia paludosa*," *Am. J. Bot.* 40:317–332 (1953).
3. J. E. Gunckel, I. B. Morrow, and A. H. Sparrow, "Variations in the floral morphology of normal and irradiated plants of *Tradescantia paludosa*," *Bull. Tor. Bot. Club* 80:445–456 (1953).
4. Å Gustafsson, "Studies on the genetic basis of chlorophyll formation and the mechanism of induced mutating," *Hereditas* 26:33–93 (1938).
5. Å. Gustafsson, "The mutation system of the chlorophyll apparatus," *Lund Univ. Årsskr.* N.F., Avd. 2, Bd. 36:1–40 (1940).
6. Å. Gustafsson, "Mutations in agricultural plants," *Hereditas* 33:1–100 (1947).
7. K. Mikaelsen, "The protective effect of glutathione against radiation-induced chromosome aberrations," *Science* 116:172–174 (1952).
8. K. Mikaelsen, "Cytological effect of chronic gamma irradiation and the protective property of certain chemicals against the radiation-induced chromosome aberrations," in *Radiobiology Symposium,* Butterworth's Scientific Publications, London, 1955, pp. 312–16.
9. K. Mikaelsen and K. Aastveit, *Studies on the Effects of Chronic Gamma Irradiation and Neutrons on Growth and Fertility in Barley and Oats.* In manuscript.
10. W. R. Singleton, "Effects of continuous gamma radiation on mutation rate in maize," *Genetics* 36:575–576 (1951).
11. W. R. Singleton, "The effects of chronic gamma radiation on endosperm mutations in maize," *Genetics* 39:587–603 (1954).
12. A. H. Sparrow, "Tolerance of *Tradescantia* to continuous exposures to gamma radiation from Co^{60}," *Genetics* 36:135 (1950).
13. A. H. Sparrow, "Somatic mutations induced in plants by treatment with X and gamma radiation," Proc. Ninth Internat. Genetics Congress, 1953, *Caryologia,* Vol. Suppl. 1954.
14. A. H. Sparrow, M. Denegre, and W. J. Haney, "Somatic mutations in *Antirrhinum* produced by chronic gamma irradiation," *Genetics* 37:627–628 (1952).

15. A. H. Sparrow and W. R. Singleton, "The use of radiocobalt as a source of gamma rays and some effects of chronic irradiation on growing plants," *Amer. Nat.* 87: 29–48 (1953).
16. L. J. Stadler, "Genetic effects of X-rays in maize," *Proc. Nat. Acad. Sci.* 14:69 (1928).
17. L. J. Stadler, "Some genetic effects of X-rays in plants," *J. Hered.* 21:3–13 (1930).
18. R. Torsell, "Mutation research in plants," *Acta Agri. Scand.* 4(3):358–642 (1954).

Chapter 26

THE CONTRIBUTION OF RADIATION GENETICS TO CROP IMPROVEMENT *

Historical

Radiation genetics had its origin in the researches of Muller and Stadler almost thirty years ago. They demonstrated that by means of X rays it was possible to induce genetic changes in animals and plants. These researches really opened a new era not only for fundamental genetics but also for the practical plant and animal breeder.

Before, the breeder was limited to the spontaneous mutations that had occurred at some time in the material with which he was working, and since in most crops there was an abundance of spontaneous mutations the plant breeder was able, by the application of genetics, to do a creditable job of plant breeding. Rapid advances were made following the rediscovery of Mendel's law in 1900 by Correns, DeVries, and Tschermak.

However, the discovery that mutations could be induced by radiation gave added impetus to plant breeding efforts. No longer was the plant breeder limited to mutations that occurred naturally. He could make them occur. Whereas a particular mutant might appear only once in a hundred years, by radiation it was possible to increase the frequency as much as a hundredfold, making it possible to induce desired mutants perhaps in one year.

Hence, evolution can be accelerated. This does not mean that the plant breeder's problems are all solved. It merely means that he can have a wealth of material from which to select. In his original paper, "The Artificial Transmutation of the Gene," Muller (Ref. 4) made the statement, "Similarly, for the practical breeder, it is hoped the method will ultimately prove useful."

The method has proved useful. Its usefulness was demonstrated first in Sweden through the brilliant researches of Nilsson-Ehle, Gustafsson, and their associates, who demonstrated in cereals that higher yields, stiffer straw, and earlier-ripening types could be produced by radiation. Considerably later,

* This chapter is taken from Geneva Conference Paper 110, "The Contribution of Radiation Genetics to Crop Improvement" by W. R. Singleton, C. F. Konzak, S. Shapiro, and A. H. Sparrow of the United States.

its usefulness was demonstrated in the United States shortly after World War II in a comprehensive experiment by Dr. Walton C. Gregory at the North Carolina Experiment Station (1955). He began by X-raying 100 lb of peanuts, *Arachis hypogaea* L., through the facilities at the Oak Ridge National Laboratory. He observed 975,000 X_2 plants and from these was able to select types that were superior in yield, better adapted to mechanical harvesting, and resistant to a serious leaf-spot disease.

The peanut breeding project of Gregory demonstrated the wisdom of large populations in a radiation breeding program. Since the rate of mutations, even under radiation, is only a few per cent and since relatively few of the induced mutants will have any practical application, it is imperative to work with large numbers.

Dr. Gregory's experiment also demonstrated that it is possible to induce mutations in a rather stable organism. The peanut has a fairly low spontaneous natural mutation rate, but mutations were abundant following radiation. Also the peanut, being self-pollinated, is not subject to natural crossing whereby new germ plasm can be incorporated. Hence, radiation has provided about the only feasible way of introducing new germ plasm into this rather stable organism and thereby increasing the genetic reservoir from which the plant breeder can extract new and desired types.

The Gamma Radiation Field at Brookhaven

Radiation genetics research at the Brookhaven Laboratory has been concerned with obtaining information regarding the effectiveness of the different

FIG. 26.1. Gamma Radiation Field at Brookhaven Laboratory. Plants growing around Co^{60} source in center.

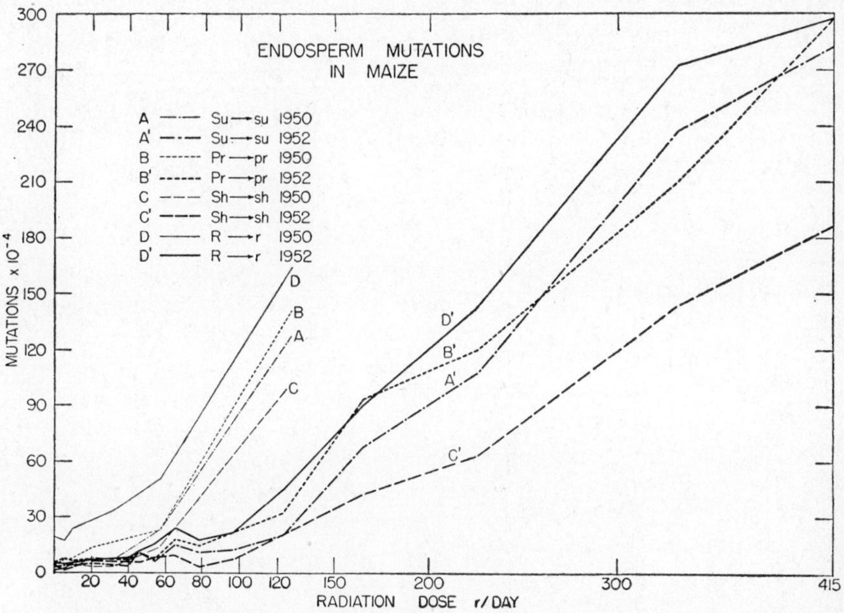

Fig. 26.2. Endosperm mutations in maize as affected by chronic gamma radiation.

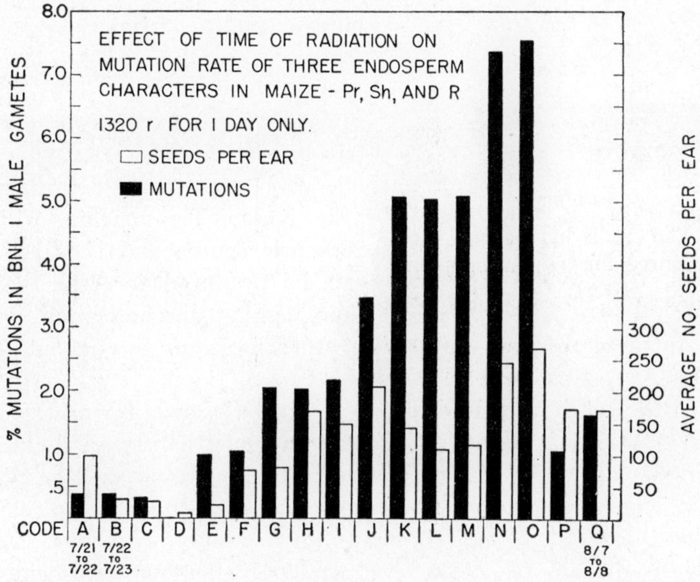

Fig. 26.3. Late stages in pollen development (N and O) more susceptible to mutations by radiation than meiosis (D) or pre-meiotic stages, (A–C).

types of radiations in producing mutations; also, how the altering of various conditions, including extra- and intracellular environment may change the effectiveness of radiation.

The effects of chronic gamma radiation on growing plants have been investigated (Ref. 8). This was made possible by placing a Co^{60} source in the center of a field with plants grown in concentric circles (Fig. 26.1). This type of radiation is effective in producing endosperm mutations in maize, mutations increasing more sharply than the radiation dose, giving (Ref. 5) a curvilinear relationship (Fig. 26.2). Also it has been found (Ref. 6) that there are extreme differences in sensitivity at different stages in the meiotic cycle (Fig. 26.3).

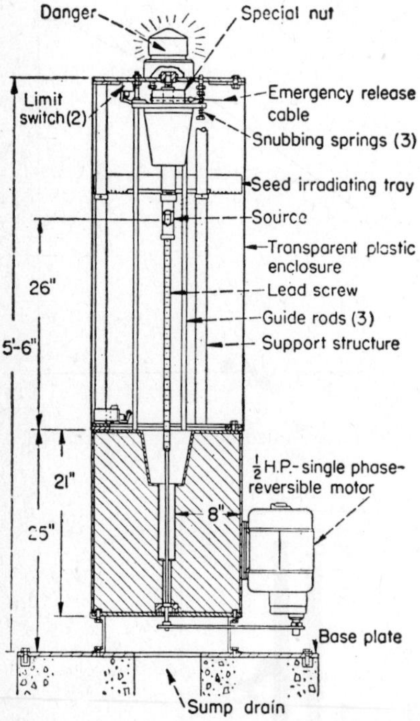

FIG. 26.4. Co^{60} radiation machine with 200 curie source for irradiating plant material.

This last finding suggests a different approach to the use of gamma radiation for the production of mutations, which can be produced as readily by giving a high exposure of one day's duration, as by exposure to a lower dose during a longer period. Sufficient radiation for mutation production can be obtained by placing plants fairly close to a small source.

With this in mind a Co^{60} gamma radiation machine has been designed by the Nuclear Engineering Department in cooperation with the Biology Department (Fig. 26.4). This machine will house a 200-curie source and it is estimated it can be produced for less than $5000, thus making chronic gamma radiation available to more research workers in other locations throughout the world (Ref. 3).

Although the sensitive period during the meiotic cycle for other plants has not been determined, it seems reasonable to assume that a period analogous to corn might be expected. If such proves to be the case it will be possible to induce mutations in the developing gametes by exposing growing plants for a short period to a Co^{60} machine similar to the one described above. For plants in which it is difficult to obtain quantities of pollen for treatment, perhaps such a technique would be best to produce mutations in the gametes. Such mutations would give rise to a whole plant carrying the mutant gene. This is in contrast to mutations induced in the somatic tissue, including seeds, in which the mutant affects only a part of the plant. This will be discussed more fully under somatic mutations.

With a plant like maize, as with some of the forest trees, where it is possible to obtain readily quantities of pollen and where mass pollinations can be made easily, it is feasible to induce mutations by exposing the mature pollen to such radiations as ultraviolet light, X rays, and either slow or fast neutrons. It has been found that slow neutrons are comparatively efficient in producing mutations in maize pollen. Mutation rates in excess of 4% per gene for endosperm characters have been obtained, with a maximum not yet reached. Different genes react differently. Mutations of the Sh_1 gene reach the maximum at a much lower dose than the other genes used, *Pr, R* and *Su*. Mutation rates in excess of 3% per gene have been obtained with X rays and ultraviolet.

FIG. 26.5. Induced resistant mutant (right) compared to parental susceptible variety, Mohawk.

SEED IRRADIATIONS

Technically mutations produced by radiation of seeds are somatic mutations (to be discussed later). However, seed irradiation is quite distinct from the radiation of growing plants for the production of somatic mutations and hence will be discussed separately.

Mutations can be produced in abundance by X rays or by slow or fast neutrons. Neutrons are considered more efficient since a more uniform effect is produced regardless of the physiological conditions of the seed.

Extensive oat irradiation experiments have shown that it is possible to produce resistance to stem rust (*Puccinia graminis avenae*) in oats (*Avena sativa* L.) by exposure of the seeds to thermal (slow) neutrons (Ref. 2). There were 48 cases of induced resistance (Fig. 26.5) among plants grown from treated seed at a rate in excess of 10%.

In another radiation experiment seeds of the Tama variety of oats, which is susceptible to Victoria blight caused by *Helminthosporium sativum* var. *victoriae,* were irradiated with thermal neutrons. From 642 R_2 progenies 10 mutations were found (Fig. 26.6). Seeds were also X-rayed (25,000 r). In this

FIG. 26.6. Oat variety normally resistant to *Helmintosporium* (left) and few resistant mutations produced by radiating susceptible variety.

experiment 3 mutations were recovered from 122 progenies. No mutations were observed in 140 control progenies.

These experiments demonstrated also that it was possible to obtain resistance for blight while retaining a type of resistance to another pathogen, crown rust, *Puccinia coronata avenae*. Rust reaction classes varying from type 1 (resistant) to type 4 (fully susceptible) have been observed in blight-resistant mutant types. The rust reaction type of the original variety may not have been recovered, but there is hope that it may be.

The result of these studies suggest that the use of mutagenic agents may make possible certain combinations of hereditary characteristics which are difficult to produce by hybridization methods and which may not yet have been found in nature. For example, resistance to certain races of crown rust is closely associated with blight susceptibility in oat varieties of Victoria parentage. This association has not been broken in regular genetic studies. However, the preliminary results reported here indicate that it may be possible with radiation to break this association and recover types not yet produced.

These preliminary experiments with inducing disease resistance in oats have been confirmed by other workers in a number of cereals and other crops. When it is realized that plant diseases cost the farmers in the United States alone an estimated 3 billion dollars a year, the importance of induced disease resistance is clear.

Somatic Mutations

Mutations in the somatic, or body, cells of the plant, in contrast to the mutations induced in the germ cells, are of considerable interest and economic importance. Many of our varieties of horticultural plants, such as the pink dogwood, Golden Delicious apples, and others, have arisen as somatic mutations. In carnations the William Sim or Red variety has given rise, by somatic mutation, to a number of others, notably White Sim, Harvest Moon, "Mamie," and Tetra Red. These derivatives of the William Sim variety, along with William Sim, constitute a fairly large percentage of all carnations grown.

Since it was found feasible to induce gametic mutations so readily, the question naturally arose whether we could induce somatic mutations. These could then be propagated asexually and new varieties established immediately.

As this appeared to be a project with practical applications it seemed wise to enlist the aid of the agricultural experiment stations who are interested in new varieties for their respective areas. A conference was called at the Brookhaven Laboratory. Representatives of the experiment stations and universities in the northeastern part of the United States were invited. They came and were enthusiastic. A cooperative project was born.

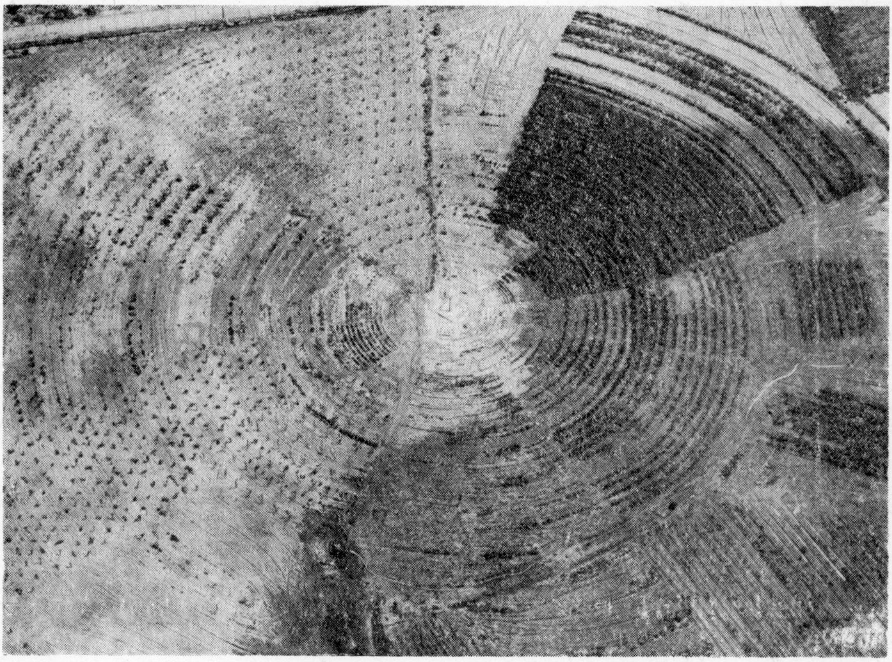

Fig. 26.7. Gamma radiation field showing orchard trees in somatic mutation program.

The original gamma radiation field was enlarged and the source increased to 1800 curies, enabling the field to accommodate a larger number of plants at areas of effective doses. Eight universities and experiment stations are growing a wide variety of horticultural crops in the field. An idea of how the "somatic" part of the field looks can be obtained from Fig. 26.7.

Although it is too soon to expect results from most of the slower-growing woody plants, positive results have been obtained with some of the herbaceous plants such as *Antirrhinum majus* L., in which tetraploid sectors were induced (Fig. 26.8). *Dahlia pinnata* Cav and *Nicotiana Sanderae* Sander have pro-

Fig. 26.8. Normal *Antirrhinum* (right) compared with induced tetraploid (4N) (left), with larger flowers.

duced mutations involving at least a whole flower. Other flowers showing smaller somatic mutations have been tabulated by Sparrow (Ref. 7).

In carnations (*Dianthus caryophyllus* L.) several propagatable somatic mutations have been produced. Eleven of 87 cuttings from plants of White Sim receiving the highest doses of radiation up to 340 r/day produced branches that had all red flowers (Fig. 26.9). The only two plants removed from the 340 r/day row later developed branches with all red flowers. Cuttings of the red-flowered type have produced plants with only red flowers. Graphs showing the dose response relationships for somatic mutations are found in Figs. 26.10–12.

It is not possible to state definitely whether there is a threshold in each case, since in some cases the control rate is below the effects produced by the low-level radiation effects. This is definitely true of the *Antirrhinum* described below, where the control is significantly below the "plateau" of low-level radiation effects. Two of the carnation graphs, for William Sim and White Sim, do not differ significantly from a straight-line relationship, while the Harvest Moon graph seems curvilinear.

FIG. 26.9. Branch with red flowers induced by radiating White Sim carnation.

Extensive somatic mutation data were obtained with the P^{RR}/p gene in maize (Fig. 26.13). More than a million kernels were observed. Most of the somatic changes were rather small, producing in many instances a very fine streak. However, for the graph presented here, only those mutations which were greater than trace were scored as mutations. In a few instances somatic mutations involved a whole kernel or a cluster of kernels. These were comparatively rare. Most mutations were small, indicating initiation rather late in the life of the ear. From the lower doses of radiation there is not a linear relationship between dose of radiation and mutations. Below 5 r/day there was no apparent effect of the radiation.

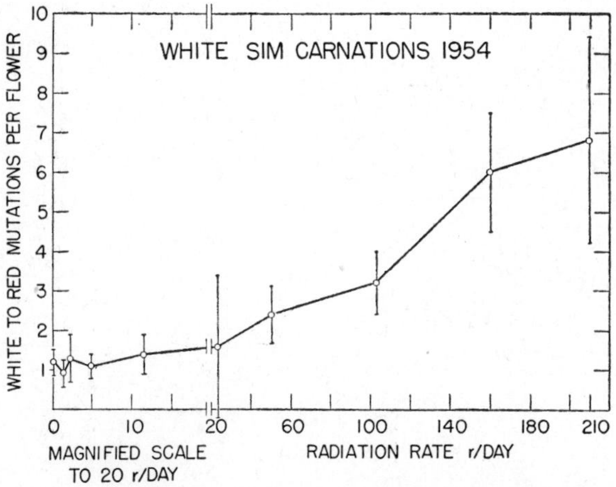

FIG. 26.10. Dose response of somatic mutations in White Sim carnations.

It was also observed that radiation injured the developing ear. From less than one to approximately 100 r/day there was an inverse linear relationship between radiation received and number of seeds per ear (Fig. 26.13).

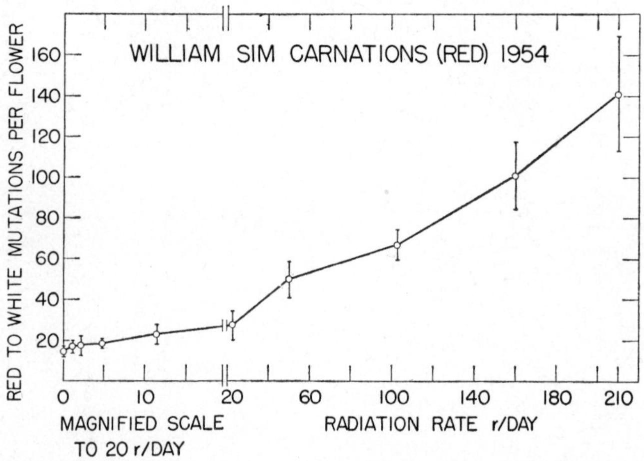

FIG. 26.11. Dose response of somatic mutations in William Sim carnations.

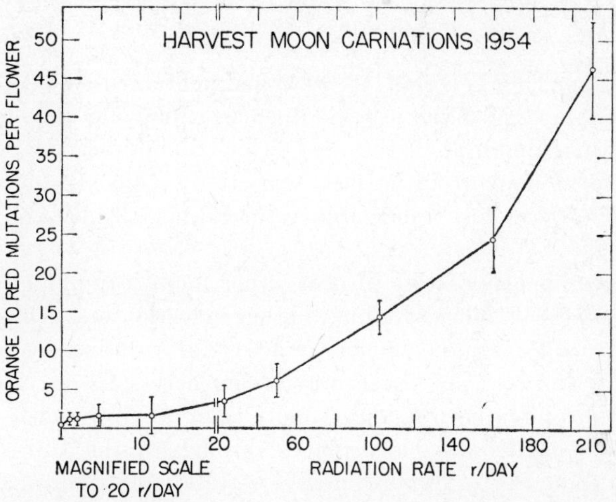

FIG. 26.12. Dose response of somatic mutations in Harvest Moon carnations.

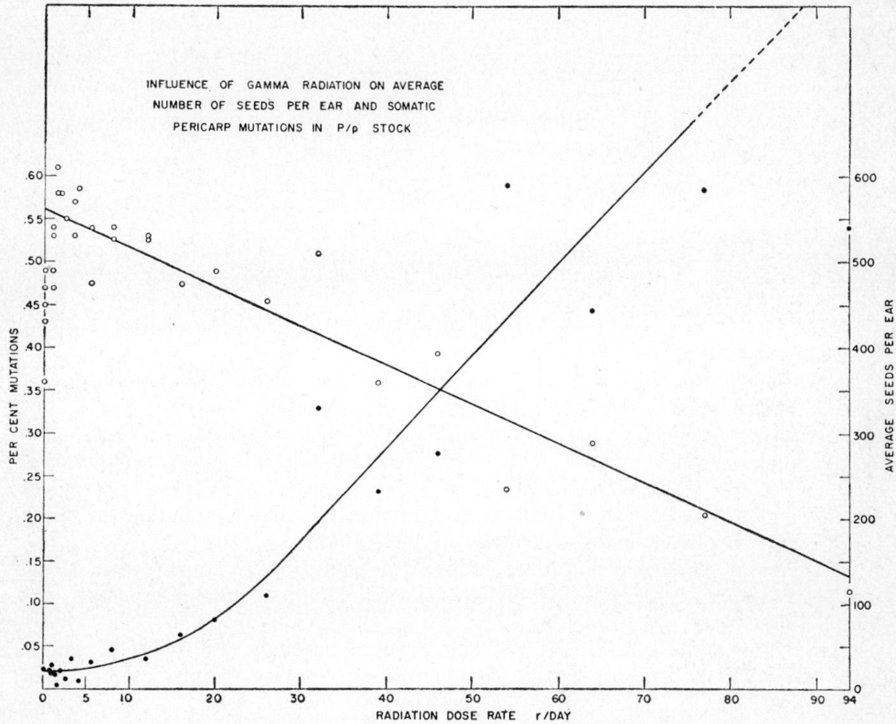

FIG. 26.13. Dose response of somatic mutations in P/p stock of maize, also reduction in seed set as a result of radiation.

Cooperative Research

It was mentioned earlier that the gamma field is used cooperatively by research workers in eight different experiment stations and universities. These workers grow their material in the field, which is maintained by the Brookhaven Laboratory. Any resulting new varieties become the property of the investigator, who takes full responsibility for taking all notes on the material growing in the field.

In addition to this cooperative project, other facilities unique to the Brookhaven Laboratory are made available to other investigators. One such facility is the thermal column of the nuclear reactor for treating seeds and scions. Also material is X-rayed for those not having access to an X-ray machine. To date radiations have been made on more than 60 crop plants for well over 100 investigators located at many places throughout the United States and Canada. It is too early to expect positive results from each of these investigators. However, workers in Minnesota are able to report positive results of disease resistance induced in plants (Dr. W. M. Myers).

Radiation as a tool in plant breeding has made a modest beginning. We feel that each succeeding year will show progress. It is impossible to envision the eventual good that may arise from the use of radiation in plant breeding; and we thoroughly agree with the late Enrico Fermi who said, "I believe truthfully, that the conquest of atomic energy may be widely used to produce not destruction, but an age of plenty for the human race."

References for Chapter 26

1. W. C. Gregory, "X-ray breeding of peanuts (*Arachis hypogaea* L.)," *Agron. J.* 47:396–9 (1955).
2. C. F. Konzak, "Stem rust resistance in oats induced by nuclear radiation," *Agron. J.* 46:538–540 (1954).
3. O. A. Kuhl, W. R. Singleton, and B. Manowitz, "A cobalt-60 field-irradiation machine," *Nucleonics* 13, No. 7:42 (1955).
4. H. J. Muller, "Artificial transmutation of the gene," *Science* 46:84–87 (1927).
5. W. R. Singleton, "The effect of chronic gamma radiation on endosperm mutations in maize," *Genetics* 39:587–603 (1954).
6. W. R. Singleton and A. L. Caspar, "Effect of time of gamma radiation on microspore mutation rate in maize," *Genetics* 39:993 (1954).
7. A. H. Sparrow, "Somatic mutations induced in plants by treatment with X and gamma radiation," Proc. Ninth Internat. Genetics Congress, 1953, *Caryologia,* Vol. Suppl. 1954.
8. A. H. Sparrow and W. R. Singleton, "The use of radiocobalt as a source of gamma rays and some effects of chronic irradiation on growing plants," *Amer. Nat.* 87:29–48 (1953).

Part VIII

FOOD STERILIZATION

Chapter 27

COLD STERILIZATION OF FOODS *

History of the Development of Radiation Sterilization

The preservation of food is one of man's greatest blessings. It has permitted him to carry the harvest over from one season to another and thus defy the ever-present law of nature according to which foods begin to spoil as soon as they are removed from her protective fold. Preservation of man's food dates back to antiquity. The largest single means of preservation today, thermal processing, dates back to the Napoleonic era, and scientific heat processing originated at the turn of the present century.

Radiation sterilization also has a history, comparatively brief but dating back, on a fundamental basis, to the discovery of ionizing radiations by Roentgen (Ref. 1) and Becquerel (Ref. 2) in 1895 and to the initial researches on the bactericidal properties of these radiations which occurred shortly thereafter (Refs. 3, 4).

The developments in radiation sources made in the 1930's and 1940's disclosed new potentialities for the beneficial uses of such types of energy sources. In 1943 research experiments were initiated at the Massachusetts Institute of Technology that were destined only a decade later to result in a multimillion dollar research effort in numerous laboratories in the United States. These investigations may lead to practical applications of ionizing radiations as a means of sterilization in the food, drug, and pharmaceutical fields. In fact, such application is already on the verge of commercial utilization in the pharmaceutical field.

From 1945 until 1948 research on the microbiological effects of ionizing radiations was limited to a small number of laboratories in the United States. In 1949 a research center in Great Britain became active in this field. In 1950, under sponsorship of the United States Atomic Energy Commission, research activity in the peacetime uses of the atom for sterilization and preservation purposes increased, and a number of excellent laboratories began applying

* This chapter is taken from Geneva Conference Paper 172, "Progress and Problems in the Development of Cold Sterilization of Foods" by B. E. Proctor and S. A. Goldblith of the United States.

their skills in their several particular disciplines toward solving the problems facing commercial utilization of radiation sterilization (Ref. 5).

Our discussion of the various phases of this development is based largely on investigations conducted by our group at the Massachusetts Institute of Technology, and citations are directed to publications that may amplify the necessarily brief factual data that can be encompassed in this limited report.

Some Results of Research Studies

Early in our own research we were able to demonstrate the ability of these radiations to destroy all types of microorganisms in every conceivable type of package, so long as the dimensions of the package did not exceed the limitations imposed by the particular type and energy level of radiations used. In addition to these findings, we have been able to demonstrate that the species of organism is the prime factor in determining the magnitude of the sterilizing dose (Refs. 6–8 and Fig. 27.1). This fact has since been confirmed by many other investigators (Refs. 9, 10).

Later research has shown the importance of environmental factors in the survival ratios of microorganisms exposed to ionizing radiations (Refs. 11, 12).

More recently we have obtained interesting data on the microbiological

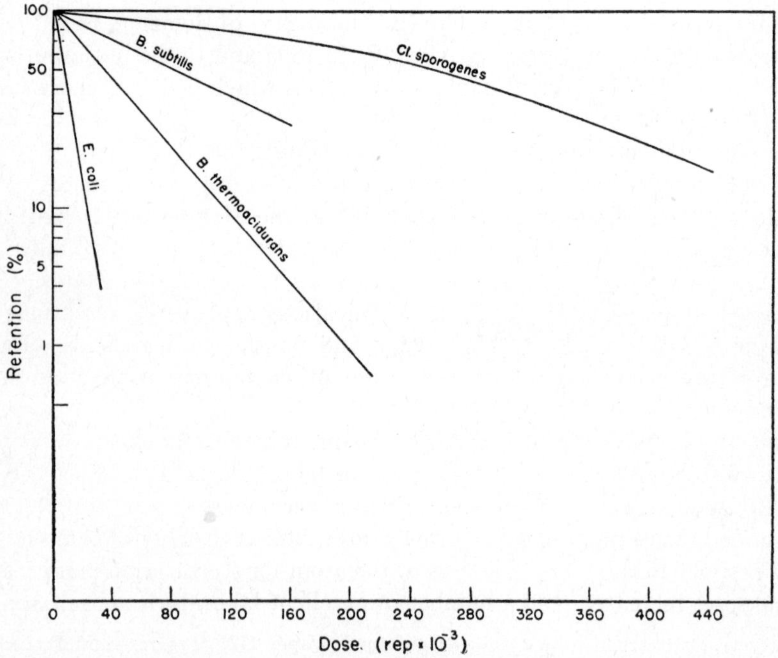

Fig. 27.1. The relative radiosensitivities of four species of bacteria (6).

factors involved in radiation sterilization. These data indicate the hazards in applying any one set formula for the use of ionizing radiations to destroy all species of microorganisms that may contaminate a food. They also demonstrate the necessity of time-consuming, tedious research under the particular experimental conditions to be encountered in processing with each particular contaminating species (Ref. 13).

The examples cited above relate to experiments conducted with *B. subtilis* spores and with *E. coli* (Refs. 14–16 and Figs. 27.2 and 27.3). The data ob-

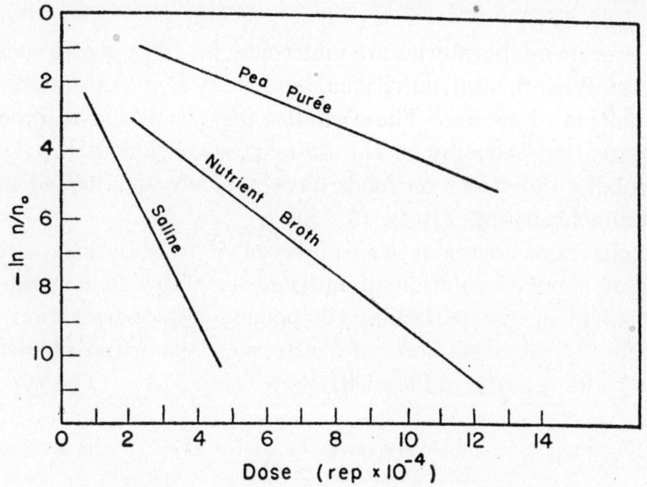

FIG. 27.2. Comparative effects of gamma radiation from cobalt-60 on saline, nutrient broth, and pea puree suspensions of *E. coli* in air atmosphere (13).

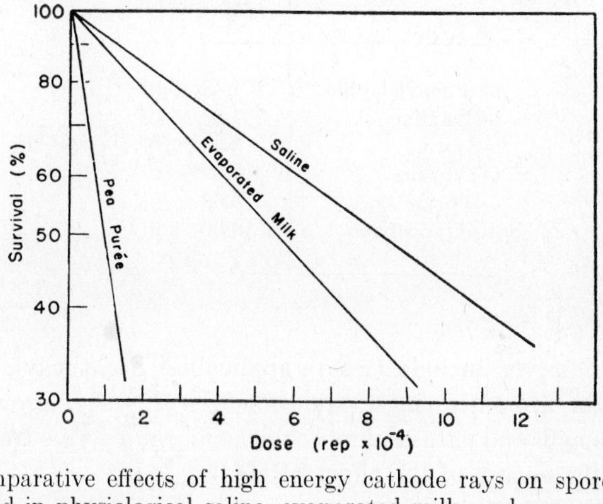

FIG. 27.3. Comparative effects of high energy cathode rays on spores of *B. subtilis* suspended in physiological saline, evaporated milk, and pea puree (13).

tained for these two organisms illustrate one of the interesting facets of the microbiological effects of radiations. The data obtained for *E. coli* might well have been predicted theoretically, from the qualitative standpoint at least. The results obtained with *B. subtilis* spores, on the other hand, were quite contrary to accepted theory and could not have been predicted.

Research on the microbiological effects of radiations may be considered to be further advanced than research on some other aspects. Studies are in progress at present on the complementary effects of heat and radiation, which may greatly reduce the radiation requirements for particular species of organisms (Ref. 17).

Studies in our own laboratories are under way on applying, to radiation processing, certain fundamental nutritional and environmental factors known to speed germination of spores. These studies may well lead to other means of obtaining increasing economy in the doses of radiation used. The chemical effects of ionizing radiations on foods have not been overlooked in these pioneering experimental years (Refs. 12, 18).

Early in our experiments it was observed that radiations affect isolated components of a mixed solution of nutrients *in vitro* to a greater or lesser degree, dependent on the particular component, its concentration in solution, its purity, and the radiation dose. *In vitro,* each particular nutrient might be classified as to its relative radiosensitivity (Table 27.1). The types of effects

TABLE 27.1. RELATIVE RADIOSENSITIVITIES OF AMINO ACIDS IN AQUEOUS SOLUTIONS [a]

Amino Acid	Specific Inactivation Dose (D_0/C), rep/gm/ml
DL-Phenylalanine	1.18×10^9
L-Histidine	1.25×10^9
L-Cystine	2.40×10^9
L-Tyrosine	2.60×10^9
L-Leucine	3.60×10^9
DL-Tryptophan	10.00×10^9

[a] From Bhatia (Ref. 19).

that might be observed include (where applicable) decarboxylation, deamination, cleavage of aromatic rings, oxidation of sulfhydryl groups, and many others. With niacin and para-aminobenzoic acid, radioactive tracers have been used to study the nature of the effect (Refs. 20, 21).

These indirect effects of radiations, which occur as the result of the absorption of large amounts of energy in the radiation sterilization process, exhibit not only the phenomena associated with dilution, rate of reaction, and temperature dependence to be expected according to the indirect action theory but also the phenomenon of protection in mixed systems (Ref. 22 and Table 27.2). The comparison given in Table 27.3, which shows that vitamins in

TABLE 27.2. EFFECT OF HIGH-VOLTAGE X RAYS ON SOLUTIONS OF U.S.P. NIACIN, U.S.P. ASCORBIC ACID, AND ON MIXTURES OF THE TWO [a]

(Dose = 125,000 r)

Solution	Concentration Before Irradiation, micrograms/ml	Retention After Irradiation, %
Niacin	50	86.0
Ascorbic acid	500	66.0
Niacin in niacin–ascorbic acid mixture	50	44.0
Ascorbic acid in niacin–ascorbic acid mixture	500	83.6

[a] After Proctor and Goldblith (Ref. 22).

TABLE 27.3. SPECIFIC INACTIVATION DOSES OF ASCORBIC ACID AND RIBOFLAVIN IRRADIATED PER SE AND IN EVAPORATED MILK

Irradiated Medium	Specific Inactivation Dose, D_0/C (r/gm/ml)
Ascorbic acid—pure solution [a]	1.19×10^9 to 3.6×10^9
Ascorbic acid—evaporated milk [b]	2.4×10^{10}
Riboflavin—pure solution [a]	2.6×10^9 to 5.2×10^9
Riboflavin—evaporated milk [b]	3.7×10^{11}

[a] From Proctor and Goldblith (Ref. 8).
[b] From Kung, Gaden, and King (Ref. 18).

pure solutions are relatively more radiosensitive than vitamins in foods, illustrates the protective nature of the complex food material (Refs. 8, 18). A further illustration (Ref. 23) of this protective action is given by the comparison in Table 27.4. These data demonstrate that nutrients may be affected by ionizing radiations, but that the effect is dependent on the phenomena mentioned above.

TABLE 27.4. EFFECT OF CATHODE RAYS (DOSE 5,700,000 r) ON AMINO ACIDS IN HADDOCK FILLETS [a]

Amino Acid	Change in Amino Acid, %	
	Loss	Gain
Phenylalanine	6.10	0.00
Tryptophan	6.92	0.00
Methionine	4.68	0.00
Cystine	0.00	0.00
Valine	0.00	6.36
Leucine	0.00	2.74
Histidine	0.00	8.11
Arginine	0.00	4.12
Lysine	4.23	0.00
Threonine	5.95	0.00

[a] From Proctor and Bhatia (Ref. 23).

Because enzymes are, for the most part, protein molecules, they might well be expected to react to ionizing radiations as do proteins or other solutes. They exhibit the same phenomena associated with dilution, temperature, and protection that are observed with proteins, amino acids, and other molecules of biological importance. As a consequence, enzymes are not completely inactivated by sterilizing doses of ionizing radiation. In fact, under a number of conditions enzymes are neither completely inactivated nor even inactivated to a desirable extent by doses of ionizing radiations 10 to 20 times the magnitude of the doses required for sterilization.

Because the spoilage of certain foods occurs partly through enzymatic action, it is usually necessary that these biological catalysts be inactivated. In most thermal processes the amount of heat used for sterilization is sufficient to inactivate the enzymes, but in radiation sterilization ("cold sterilization") some other means for enzyme inactivation could be used to advantage. The possibility of applying high-frequency electronic heating is intriguing. Such application may well prove to be feasible, since it is a rapid, efficient, and satisfactory method for this purpose. The properly controlled use of microwave heating results in even heating throughout the mass of food and thereby avoids overheating of the exterior surface while producing only the minimal quantity of heat needed for enzyme inactivation in the interior of the food mass.

Problems in Radiation Sterilization

Some indirect effects of ionizing energy have given rise to the greatest problems existing today in radiation sterilization, namely, changes in color, texture, and flavor of irradiated foods. The fundamental studies mentioned above have suggested several technological means of obviating these undesirable side-reactions. These include (*a*) irradiation of food in the frozen state, (*b*) irradiation in inert atmospheres, and (*c*) addition of free-radical acceptors. It may be that all three of these technological means are really one and the same fundamentally, that is, means of reducing the indirect effect of radiation, leaving only the direct effect for destroying microorganisms.

Fundamental investigations are being conducted today to study the basic changes that take place in food materials when irradiated. The premise is made that once the type or types of changes in complex food materials are known, the possibility of finding means of avoiding or minimizing these should logically follow.

In addition to the chemical changes occurring in the protein components of irradiated foods, modifications in the structure of carbohydrates are observed. The changes in lipoidal material have been studied by several investigators, with interesting results (Ref. 30). The data obtained indicate the complex interaction that takes place among the many components which make up foods. The basic effect of ionizing radiations responsible for destruction of nutrients is probably also the simultaneous cause of changes in flavor, and any means of reducing the one effect will be a means of reducing the other effect.

These same indirect effects have also given rise to speculation as to whether any compound of a toxic nature might be produced by these radiations. The United States Food and Drug Administration has given its views on the required testing procedure (Ref. 24). This testing is not easy or inexpensive, and details of the procedure are frequent subjects of controversy. For instance, it has been stated that laboratory animals should be fed with irradiated food material in such an amount that a positive answer (if there can be one) will be obtained.

In other words, to determine the factor of safety, it is stipulated that one should ascertain what irradiation dose will produce changes of a toxic nature in foods. The procedure suggested to accomplish this is to overirradiate a food by a dose ten times, for example, as great as the sterilizing dose (Ref. 24). Such a procedure, however, may cause certain fallacious interpretations. For instance, let us assume that compound B is a toxic compound produced by irradiation from compound A (the precursor compound) in a food. Increasing the irradiation dose may produce more of compound B, but it may also have an entirely different effect, that of producing compound C from compound B. Compound C may be entirely innocuous. Such a combination of circumstances might result in a false and misleading sense of security. This example points out the difficulties in this particular problem alone.

In two toxicity studies that have been published to date (Refs. 25, 26), evidence has been lacking that ionizing radiations produce compounds of a toxic nature in foods. Destruction of nutrients has been noted, but this occurs even with conventional heat processing and is not unexpected.

Research to study this important point thoroughly is being conducted today in many laboratories in the United States under sponsorship of the Quartermaster Corps of the Department of the Army and with the scientific supervision of the Office of the Surgeon General. Not only are animal feeding studies and chemical studies under way, but also feeding studies with humans.

A problem closely related to that of wholesomeness or toxicity of irradiated foods is that of a possibly induced radioactivity in irradiated foods. This may be stated theoretically to be a function of the type of radiation to be used, the energy level, the cross section of the precursor element, and its relative isotopic abundance. With fast electrons or gamma rays, induced radioactivity should theoretically not occur below energies of 8 to 10 million electron volts. Studies are in progress in our laboratories at the Massachusetts Institute of Technology and elsewhere to ascertain the particular energy level of fast electrons necessary to produce induced radioactivity, the particular species of isotope, and the quantity of the isotope (if any) produced.

Recent research on the sterilization of milk in these laboratories has indicated that many of the undesirable changes taking place in irradiated foods occur almost immediately. Therefore, any efforts to avoid these undesirable changes must be made during or before the irradiation process. This research has resulted in a technique for obtaining an organoleptically acceptable milk product treated by ionizing radiations in doses up to 10^6 rep (roentgen-equivalent-physical).

Concurrent with the developments just mentioned, great strides have been made in the design, fabrication, and utilization of radiation sources. Multikilowatt sources of fast electrons are now available at increasing energy levels, and kilocurie radioisotope sources have been developed. Designs for reactors for this purpose, as well as megacurie radioisotope sources, are now in fabrication. All this has been partly instrumental in making increases possible in the research effort in this field.

There are problems relating to the use of these sources of radiation, however, just as there are problems relating to irradiated foods. For example, there is a continually recurring economic loss in the use of a radioisotope source because of its nuclear disintegration and the necessity for replenishment of the source.

Theoretically it would be most efficient if this nuclear energy could at all times be absorbed by foods requiring sterilization. Since many food plants are widely dispersed in location and their operation is often highly seasonal in nature, this is not possible. Plants that operate twenty-four hours daily throughout the year are in the minority, and many now work only one eight-hour shift per day, five days a week. Even from the standpoint of location alone, it would be difficult for such plants to tie in with projected future nuclear power stations, which are unlikely to be established in the most flourishing rural agricultural regions.

The use of particle accelerators presents problems as well. These relate to the tremendous amount of development necessary to produce machines of high output reliably delivered over extended periods of time during the harvest season.

Packages also present certain problems, because during storage a food material can be of no better quality than the functional properties of the package in which it is contained. Plastic containers have been suggested for use in the radiation sterilization of foods and drugs. Some of these containers require study to determine the effect of the radiations on them. It is known, for example, that certain plastics degrade on irradiation, whereas others crosslink. The question arises whether these changes affect the functional properties of the containers. In studies conducted in our laboratories on certain plastics, no such effects have been observed. Further study is necessary with other plastics.

Because, in general, the penetrability of radiations into foods is related to the product of thickness and density of the packaging material, it is obvious that a metal container as thin and of as low a density as possible is desirable. The metal containers designed for use in this field of research need not be so rigid as those used for thermal processing, because they need not withstand the heavy pressures of retorting. Hence savings in metal containers may result.

Today the accelerated research on food sterilization in the United States, alluded to above, has made it possible for many research investigators skilled in their several disciplines to become associated with those already active in the field of radiation sterilization and to participate in solving many of the important and difficult problems facing radiation sterilization. Over thirty research laboratories throughout the United States are now active in this field. Both food and pharmaceutical industrial laboratories are also playing an important role, and the outlook for more complete understanding and possible utilization of radiation sterilization is improving.

Applications of Radiation Sterilization

Certain limited applications appear today to be possible. For example, it seems possible that pharmaceuticals such as ointments, antibiotics, and other products of microbial metabolism that are heat labile and certain parenterally administered products such as steroidal compounds may be sterilized by ionizing radiations. Because pharmaceuticals or mixtures thereof are relatively simple compounds as compared with foods, they lend themselves more readily to tests for chemical changes induced by ionizing radiations. Also the changes produced in pharmaceuticals by irradiation are less in degree than those in foods. These facts, combined with the relatively greater cost of pharmaceuticals, the economic structure of drug distribution, and the smaller over-all mass of a drug per unit package cause us to predict again that the first industrial application of radiation sterilization may be with pharmaceuticals (Ref. 31).

This research has also resulted in another interesting and important application of ionizing radiations, namely, the sterilization of segments of human aortae for subsequent surgical transplant in persons suffering from coarctation of the aorta. Although this is a relatively small-volume application of ionizing radiations, it is one of great importance and one that has already been of great value in saving lives.

Another promising application of radiation sterilization is in the field of surgical sutures. Irradiation will do away with separate heat sterilization of suture and ampoule and aseptic handling of the suture through the filling stage. This will result in a better product with respect to tensile strength.

Ionizing radiations may also be used for the sterilization of the surface of certain foods, to increase shelf life. Such foodstuffs as comminuted meats in casings may have their shelf life greatly increased if cathode rays of low energy level are used to reduce their surface contamination.

The research of Doty and Wachter (Ref. 27) has pointed to the fivefold increase in shelf life of meats irradiated with low doses of gamma radiation (less than 100,000 rep).

Research at the Department of Food Technology of the Massachusetts Institute of Technology has shown that by the application of free-radical acceptors, the irradiation dose may be increased to 800,000 rep without detectable side reactions, and the shelf life of irradiated comminuted meats at 36° to 40° F may be increased from five to seven days for the control sample to three months for the irradiated product. Similar promising results in preservation have been obtained with other meat products.

It has been found (Ref. 28) that insects can be destroyed by relatively low doses of ionizing radiations (less than 50,000 rep). This finding points out one of the more promising applications of ionizing radiations and one that may be a means of saving many millions of dollars now lost annually because of destruction of cereals by insects. The relatively low dose of radiation required for insect eradication in grain products results in a comparably reduced cost of preservation.

The recent work of Sparrow and Christensen (Ref. 29) has shown that ionizing radiations successfully inhibit sprouting of potatoes at low doses (approximately 20,000 roentgens) and thus the storage life of these potatoes is increased by many months. Whether this method can compete with chemical means of inhibition is as yet unknown.

Summary

Summing up the activities of the decade now drawing to a close, one might say that the horizons have been scanned in respect to the many factors relating to a possible new method of food preservation using ionizing radiations and a method providing an extension of food storage life for perishable foods. These accomplishments may have a pertinent influence on the food processes of the future.

That food, pharmaceuticals, and tissues may be adequately sterilized by ionizing radiations is now an established fact, and the levels and types of radiation that may accomplish this sterilization are known.

For this method to be successfully applied to a wide variety of foods, extended research will be required into many specific and fundamental factors, largely chemical in character, which are severally and simultaneously affected by the extent and magnitude of radiation exposure. This is because the concomitant results of other changes, such as changes in flavor, color, and texture which sometimes occur during the sterilization period, may necessitate modification by some additional treatment.

On the assumption that the above difficulties may be alleviated as the result of sufficient application of research and subsequent process control, economic establishment of the available radiation sources is obviously necessary if these sources are to be put to optimum use. Because the processing of fruits, vegetables, and fish in the United States is highly seasonal and the processing plants are widely separated geographically, atomic radiation sources would be used only a relatively small portion of the year and thus would be unused and depreciate during the remainder of the year. The processing of meats and dairy products is not limited to the same extent, although the quantities of raw materials for these products may vary greatly during different parts of a given year.

If research is continued at present levels toward solution of the problems cited above, it would seem conservative to predict the initiation of commercial processes for irradiation sterilization of some pharmaceuticals, foods, tissues, and packaging materials in the present decade. To accomplish this will require the best type of scientific cooperation, diligence, and interchange of ideas.

References for Chapter 27

1. W. C. Roentgen, *Ann. Physik* 64:1 (1898).
2. H. Becquerel, *Compt. rend.* 122:501 (1895).
3. G. Pacinotti and V. Porcelli, cited by R. E. Buchanan and E. I. Fulmer, *Physiology and Biochemistry of Bacteria,* Williams & Wilkins, Baltimore, 1898, Vol. 2, p. 188.
4. S. C. Prescott, "The effect of radium rays on the colon bacillus, the diphtheria bacillus, and yeast," *Science,* N.S. 20:246–248 (1904).
5. J H. Hayner and B. E. Proctor, "Investigations relating to the possible use of atomic fission products for food sterilization," *Food Tech.* 7:6–10 (1953).
6. S. A. Goldblith, "Preservation of foods by ionizing radiations," *J. Am. Diet. Assoc.* 31:243–249 (1955).
7. B. E. Proctor and S. A. Goldblith, 'Food processing with ionizing radiations," *Food Tech.* 5:376–380 (1951).
8. B. E. Proctor and S. A. Goldblith, *A Critical Evaluation of the Literature Pertaining to the Application of Ionizing Radiations to the Food and Pharmaceutical Fields,* Tech. Rept. No. 1 on Contract AT(30–1)-1164 with U.S. Atomic Energy Commission, Jan. 1, 1952 (NYO 3337).
9. B. H. Morgan and J. M. Reed, "Resistance of bacterial spores to gamma irradiation," *Food Research* 19:357–366 (1954).
10. C. B. Denny, C. W. Bohrer, and J. M. Reed, *Investigation of Gamma Sterilization,* Research Report No. 3-54, Nat. Canners Assoc., Washington, D. C. Final report for July 1, 1953 through June 30, 1954 on Contract AT(30–1)-1567 with U.S. Atomic Energy Commission, Oct. 1954.
11. G. E. Stapleton and A. Hollaender, "Mechanism of lethal and mutagenic action of ionizing radiations on *Aspergillus terreus,*" *J. Cell. Comp. Physiol.* 39 (Suppl. 1):101–113 (1952).
12. S. A. Goldblith, B. E. Proctor, S. Davison, E. M. Oberle, C. J. Bates, B. Kan, O. A. Hammerle, and B. Kusmierek, "How processing conditions affect microorganism, radioresistance," *Nucleonics* 13(1):42–45 (1955).
13. S. A. Goldblith and B. E. Proctor, "Radiation sterilization: Review of status and problems of radiation preservation of foods and pharmaceuticals," *J. Agric. Food Chem.* 3:253–256 (1955).
14. B. E. Proctor and S. A. Goldblith, *Bactericidal Effects of Ionizing Radiations on E. coli, B. thermoacidurans, and Cl. sporogenes as Influenced by Different Conditions of Atmosphere, Medium and Physical State in Which Samples Are Irradiated.* Report on Contract AT(30–1)-1164, U.S. Atomic Energy Commission, June 30, 1954 (NYO 3344).
15. B. E. Proctor and S. A. Goldblith, *Bactericidal Effects of Ionizing Radiations as Influenced by Different Conditions of Atmosphere, Medium and Physical State in Which Samples Are Irradiated.* Report on Contract AT(30–1)-1164, U.S. Atomic Energy Commission, Jan. 1, 1954 (NYO 3342).
16. W. C. Miller, Jr., *Study of Environmental Factors Modifying Radiation Sensitivity of Bacillus subtilis spores,* S.M. thesis, Dept. of Food Technology, Massachusetts Institute of Technology, 1954.
17. L. L. Kempe and J. T. Graikoski, *Combined Effects of Heat and Radiation in Food Sterilization.* Paper A21, presented at 55th General Meeting, So. Am. Bact., New York, May 9, 1955.
18. H. C. Kung, E. L. Gaden, Jr., and C. G. King, "Vitamins and enzymes in milk: Effect of gamma-radiation on activity," *J. Agric. Food Chem.* 1:142 (1953).

19. D. S. Bhatia, *Chemical and Physical Effects of Supervoltage Cathode Rays on Amino Acids in Foods and in Aqueous Solutions,* Sc.D. thesis, Dept. of Food Technology, Massachusetts Institute of Technology, 1950.
20. S. A. Goldblith, B. E. Proctor, J. R. Hogness, and W. H. Langham, "The effect of cathode rays produced at 3000 kilovolts on niacin tagged with C^{14}," *J. Biol. Chem.* 179:1163–1167 (1949).
21. M. Corson, S. A. Goldblith, B. E. Proctor, J. R. Hogness, and W. H. Langham, "The effect of supervoltage cathode rays on *P*-aminobenzoic acid and anthranilic acid labeled with C^{14}," *Arch. Biochem. Biophys.* 33:263–269 (1951).
22. B. E. Proctor and S. A. Goldblith, "Effect of high-voltage X-rays and cathode rays on vitamins (niacin)," *Nucleonics* 3(2):32–43 (1948).
23. B. E. Proctor and D. S. Bhatia, "Effect of high-voltage cathode rays on amino acids in fish muscle," *Food Tech.* 4:357–361 (1950).
24. A. J. Lehman and E. P. Laug, "Evaluating the safety of radiation-sterilized foods," *Nucleonics* 12(1):52–54 (1954).
25. C. E. Poling, W. D. Warner, F. R. Humburg, E. F. Reberl, W. M. Urbain, and E. E. Rice, "Gross reproduction, survival, and histopathology of rats fed beef irradiated with electrons," *Food Research* 20:193–214 (1955).
26. T. D. Luckey, M. Wagner, J. A. Reyniers, and F. L. Foster, Jr., "Nutritional adequacy of a semi-synthetic diet sterilized by steam or by cathode rays," *Food Research* 20:180–185 (1955).
27. D. M. Doty and J. P. Wachter, "Influence of gamma radiation on proteolytic enzyme activity of beef muscle," *J. Agric. Food Chem.* 3:61–63 (1955).
28. B. E. Proctor, E. E. Lockhart, S. A. Goldblith, A. V. Grundy, G. E. Tripp, M. Karel, and R. C. Brogle, "The use of ionizing radiations in the eradication of insects in packaged military rations," *Food Tech.* 8:536–540 (1954).
29. A. H. Sparrow and E. Christensen, "Improved storage quality of potato tubers after exposure to Co^{60} gammas," *Nucleonics* 12(8):16–17 (1954).
30. D. A. Lang, *Effects of Ionizing Radiations on Lipids,* Ph.D. thesis, Dept. of Food Technology, Massachusetts Institute of Technology, 1953.
31. B. E. Proctor and S. A. Goldblith, "Electromagnetic radiation fundamentals and their applications in food technology," *Advances in Food Research* 3:188 (1951).

Chapter 28

RADIATION CONTROL OF TRICHINOSIS *

The control of trichinosis, a prevalent and typical food-borne helminthic disease, through the irradiation of the larvae encysted in meat, has been under study for several years. The effectiveness of X rays, gamma rays from cobalt-60, and gamma rays from mixed fission products have been evaluated and the results are summarized. In addition, the effectiveness of the newly available pure cesium-137 gamma rays for breaking the trichinosis cycle has been successfully proven. The results of the study demonstrate the use of ionizing radiation as a means of eradicating this particular food-borne helminthic disease and also suggest the use of radiation as a new approach to the world-wide problem of control of all food-borne helminthiases.

Some Important Food-borne Helminthic Diseases

Table 28.1 lists the principal helminthic diseases of man, the causative organisms of each, the form in which the parasite is ingested by man, and the incidence and geographic distribution of the parasite or disease. All the diseases listed are contracted as a result of eating raw or inadequately cooked foods. Thus, trichinosis, beef or pork tapeworm infections, pork cysticercosis, and hydatid disease are acquired as a result of eating raw or undercooked infected meats; fish tapeworm infection and clonorchiasis (Chinese liver fluke), through eating raw infected crabs or crayfish; fascioliasis (sheep liver fluke), fasciolopsiasis (intestinal fluke), and ascariasis (in some cases), through eating vegetables or raw fruit contaminated with infected soil.

In a number of countries of the Orient, the use of night soil as fertilizer is an age-old, widespread, and firmly established practice. This practice is responsible for contamination of fresh fruits and vegetables, leading to infestation with sheep liver fluke or intestinal fluke.

In each case thorough cooking could destroy the parasite and break the disease cycle. However, the natural food flavor and texture preferences and

* This chapter is taken from Geneva Conference Paper 225, "Prevention of Human Helminthic Diseases by Gamma Radiation of Food, with Particular Reference to Trichinosis" by H. J. Gomberg and S. E. Gould of the United States.

the established eating habits of the population groups involved has made this approach ineffective. It may, however, be possible to break these disease cycles by irradiating the food with doses which, while preventing parasite growth and reproduction, do not affect the flavor and texture of the foods concerned.

This principle of breaking a parasitic disease cycle by inhibiting continued reproduction has been rather completely explored in the case of *Trichinella spiralis,* the causative agent of trichinosis.

The Trichinosis Cycle

The cycle of trichinosis (Ref. 1) takes approximately the following course. On ingestion of meat containing encysted trichina larvae (*Trichinella spiralis*) the muscle fibers and cyst walls that enclose the parasites are digested, liberating the larvae. In the small intestine, the larvae mature in two to four days (Ref. 2) to adult worms and then copulate. Beginning with the sixth day (Ref. 2) gravid females partially embedded in the mucosa of the intestine give birth to young larvae of the second generation. The larvae enter the blood stream and are carried into the general arterial circulation. The young larvae leave the capillaries and penetrate skeletal muscle fibers and encyst there. At this point, the course in man ends, although the encysted parasites may remain alive in a latent state for the rest of the life of the infected person. In hogs, however, the parasitic cycle is continued when inadequately treated meat of a trichinous pig is eaten by another pig. A secondary avenue lies through consumption of infected meat by rats, which then become infected and in turn are eaten by pigs.

Methods of Control

Man may be directly protected by killing the trichina larvae in the meat by adequate cooking, by suitable refrigeration, or by ionizing radiation. Adequate cooking which raises the temperature of all parts of the meat to 137° F will kill the larvae, but is a precaution that must be taken by the consumer and is not suited for public health control measures. Holding the meat at 5° F (temperature of meat, not of refrigerator) for 20 days or sharp chilling to −35° F for a few minutes will also kill the trichinae (Ref. 3), but this method of control is expensive and results in changes in the meat texture.

Irradiation of Meat for the Control of Trichinosis

From our work to date on breaking the trichinosis cycle through the use of radiation, the effects produced by irradiation of trichinous meat may be summarized as follows.

TABLE 28.1. SOME IMPORTANT HELMINTHIC DISEASES OF MAN ACQUIRED THROUGH FOOD

Causative Organism	Form in Which Ingested by Man	Disease Produced in Man	Natural Host	Incidence and Geographic Distribution
Trichinella spiralis	Cysts in raw or undercooked pork.	Trichinosis.	Hog. (Rat, man, carnivorous animals.)	Cosmopolitan. In 16% human autopsies in U.S. (p. 370).[a] In U.S. 1.5% of all hogs are infected (p. 370).[a] 27.8 million people.[b]
Clonorchis sinensis	Metacercariae in fresh-water fish.	Clonorchiasis (Chinese liver fluke).	Man, dog, cat, hog. (Snails and fish are intermediate hosts.)	Japan, China, Formosa, Korea, French Indochina. "Disease is equally prevalent in poor and the prosperous, in illiterate as well as the educated," because of custom of eating raw fish (p. 650).[a] 19 million people.[b]
Ascaris lumbricoides	Ova in raw vegetables. (Principal mode of transmission is infected soil.)	Ascariasis. (Intestinal obstruction in children, pneumonia.)	Hog, man.	Cosmopolitan. In surveys of U.S. Southern States, 1% to 60% in children (p. 447).[a] In Shantung, China, 23% to 44% in children. 644.4 million people.[b]
Fasciolopsis buski	Metacercariae on raw fruit or edible water plants.	Fasciolopsiasis (intestinal fluke).	Man, hog. (Snail is intermediate host.)	Central and South China. 10 million people.[b]
Taenia saginata	Cysticerci in meat.	Beef tapeworm.	Man. (Cattle are intermediate hosts.)	Cosmopolitan. 0.38% of persons infected in Peiping, China. Incidence in cattle: U.S. 0.37%, Holland 3.3% to 5.6%, S. Africa 5% (p. 578).[a] 38.9 million people.[b]

Taenia solium	Cysticerci in pork.	Pork tapeworm; cysticercosis in tissues.	Man. (Hog is intermediate host.)	Cosmopolitan. Incidence of tapeworm in man: Russia 0.2% to 1.5%; incidence of cysticercosis in man: 2.9% Mexico (p. 575).[a] Incidence in hogs: S. Africa 25%, Mexico 4.3%. 2.5 million people.[b]
Diphyllobothrium latum	Plerocercoid larva in fresh-water fish.	Fish tapeworm anemia resembling pernicious anemia.	Man, dog, cat. (Fish is intermediate host.)	N. America, Europe, Tropical Africa, N. Asia, S. Asia. In Finnish Army 14.5% (p. 548).[a] About 20% of pickerel and pike in Lake Winnipeg and Lake Manitoba are infected (p. 549).[a] 10.4 million people.[b]
Paragonimus westermani	Crabs or crayfish containing metacercariae.	Paragonomiasis (Oriental lung fluke).	Man, dog, cat, many other animals. (Snail and crab or crayfish are intermediate hosts.)	Japan, Korea, Manchuria, Formosa, China, other countries of Orient; Africa, S. America. Adult parasites have been found in mammals in Canada and many states of U.S. 3.2 million people.[b]
Fasciola hepatica	Encysted cercaria in water cress, lettuce.	Fascioliasis (sheep liver fluke.)	Man, sheep. (Snails are intermediate hosts.)	Cosmopolitan, in sheep-raising areas of world. Incidence in Estonia 57% in sheep, 72% in cattle; in Transcaucasia, 48% in sheep; in Southern Texas, 21% to 32% of rabbits, reservoir hosts for new infections (p. 629).[a] 100,000 people.[b]

[a] Page numbers have reference to D. L. Belding, *Textbook of Clinical Parasitology*, 2nd ed., New York, Appleton-Century-Croft.
[b] N. R. Stoll, "This wormy world," *J. Parasitol.* 33:15 (1947).

1. Killing the trichina larvae in the meat. This can be done by exposing the meat containing the trichinae to about 1,000,000 rep (Refs. 4, 5, 6). This value applies for 250-kv X rays, and for cobalt-60 gamma radiation of about 1.3 Mev energy. However, a radiation dose this large produces highly objectionable flavor changes, making the pork rancid and sour (Ref. 7).

2. Inhibiting maturation of the trichina larvae. The primary danger from trichinosis occurs as a result of multiplication of trichinae in the intestinal tract, about 1500 young being produced for every female adult worm. Any measure which prevents reproduction can be effective in trichinosis control. About 18,000 rep of cobalt-60 gamma radiation (Refs. 8, 6) or 15,000 rep of 250-kv X rays (Ref. 5) will inhibit the development of the larvae that are liberated from the meat by digestion. These dose levels produce no detectable change in flavor, texture, or appearance of the pork (Ref. 8).

3. Sexual sterilization of the trichinae. Adult female trichinae, developing from larvae which were exposed to 15,000 rep of cobalt-60 gammas (Refs. 8, 6) or 5000 rep of 250-kv X rays (Ref. 5) are completely sterile. The complete degeneration of the reproductive system is apparent in female worms developed from irradiated encysted larvae.

While the evidence is not yet conclusive, it appears that the males have likewise been rendered sterile by these doses of radiation. This sterilizing effect is significant, since for purposes of trichinosis control, it can serve as a second line of protection in processing intended to inhibit maturation of the larvae.

To judge by these results, a dose of 30,000 rep of cobalt-60 gammas, or its equivalent, delivered to pork will serve as a method of preventing and eradicating trichinosis.

Secondary Considerations in Controlling Trichinosis by Meat Irradiation

1. Enteric phase of trichinosis. While the muscular phase of trichinosis is usually the most serious, in the case of infection with large numbers of larvae, a violent enteric reaction may be produced, sometimes with fatal results. In tests, 12,000 larvae fed to rats weighing between 200 and 250 gm killed 12 out of 12 animals within 24 days (Ref. 9). A similar group fed 12,000 larvae which received 10,000 rep survived the test period of 31 days, although 6 developed a transient diarrhea. In a third group of 10, fed 12,000 larvae which received 18,000 rep, all survived with no diarrhea and continued to show weight increase during the test period. A fourth group, fed 24,000 larvae which received 18,000 rep, fared as well as the third group.

It is expected that consumption of pork containing larvae exposed to 30,000 rep will not result in any intestinal disturbance.

2. Immunity. A small dose of trichina larvae, producing a mild infection, can also produce immunity to further reinfection (Ref. 9). It has been found that experimental feeding of larvae exposed to 10,000 rep of cobalt-60 gamma rays will produce a marked immunity, although not one as effective as that produced by unirradiated larvae. An exposure to 18,000 rep destroys the ability of larvae to induce immunity.

3. Selection of a radiation-resistant strain of trichina larvae. When doses approaching but not equal to the total sterilizing dose of radiation are applied, the progeny of the unaffected trichinae appear to breed normally. These were tested (Ref. 9) to determine if the young produced by trichinae which were not sterilized were in any sense more resistant to radiation than others not so selected. There is no indication as yet that the young produced by the survivors of the sterilization process differ in radiation resistance from the controls.

4. Induced radioactivity. There is no induced radioactivity in food treated with cobalt-60 gamma rays or radiation of lower energy. Thresholds for photo-neutron reactions which would result in induced radioactivity occur at much higher energies (Ref. 10).

Among the secondary effects examined, there is no indication of deleterious side effects which would contraindicate on medical or biological grounds the use of radiation as a means of trichinosis control.

Effectiveness of Cesium-137 Gamma Rays for Irradiation of Pork

While the tests run to determine the effectiveness of cobalt-60 and 250-kv X rays in breaking the trichinosis cycle yielded positive results, it was felt that definitive tests using cesium-137, the principal long-lived gamma-emitting fission product, as the radiation source were necessary. Arrangements were made with Dr. Marshall Brucer of the Oak Ridge Institute for Nuclear Studies for the use of the cesium therapy unit being installed at the medical division of the Institute. It was possible to use the unit before its installation for human therapy use was quite completed; and so the first irradiation tests using a pure cesium-137 radiation source were made at Oak Ridge on March 25, 1955.

Cesium-137 irradiation technique and dosimetry. A ham weighing about 30 lb after being trimmed of the fat, which came from a hog which had been fed 10,000 trichina larvae about 21 months before sacrifice, was transported to Oak Ridge by air. The ham formed a roughly rectangular meat slab which, when irradiated, was about 11½ in. tall, 15 in. long, and 6 in. wide. It was centered under the cesium irradiator, which was adjusted so that the radiation emerged vertically downward. The top surface of the meat was 27¾ in. below the cesium source within the irradiator housing. Dosimeters consisting of small glass vials containing ferrous sulfate were inserted in the meat at intervals, starting ½ in. below the top surface and working down. The dosimeters were placed chiefly in the cone of direct radiation, although a few were placed in peripheral positions. The relative position of all elements is shown in Fig. 28.1. The radiation cone had previously been monitored with a

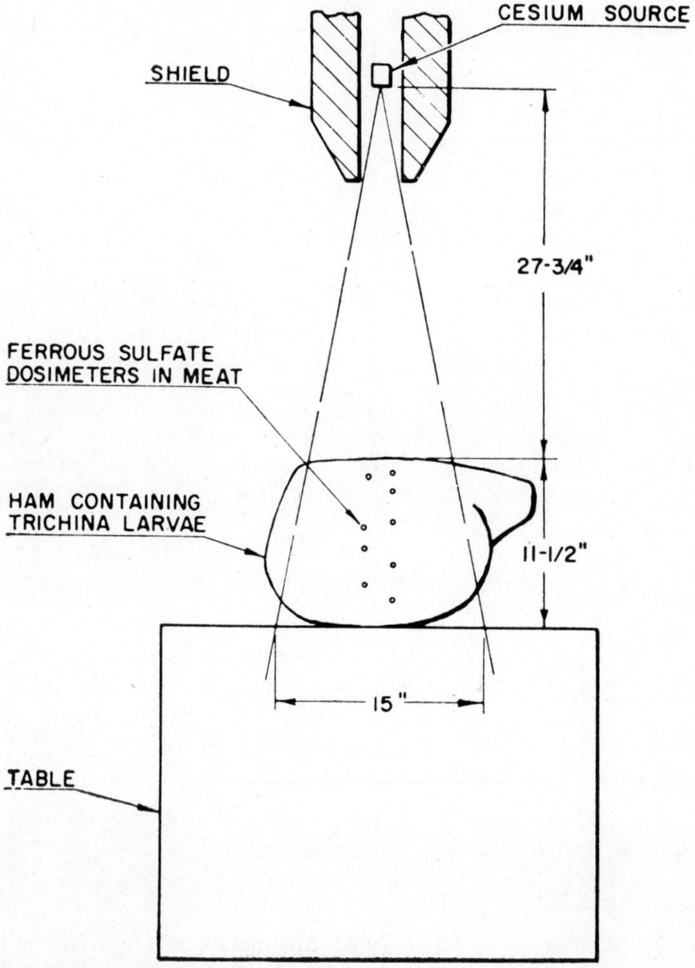

FIG. 28.1. Arrangement of Cs^{137} source and meat during irradiation.

Victoreen roentgen rate meter and the relative position of the irradiator and meat chosen so that the center of the top surface received 11.5 r per min.

Irradiation continued for 23 hr and 40 min. The radiation source was then closed and the meat and dosimeters returned to Ann Arbor. Radiation dose was measured, using a Beckman DU spectrophotometer, after return to Ann Arbor. Checking control dosimeters which had been left behind with control dosimeters which had been carried along, no difference was found. The indicated doses were then accepted as correct. The location of the dosimeter in the meat and the corresponding dose are shown in Fig. 28.2. The variation in dose with depth in the meat has been separated into that caused by geometry (inverse square law) and that caused by geometry plus absorption. The half-value layer for cesium-137 radiation in meat as found from these data is 6¾ in.

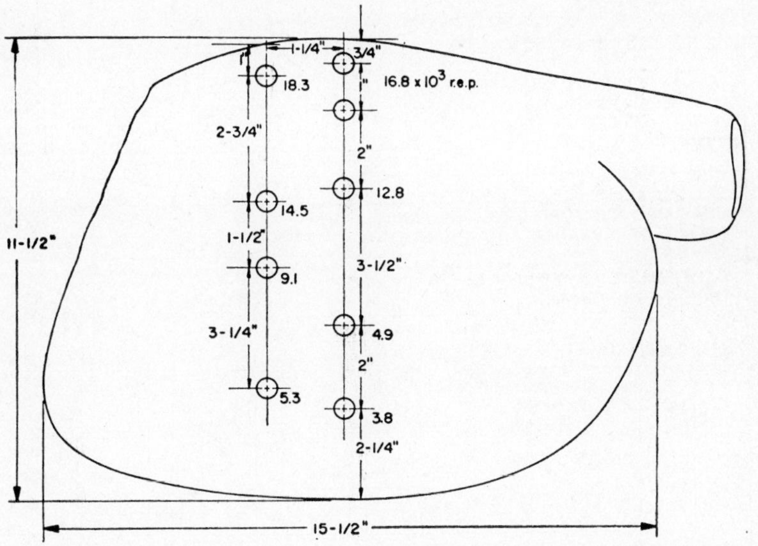

Fig. 28.2. Location of ferrous sulfate dosimeters and corresponding dose recorded. Readings indicate center line of meat and radiation beam did not coincide.

As in previous experiments, using cobalt-60 gamma rays (Ref. 8) small blocks of meat were cut from sites surrounding the dosimeters. It was assumed that on the average all the trichina larvae in the meat in a single small block had received the same dose. The larvae in each sample block were recovered by digestion of the meat and then 5000 larvae fed by stomach tube to each of a number of test rats. In addition, control specimens of meat taken from unirradiated portions of the same hog carcass were processed and the recovered larvae fed to rats intended to serve as controls on the infectivity of the larvae.

Cesium irradiation results. The results have been evaluated in two ways: (1) the ability of the radiation to sexually sterilize the female larvae, and (2) the dose needed to prevent the larvae from remaining in the intestine of the test animal in the presence of normal or violent intestinal clearance. The results are summarized in Figs. 28.4, 28.5, and 28.6.

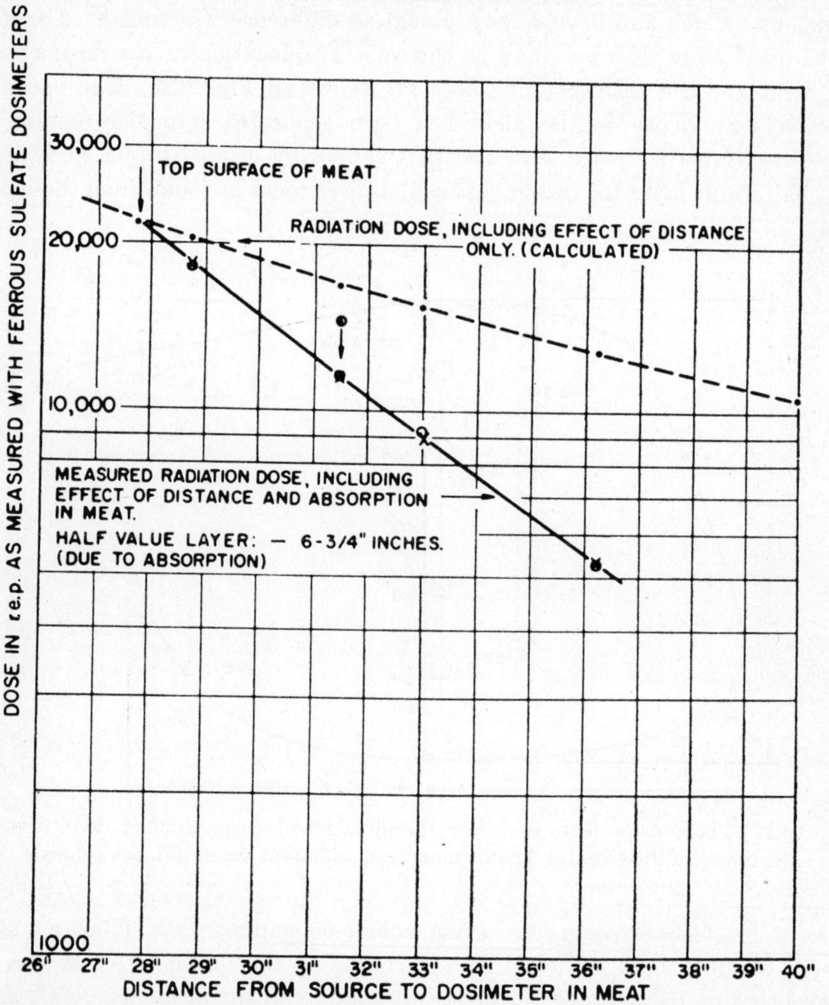

FIG. 28.3. Dose of Cs^{137} gamma radiation at varying depth in irradiated meat as measured with ferrous sulfate dosimeters and calculated dose in air at same points for orins Cs^{137} irradiator.

(1) In Fig. 28.4, determination was made of the percentage of adult female trichinae with young, recovered from the gut of test rats 6 days after feeding irradiated larvae. This curve is quite similar to those found previously for cobalt-60 and waste fission product irradiation (Ref. 8). The points are scattered somewhat more than previously, but the trend and cut-off point are clear. About 14,000 rep from cesium-137 apparently sterilizes more than 99% of the female trichinae larvae.

(2) A more significant test is the measure of the number of second-gen-

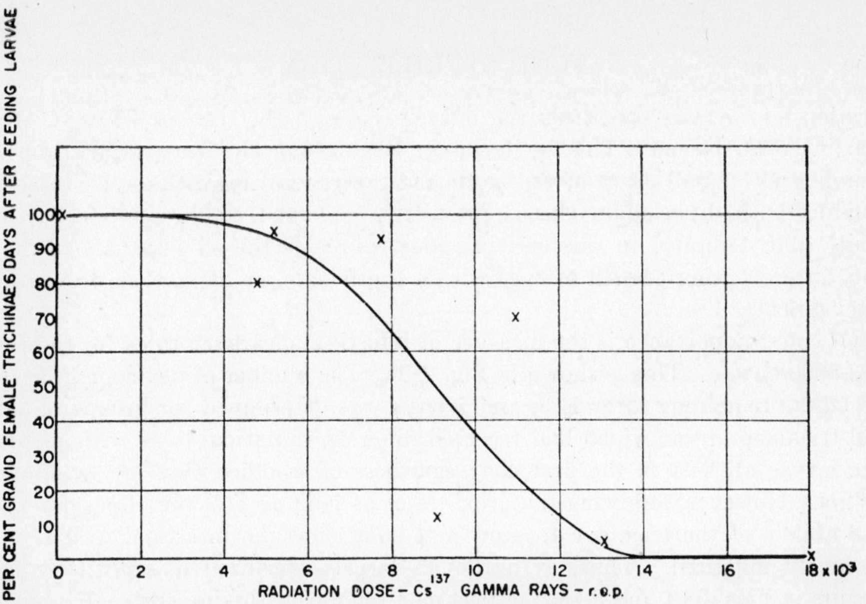

FIG. 28.4. Sexual sterilization of trichinae by irradiation of the immature larvae, as measured by examination for maturing eggs in 100 adult female trichinae recovered from intestine of test rat 6 days after feeding irradiating larvae.

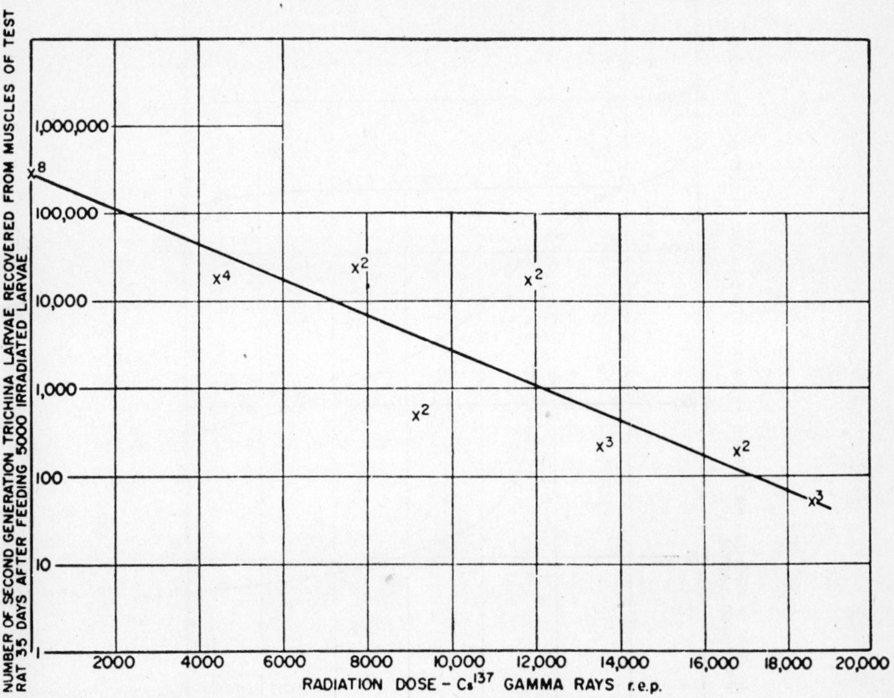

FIG. 28.5. Sexual sterilization of trichinae by irradiation of the immature larvae, as measured by reduction in number of encysted second generation larvae found 35 days after feeding 5000 irradiated larvae. Numbers at test points show number of rats used to obtain larvae count.

eration larvae recovered from the muscle tissue of the test rat 35 days after feeding 5000 larvae. This is shown in Fig. 28.5. The very rapid drop in number with small doses of radiation and the general logarithmic decrease in number with dose, again checks previously observed results from cobalt-60 quite well. Implied in this and previous results is the effectiveness of "one hit" in preventing growth of eggs within the female, or at least in preventing encystment.

Of equal importance is the measure of intestinal clearance rates for the rats fed 5000 larvae. This is shown in Fig. 28.6. The number of developing larvae or adults remaining after 1, 3, and 7 days was determined for groups of rats fed trichinae larvae which had received different radiation doses. About half the larvae are lost in the first day regardless of whether they are irradiated or not. However, following radiation doses as high as 11,900 r of cesium-137, the ability of the trichinae to grow and hang onto the intestinal wall is not seriously impaired. Thus, even though largely sterilized as shown by the results in Figs. 28.4 and 28.5, the trichinae remain in the intestine. However,

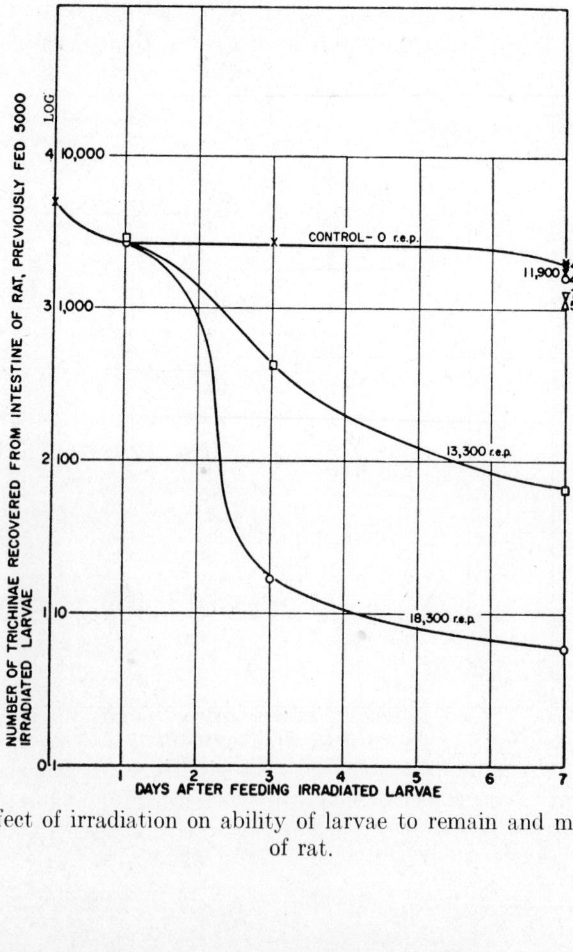

FIG. 28.6. Effect of irradiation on ability of larvae to remain and mature in intestine of rat.

beyond about 12,000 r the ability to grow or at least to remain attached to the intestinal wall is rapidly lost. At 18,300 r, practically all the larvae have been cleared out (or digested) in 7 days. These results correlate directly with the observed disappearance of the enteric phases of the disease when larvae receiving 18,000 r cobalt-60 radiation are fed even in massive numbers (Ref. 9).

It would appear that except for a reduction in half-value layer from about 8 in. to 6¾ in. in meat, the cesium-137 gamma rays are equally as effective as cobalt-60 in breaking the cycle of trichinosis.

Techniques of Irradiation

Based on the trichinosis results, the general use of irradiation as a method of parasitic disease control appears worthy of consideration. While the dose to be used to inhibit development of other parasites remains to be determined, the general problems of mass food-handling applicable to pork will be characteristic of most other foods. The effectiveness of food irradiation for large-scale use anywhere in the world will be governed by the technical feasibility of setting up the necessary facilities, not only in large technically advanced communities, but also in less well developed areas, even those without basic utilities such as electricity and running water.

Two basic sources of high-energy radiation are available: (1) electric converters such as X-ray machines and particle accelerators, and (2) radioactive materials, particularly by-products of nuclear reactors such as cesium-137. Small nuclear reactors could also be used, but for the application proposed here, they do not appear sufficiently simple or inexpensive, and also because they involve fissionable material, their use may not be politically feasible at this time.

In the case of trichinosis, the simplest mode of treatment would be the irradiation of whole animal carcasses. The thorough penetration of all parts requires the use of fairly high-energy gamma rays. Radiations from cobalt-60 and from cesium-137 have been demonstrated as capable of providing adequately uniform penetration of all sections of the split carcass of a 450-lb hog (Ref. 8). Using a 1- to 3-million-volt X-ray machine, similar results would be obtained.

However, the technical complications of operation of such a system would be great. One great advantage of the radioactive sources is that they require no external power supply. However, they cannot be turned off and so must be adequately shielded at all times.

The safe operation of a food irradiation unit energized by a radioactive source can be assured by indoctrination of semiskilled personnel in safety procedures and simple methods of radiation monitoring. No complex ultra-high-voltage power system, radio frequency generator, or vacuum systems are involved.

A facility for processing pork carcasses or other foods for parasite control can be built from simple material. Given the radiation source, the rest of the facility, such as the conveyor belt, source storage pit, handling system and shielding, could be built and improvised from materials available almost anywhere. Central facilities of a permanent nature in urban and population centers could serve large groups and have high handling capacities. Facilities to handle 2000 hog carcasses per day for operation in conjunction with a modern American style packing plant have been studied and designed (Ref. 11).

Smaller units with lower capacities, and even portable or mobile plants which could serve a number of communities on a rotating basis when slaughtering is in progress, are technically feasible, although no detailed design studies have been made. Where fish or fresh produce are to be irradiated, simple conveyor-belt systems could carry the food past radiation sources. A recent study of a potato irradiation plant showed that bulk food items could be irradiated with 10,000 rep from cesium-137 at the rate of 270 bushels per hr, using a 150,000-curie source made up of 20 individual rods (Ref. 12).

Sufficient radioactive material for many radiation sources is already in existence in the waste storage pits of chemical processing plants of the United States Atomic Energy Commission. Similar supplies exist in other parts of the world wherever large-scale reactor operation is in progress. And as the use of atomic energy for electric power generation expands, suitable radioactive supplies will be even more abundant.

Costs of Plant Construction and Operation

In a detailed cost study of a plant designed to process 2000 hog carcasses per day and allowing costs for the radioactive source which are believed to reflect the developing technology of chemical processing, it was found that the meat could be treated at 30,000 rep for less than ¼ cent per lb (Ref. 11).

Processing the meat (or other food product) in a more compact package than the split and spread carcass could lead to further reduction of cost. On the other hand, increased costs of the radioactive sources could raise the price of processing. The potato irradiation plant (Ref. 12) mentioned previously would operate at a cost of $0.066 per bushel or about $3.91 per ton of bulk food, provided it operated 16 hr a day, 260 days a year. Smaller plants may in general be expected to operate at somewhat higher costs than the reference design unit. The construction and test operation of a pilot plant could contribute substantially to our information in this and other areas.

Conclusions and Recommendations

The technical feasibility of breaking the trichinosis cycle by irradiation of pork has been established. No undesirable secondary effects from this method of processing have been discovered. Present estimates, based on admittedly meager data on costs of radioactive sources, indicate that the cost of such processing can be a tiny fraction of the value of the product.

In addition, the principle of using moderate amounts of radiation to break the life cycle of parasites by treating food or other materials they infest appears to be well established. Additional applications, particularly where uncooked food such as fish is involved, are worth consideration. If the potential for growth of the parasite when lodged in the intermediate host is destroyed, the disease can be eliminated.

Radiation doses of the size previously shown to be capable of inhibiting development of the parasite (*Trichinella spiralis*) do not affect the taste or texture of the food so treated. Thus, the normal eating habits and dietary preferences of population groups need be altered to only a minor degree.

Further work along the following lines is either under way in our laboratories or is recommended for consideration by the cognizant authorities:

1. Determination of the sensitivity to radiation of the parasites carried by food to man or his essential livestock.
2. Establishment of the technical feasibility of carrying out the processing where it is needed.
3. Possible erection of pilot-plant facilities in the world areas where need for control of food-borne parasitic disease is greatest. The implementation of this recommendation might best come through consultation with the World Health Organization and other cognizant United Nations groups. It is planned to investigate this aspect of the problem in more detail.

References for Chapter 28

This is not intended to be an exhaustive reading list. It is to serve merely as a guide to the literature from which the results quoted have been taken.

1. S. E. Gould, H. J. Gomberg, and F. H. Bethell, "Control of trichinosis by gamma irradiation of pork," *J. Amer. Med. Assoc.* 154:653–658 (1954).
2. S. E. Gould, H. J. Gomberg, F. H. Bethell, J. B. Villela, and C. S. Hertz, "Studies on *Trichinella spiralis*, Parts I and II," *Amer. J. Pathol.*
3. D. L. Augustine, "Low-temperature treatment of pork," *Proc. First National Conference on Trichinosis,* Public Health Reports, 68:419, 1953.
4. J. E. Alicata, "Effects of roentgen radiation on *Trichinella spiralis*," *J. Parasitol.* 37(5):491–501 (1951).
5. S. E. Gould, J. G. Van Dyke, and H. J. Gomberg, "Effects of X-rays on trichina larvae," *Amer. J. Pathol.* 29:323–337 (1953).
6. H. J. Gomberg and S. E. Gould, "Effect of irradiation with cobalt-60 on trichina larvae," *Science* 118:75–77 (1953).
7. L. E. Brownell *et al., Utilization of Gross Fission Products,* Progress Report #3, COO-91. U.S. Atomic Energy Commission, 1952.
8. H. J. Gomberg, S. E. Gould, J. V. Nehemias, and L. E. Brownell, "Using cobalt-60

and fission products in pork irradiation experiments, *Nucleonics* 12(5):38–42 (1954).

9. S. E. Gould, H. J. Gomberg, F. H. Bethell, J. B. Villela, and C. S. Hertz, "Studies on *Trichinella spiralis*, Parts III, IV, and V," *Amer. J. Pathol.* 31:933–963, 1955.
10. W. W. Meinke, "Does irradiation induce radioactivity in food?" *Nucleonics* 12(10): 37–39 (1954).
11. H. J. Gomberg, S. E. Gould, J. V. Nehemias, and L. E. Brownell, "Design of a pork irradiation facility using gamma rays to break the trichinosis cycle," *Chem. Eng. Progress Symp.*, Series 13, 50:89–104 (1954); "Design fission irradiator to break trichinosis cycle," *Food Engineering* 78–80 (1954).
12. L. E. Brownell *et al.*, *Utilization of Gross Fission Products*, Progress Report #7, U.S. Atomic Energy Commission Contract AT 11-1-162, University of Michigan, Ann Arbor, 1955.

GENEVA CONFERENCE PAPERS RELEVANT TO NUCLEAR RADIATION IN FOOD AND AGRICULTURE

Paper Number	*Volume and Page* *	*Author and Title*	*Country*
10	12:75	J. W. T. Spinks. Studies of special problems in agriculture and silviculture by the use of radioisotopes.	Canada
82	11:71	A. M. Brues. Commentary on the modes of radiation injury.	U.S.A.
83	11:256	T. N. Tahmisian. Studies on the biological basis of radiosensitivity.	U.S.A.
87	11:260	G. D. Adams *et al.* Relative biological effectiveness.	U.S.A.
89	13:132	R. S. Stone. Maximum permissible exposure standards.	U.S.A.
90	12:245	C. L. Comar. Radioisotopes in animal physiology and nutrition-mineral metabolism.	U.S.A.
91	12:313	C. Adrian and M. Hogben. The mechanism of gastric acid secretion as revealed by radioisotopes.	U.S.A.
93	12:292	M. Kleiber *et al.* Isotopes in research on animal nutrition and metabolism.	U.S.A.
94	11:273	R. E. Zirkle *et al.* Use of partial-cell irradiation in studies of cell division.	U.S.A.
95	12:306	E. P. Reineke and H. A. Henneman. Use of radioactive iodine (I^{131}) and thyroxine to determine the thyroid hormone secretion rate of intact animals.	U.S.A.
96	12:281	B. M. Tolbert *et al.* Respiratory carbon-14 patterns and physiological state.	U.S.A.
97	11:283	S. A. Gordon. Studies on the mechanism of phytohormone damage by ionizing radiation.	U.S.A.
100	12:170	L. P. Miller and S. E. A. McCallan. Use of radioisotopes in tracing fungicidal action.	U.S.A.
101	12:40	R. S. Caldecott. Ionizing radiations as a tool for plant breeders.	U.S.A.
103	12:60	W. M. Myers *et al.* Resistance to rust induced by ionizing radiations in wheat and oats.	U.S.A.
104	12:89	L. A. Dean. Applications of radioisotopes to the study of soils and fertilizers: A review.	U.S.A.

* Volume and page number refer to the 16-volume series, *Peaceful Uses of Atomic Energy: Proceedings of the International Conference in Geneva, August, 1955*, published by Columbia University Press, New York, 1956.

Paper Number	*Volume and Page*	*Author and Title*	*Country*
105	12:144	J. E. Kuntz and A. J. Riker. The use of radioactive isotopes to ascertain the role of root-grafting in the translocation of water, nutrients, and disease-inducing organisms among forest trees.	U.S.A.
106	12:138	H. B. Tukey *et al.* Utilization of radioactive isotopes in resolving the effectiveness of foliar absorption of plant nutrients.	U.S.A.
107	12:48	W. C. Gregory. The comparative effect of radiation and hybridization in plant breeding.	U.S.A.
108	12:203	H. C. Harris. Radioactive sulfur absorbed by the peanut fruit in comparison to the root.	U.S.A.
109	12:159	I. Stewart and C. D. Leonard. Use of isotopes for determining the availability of chelated metals to growing plants.	U.S.A.
110	12:25	W. R. Singleton *et al.* The contribution of radiation genetics to crop improvement.	U.S.A.
111	12:157	A. S. Crafts. Use of labeled compounds in weed research	U.S.A.
112	12:98	E. Epstein and S. B. Hendricks. Uptake and transport of mineral nutrients in plant roots.	U.S.A.
113	12:223	J. A. Vomocil. *In situ* measurement of soil bulk density.	U.S.A.
114	12:216	R. C. Bushland *et al.* Eradication of the screw-worm fly by releasing gamma-ray–sterilized males among the natural population.	U.S.A.
115	12:117	G. Burr *et al.* Uses of radioisotopes by the Hawaiian sugar plantations.	U.S.A.
172	15:245	B. E. Proctor and S. A. Goldblith. Progress and problems in the development of cold sterilization of food.	U.S.A.
173	15:265	D. M. Doty *et al.* Basic studies relating to the use of ionizing radiations for mean processing.	U.S.A.
174	15:269	R. G. H. Siu *et al.* Research in the United States on the radiation sterilization of foods.	U.S.A.
175	15:258	L. E. Brownell and J. J. Bulmer. Sterilization of medical supplies with gamma radiation.	U.S.A.
178	10:417	A. Baird Hastings. The use of isotopes in biochemical and medical research.	U.S.A.
225	15:251	H. J. Gomberg and S. E. Gould. Health protection against food-borne parasitic diseases with particular reference to control of trichinosis.	U.S.A.

Paper Number	*Volume and Page*	*Author and Title*	*Country*
234	11:400	H. J. Muller. How radiation changes the genetic constitution.	U.S.A.
235	11:382	W. L. Russell. Genetic effects of radiation in mice and their bearing on the estimation of human hazards.	U.S.A.
236	11:188	J. W. Gowen. Genetic differences in longevity and disease resistance in animals under irradiation.	U.S.A.
238	11:377	B. Wallace. The genetic structure of Mendelian populations and its bearing on radiation problems.	U.S.A.
239	11:266	E. L. Powers. Radiation effects on cells: Genetics or Physiology?	U.S.A.
256	11:105	J. W. Gowen. Effects of whole body exposure to nuclear energy on life span and efficiency throughout life.	U.S.A.
257	11:184	G. W. Casarett and J. B. Hursh. Effects of daily low doses of X-rays on spermatogenesis in dogs.	U.S.A.
259	12:347	M. Calvin and J. A. Bassham. The photosynthetic cycle.	U.S.A.
266	12:52	A. H. Sparrow and J. E. Gunckel. The effects on plants of chronic exposure to gamma radiation from radiocobalt.	U.S.A.
269	12:334	J. M. Siegel. The application of carbon-14 to studies on bacterial photosynthesis.	U.S.A.
271	12:286	A. Dorfman and S. Schiller. The metabolism of mucopolysaccharides.	U.S.A.
274	12:377	N. J. Scully *et al.* The biosynthesis of C^{14}-labeled plants and their use in agricultural and biological research.	U.S.A.
275	12:266	C. Blincoe and S. Brody. Use of I^{131} in the study of the influence of climatic factors on thyroid activity and productivity of livestock.	U.S.A.
314	14:68	A. F. Rupp. Large-scale production of radioisotopes.	U.S.A.
315	14:128	A. F. Rupp. Methods of handling multikilocurie quantities of radioactive materials.	U.S.A.
371	11:244	H. Marcovich. The problem of the biological action of low doses of ionizing radiation.	France
373	12:324	J. P. Aubert and G. Milhaud. C^{14}-labelled glucose and ethanol metabolism in baker's yeast.	France
379	12:311	F. Morel and A. Combrisson. Kinetics of the distribution of radio-sodium in rabbits during experimental hypothermia.	France

Paper Number	*Volume and Page*	*Author and Title*	*Country*
380	12:221	G. Barbier and P. Quillon. A study by isotopic dilution of the solution of slightly soluble phosphates in the presence of an anion exchanger.	France
381	12:211	R. Ortavant. Study of the spermatogenesis in domestic animals using P^{32} as a tracer.	France
382	12:200	Y. Demarly. Some notes on the pollination of alfalfa.	France
449	11:384	T. C. Carter. The genetic problem of irradiated human populations.	United Kingdom
459	12:364	H. K. Porter and J. Edelman. Some aspects of sucrose metabolism in plants.	United Kingdom
460	12:103	R. Scott Russell *et al.* Factors affecting the availability to plants of soil phosphates.	United Kingdom
618	12:3	A. L. Kursanov. The utilization of radioactive isotopes in biology and agriculture in the USSR.	U.S.S.R.
688	12:257	K. S. Zamychkina and D. E. Grodzensky. The role of radioactive isotopes in investigating the physiology and biochemistry of digestion.	U.S.S.R.
694	12:109	V. M. Klechkovski. The use of tracer atoms in studying the application of fertilizers.	U.S.S.R.
695	12:118	A. V. Sokolov. Determination of assimilation of soil phosphates and fertilizers by means of radioactive isotopes of phosphorus.	U.S.S.R.
696	12:165	A. L. Kursanov. Analysis of the movement of substances in plants by means of radioactive isotopes.	U.S.S.R.
697	12:340	A. A. Nichiporovich. Tracer atoms used to study the products of photosynthesis as depending on the conditions in which the process takes place.	U.S.S.R.
698	12:130	I. N. Antipov-Karatayev. Application of the isotope method to the study of absorption of electrolytes by soils in connection with land improvement.	U.S.S.R.
699	12:149	A. M. Kuzin. Utilization of ionizing radiations in agriculture.	U.S.S.R.
700	12:368	S. I. Kuznetsov. Application of radioactive isotopes to the study of processes of photosynthesis and chemosynthesis in lakes.	U.S.S.R.
701	12:185	K. A. Gar and R. Y. Kipiani. Research by means of radioactive isotopes concerning penetration into and residues of phosphoorganic insecticides in plants.	U.S.S.R.
715	12:358	T. N. Godnev and A. A. Shlik. C^{14} in the study of the biosynthesis of chlorophyll.	Byelorussian S.S.R.

Paper Number	*Volume and Page*	*Author and Title*	*Country*
716	12:123	O. K. Kedrov-Zikhman. Co^{60} and its part in studying the role of cobalt as a microelement in plant nutrition.	Byelorussian S.S.R.
780	12:10	R. A. Silow and M. E. Jefferson. The peaceful uses of atomic energy in food and agriculture.	U.N.
793	12:31	L. Ehrenberg *et al.* The production of beneficial new hereditary traits by means of ionizing radiation.	Sweden
874	12:278	A. S. Paintal. A method of recording the circulation times in the cat.	India
889	12:208	N. Mikaelsen *et al.* Effects of gamma-rays on growth and sprouting in carrots and potatoes during storage.	Norway
890	12:34	K. Mikaelsen. Genetic effects of chronic gamma-radiation from Co^{60} in plants.	Norway
899	11:209	L. H. Gray. The effects of ionizing radiations on biological systems.	United Kingdom
904	11:219	A. R. Gopal-Ayengar. Cytological and cytochemical effects of radiation in actively proliferating biological systems.	India
908	12:252	H. H. Ussing. Isotopes in permeability studies.	Denmark
1040	12:184	S. Mitsui. The importance of isotopes in agriculture.	Japan
1042	12:46	K. Murati and D. Moriwaki. Genetic effects induced in plants.	Japan
1046	12:298	R. Sasaki. Studies on physiology of lactation.	Japan
1047	12:275	R. Sasaki. Studies on the metabolism of calcium and phosphorus in the laying hen.	Japan
1048	12:326	Y. Okada *et al.* Studies on pearl formation mechanism by radio-autography.	Japan
1049	12:87	S. Mitsui. Studies on the plant nutrition, fertilizer and soil by use of radioisotopes.	Japan
1067	12:330	R. Sasaki. Some observations on the biological influences of radioactive isotopes upon the physiological functions.	Japan
1079	12:214	H. B. D. Kettelwell. Labelling locusts with radioactive isotopes.	United Kingdom

SUBJECT INDEX

NAME INDEX